A move towards more flexible, sustainable agricultural practices is being seen increasingly as the way to address or av [illegible] lems associated with existing, predomina [illegible] ook examines the implications of adop [illegible] ural practices, both at the level of indivi [illegible] cale agro-ecosystems such as water catc [illegible] on human and social aspects, rather than [illegible] ons, focusing on the learning process nece [illegible] d, in turn, on the facilitation of that learning through participatory approaches and appropriate institutional support and policy structure.

Following a general introduction by the editors, a range of case studies from around the world presents examples of attempts to find sustainable solutions to problems comprising complex conflicts of interest. In the final chapter the editors summarize the book's findings, concluding that the successful facilitation of sustainable development demands a new perspective, fundamentally different from that needed to manage change in conventional agricultural systems. This perspective is based on an understanding of the underlying social interactive dynamics of those involved and an appreciation of the importance of the learning process through which change can be mediated.

Facilitating Sustainable Agriculture

Facilitating Sustainable Agriculture

Participatory learning and adaptive management in times of environmental uncertainty

N.G. RÖLING and
M.A.E. WAGEMAKERS

Department of Communication and Innovation Studies
Wageningen Agricultural University
The Netherlands

PUBLISHED BY THE PRESS SYNDICATE OF THE UNIVERSITY OF CAMBRIDGE
The Pitt Building, Trumpington Street, Cambridge, United Kingdom

CAMBRIDGE UNIVERSITY PRESS
The Edinburgh Building, Cambridge CB2 2RU, UK http://www.cup.cam.ac.uk
40 West 20th Street, New York, NY 10011–4211, USA http://www.cup.org
10 Stamford Road, Oakleigh, Melbourne 3166, Australia
Ruiz de Alarcón 13, 28014 Madrid, Spain

First published 1998
First paperback edition 2000

Typeface Bembo MT 10/12pt *System* QuarkXPress™ [SE]

A catalogue record for this book is available from the British Library

Library of Congress Cataloguing in Publication data

Social learning for sustainable agriculture / [edited by] N.G. Röling
and M.A.E. Wagemakers.
p. cm.
Includes bibliographical references.
ISBN 0 521 58174 5 (hardback)
1. Sustainable agriculture 2. Agriculture–Technology transfer.
3. Agricultural extension work. I. Röling, Niels G.
II. Wagemakers, M.A.E. (Maria Annemarie Elisabeth),
S494.5.S86S635 1998 97-11992 CIP
630–dc21

ISBN 0 521 58174 5 hardback
ISBN 0 521 79481 1 paperback

Transferred to digital printing 2004

Contents

About the authors

Noelle Aarts graduated from the University of Nijmegen in the Netherlands in cultural anthropology, producing an MSc thesis on the tension between 'western' and local knowledge among a group of indigenous people in Venezuela. She is currently engaged in research for a PhD at the Department of Communication and Innovation Studies at the Agricultural University, Wageningen, on issues relating to communication between the Dutch government and other groups involved in nature and nature-related policies, with particular interest in aspects of interpretation in relation to policy acceptance.

Abraham Blum was born 1928 in Basel, Switzerland. After an agricultural apprenticeship, he immigrated to Israel, where he was a member of a Kibbutz, and later responsible for the orchards in an agricultural school. In 1966, after studying Agriculture (in Israel) and Education (in Chicago), he joined the Israeli Ministry of Education as Director of National Curriculum Projects in Agriculture and Environmental Studies. He was awarded a PhD for studies in this area. He founded the Extension Centre at the Faculty of Agriculture of the Hebrew University and currently is the incumbent of the Haim Gvati Chair in Agricultural Extension there. His main interests are in agricultural knowledge systems and rural development.

Jouke Boerma was born in 1960, in Petaling Jaya, Malaysia, and graduated in tropical land use and the management of natural resources in Wageningen. She has worked on negotiation processes in natural resource use and has examined the use of 'platforms' for this purpose. She has carried out research on tourism and safety in Amsterdam for a police station in the Red Light District. The practical sides of human communication were learned working for a period as a ticket clerk on the Dutch Railways. Jouke is currently engaged in setting up a therapeutic farm for mentally and/or physically disabled children in Sucre, Bolivia, most of whom live in an orphanage for the handicapped. The aim is to teach children farm work so they might return to their rural families.

Andrew Campbell works for the Ministry of the Environment in Canberra Australia as director of the Sustainable Land and Vegetation Branch of Environment Australia. He graduated in forestry at the University of Melbourne, and has a Masters degree from the Agricultural University Wageningen, the Netherlands. He is presently writing a doctorate on policies for rural sustainability on a community scale, based on case studies in Australia and southern France. He also manages, with the help of a neighbour, the family wool-producing farm in south-eastern Australia. He was instrumental in developing the Landcare movement in Australia and was Australia's first National Landcare Facilitator. His major publications are *Landcare: Communities Shaping the Land and the Future, 1994*; and *Planning for Sustainable Farming*, 1991.

Gelia Castillo is a rural sociologist and Professor Emeritus at the University of the Philippines at Los Banos. She has published widely on issues concerning the human aspects of agricultural research, rural development, poverty, participatory development, social dimensions of science and technology, research and partnerships, and gender issues. She has served, and continues to be involved in, numerous national and international boards, committees, review and advisory panels concerned with rural development, agricultural and health research, development policy etc. In recognition of her contribution, she has been awarded a number of honorary doctorates.

John Fisk is consultant to the W.K. Kellogg Foundation, Battle Creek, Michigan, on Food Systems and Rural Development programming. He is also a C.S. Mott Fellow in sustainable agriculture and receives a doctorate in crop and soil science at Michigan State University in East Lansing in 1997.

Elske van de Fliert was born in 1960, graduated in biology from the University of Utrecht, the Netherlands, and has a PhD in Agricultural Sciences from the Agricultural University Wageningen. Her published thesis (1993) is on the results of a field evaluation of the Indonesian IPM programme in irrigated rice in Central Java, carried out while working as an associate expert for the FAO. She participated in a baseline study of FAO's programme on IPM in irrigated rice in Sri Lanka (van de Fliert & Matteson, 1989, 1990) and worked for three and a half years as an FAO associate expert in the Indonesion IPM programme on a field evaluation study and on the development of rat IPM. She lives in Yogyakarta, Indonesia, and is currently a lecturer and freelance development consultant.

Alexander Gerber was born in 1966. He served two years apprenticeship on a bio-dynamic farm beside Lake Constance. Between 1989 and 1994 he studied Agricultural Science (1989–94) at the University of Hohenheim, Stuttgart Germany, majoring in eco-farming and sustainable agriculture. At present he is working in research on training, agricultural knowledge systems, eco-farming and sustainable agriculture at the University of Hohenheim in the Department of Agricultural Communication and Extension, also working as a freelance moderator and trainer.

Gus Hamilton is a Principal Extension Agronomist in southern Queensland, Australia with experience in the development and implementation of participatory action learning for improving service delivery. He has been involved in developing new tools to assist farmers in exploring and improving their farming systems. His current interest is in curriculum-based learning activities with farmer groups, providing a key role in developing information systems that support farmer learning and improve client access to information, and in developing research and extension methods that empower clients to become self-directed learners.

Oran Hesterman is Program Director for Food Systems and Rural Development programming at the W.K. Kellogg Foundation, Battle Creek, Michigan. In this role, he provides leadership to the integrated farming systems (IFS) initiative and supports development of related initiatives. Previously, Dr Hesterman researched and taught forage cropping systems management, sustainable agriculture, and leadership development in the Crop and Soil Sciences Department at Michigan State University in East Lansing. He received his doctorate in agronomy from the University of Minnesota in St Paul.

Volker Hoffmann was born in 1947. He worked in agricultural practice in southern Germany and France, studied agricultural economics at the University of Hohenheim, and was later awarded a PhD in the Social Sciences at the same university. He worked as Professor of Communication in Horticulture at Hannover 1991–1993, and is presently Professor of Agricultural Communication and Extension at Hohenheim.

Janice Jiggins is Professor of Human Ecology at the Swedish University of Agricultural Sciences, Uppsala, and Visiting Professor at the University of Guelph, Ontario. She has lived and worked for over 20 years in South Asia and Sub-Saharan Africa, Australia and New Zealand, mainly on problems of rural service provision, agricultural research management, and extension. Farmers' participation in agricultural research and technology development, resource users' management of natural resources, and people-centred fertility management, are key areas of concern. Her recent book, *Changing the Boundaries*, examines women's perspectives on population issues and the environment.

Alex Koutsouris graduated in Agricultural Economics at the Agricultural University of Athens, took his MSc degree in Agricultural Extension, University College of Dublin and PhD in Agricultural Extension and Education at the Agricultural University of Athens. He is currently working with the Development Agency of Karditsa, Greece, as Scientific Manager of Local Development Projects, and co-operates with the Unit of Agricultural Extension, Agricultural University of Athens in research and in teaching (1995/1996). He has participated in an academic and organizing capacity at various National and European Conferences and participated in the main teams of National and European research projects (CAMAR, FAIR, TEMPUS, etc.). His main interests are in rural development, extension and training.

Dimitrios Papadopoulos graduated in Agricultural Economics at the Agricultural University of Athens and holds an MSc degree in Social Research (MAKS) from the Agricultural University at Wageningen, the Netherlands. He is a PhD candidate in the Agricultural University of Athens in the field of Agricultural Extension and Sociology. His main interests are on topics related to sociology of organizations and agricultural extension.

Jules Pretty is a Fellow of the International Institute for Environment and Development, an independent policy research institute based in London. He is the former Director of the Sustainable Agricultural Programme at IIED, where he has since 1986 been engaged in a wide range of collaborative research, training and capacity building activities, both at local and policy levels. He has run training courses in participatory methodologies for many institutional contexts in a wide range of countries. He has published widely, and his books include *Regenerating Agriculture: Policies and Practice for Sustainability and Self-Reliance* (1995: published by Earthscan Publications, London; National Academy Press, Washington DC and Vikas Publishers, Bangalore); he co-authored *Unwelcome Harvest: Agriculture and Pollution* (1991); and the co-authored *A Trainers' Guide for Participatory Learning and Action* (1995). Jules Pretty has worked in Europe, USA, East and West Africa, South and South East Asia, Central America and the Pacific. Jules Pretty is currently on secondment from IIED to the University of Essex as Visiting Professor.

Jet Proost, born in 1957, is Assistant Professor at Wageningen Agricultural University in the Department of Communication and Innovation Studies. She specializes in the fields of environmental issues in agricultural extension and management aspects of extension services. She has undertaken various research projects and advisory tasks for the Ministry of Agriculture and the European Union. A PhD thesis is in preparation on farmers' study groups and their role in research and extension. Her background is in home economics and rural sociology. Her work over the past 12 years in West Africa and the Netherlands has created an interest in the role and functioning of farmers' organizations in agricultural knowledge and information systems.

Niels Röling, born in 1937, is a professor at Wageningen Agricultural University in the Netherlands, with an MSc in Rural Sociology from Wageningen and a PhD in Communication from Michigan State University. His research interest has moved from the diffusion of innovations (Röling *et al.*, 1976), via targeting technology development for poverty alleviation (Röling, 1988), and agricultural knowledge systems analysis (e.g. Röling, 1990), to a constructivist approach to sustainable natural resource management (e.g. Röling, 1993). He is currently part of a research programme on 'Knowledge Systems for Sustainable Agriculture' which focuses on the facilitation of social learning, on the linkage between 'soft' platforms for decision making and 'hard' ecosystems and on the institutional and policy conditionalities for learning sustainable agriculture at the farm and higher agro-ecosystem levels.

Michel Roux, born in 1956, Switzerland, studied agronomy at the Swiss Federal Institutes of Technology (SIT) and has a PhD in rural economy and sociology. Since 1984 he has worked in training and research activities at the Agricultural Advisory Centre (LBL), a private non-profit institution contracted by government to support the cantonal and private extension services in the German and Raetoroman part of Switzerland. He lectures in rural sociology at the Swiss College of Agriculture, and is a member of the Board of the Swiss Association of Rural Economics and Rural Sociology and the Swiss Academic Society for Environmental Research and Ecology. He is presently engaged in action research on platforms for resource use negotiations in landscapes and countryside under the priority programme 'Environment' of the Swiss National Science Foundation.

Nadet Somers studied social anthropology at the University of Amsterdam, and completed her PhD at the Department of Communication and Innovation Studies of the Agricultural University, Wageningen, the Netherlands. She has worked for several years at the Agricultural Economics Institute in the Hague, working on a wide range of topics such as prospects for small-scale farmers, possibilities and impediments for integrated arable farming, and societal functions of agriculture in the 21st century. She is a freelance socio-economic researcher and writer.

Thomas Thorburn is a program director for the W.K. Kellogg Foundation, Battle Creek, Michigan, with major responsibility for food systems, rural development, and water resource programs. Prior to joining the Foundation, he was an agricultural program leader for the Minnesota Extension Service at the University of Minnesota in St Paul. Dr Thorburn received his doctorate in adult and continuing education from Michigan State University in East Lansing.

Annemarie Wagemakers, born in 1964, graduated from the Wageningen Agricultural University in human nutrition with a special interest in extension, health, and developing countries. She has worked as research assistant in the Department of Communication and Innovation Studies in the field of women's health and extension, and women's re-entry into the labour process. She is currently working part-time with Professor Niels Röling on sustainable agriculture, and part-time for Aletta, the Dutch National Centre for Women's Health Care. Her task is to integrate the principles of women's health care into the regular national and municipal health extension organizations.

Mathieu Wagemans, born in 1947, is employed by the Dutch Ministry of Agriculture, Nature Management and Fisheries. He has a special interest in questions of planning, the introduction of participatory methods in policy development, and organizational development. He has been, and remains, engaged in the process of transforming the Dutch Ministry, from a prescribing to a facilitating policy context. At local level he is politically active as an elected Member of the

Community Council. His thesis dealt with rational planning models in public service bureaucracies and focuses in particular on the question of why application so often proves problematic. At international level he has been involved as consultant in various projects in China, Russia, Moldova and Vietnam, mainly concentrating on public administration reform at national, regional and local level.

Willem van Weperen graduated in animal production and agricultural education, and completed his MSc degree at the Agricultural University Wageningen. He has field experience in extension work in Asia and Africa in various dairy development projects, and has taught in the field of animal production in the Netherlands. He has worked as a freelance researcher at the Department of Communication and Innovation Studies in Wageningen and is currently a freelance consultant in agricultural extension. His principal interest is in participatory action research relating to agricultural innovation and farmers' learning processes in the shift towards sustainable farming practices.

Cees van Woerkum, born in 1947, is Professor of Extension Science at the Department of Communication and Innovation Studies at the Agricultural University in Wageningen, the Netherlands. His field is sociology, especially the sociology of mass communications. Up to 1980 his primary research interests were in constructing working plans for extension via mass media, moving then to the topic of 'extension as a policy instrument'. His major publications are on the concept of communication in sociology, on a working plan for mass media and on written communication. He is currently working in the field of agriculture, environment and health.

James Woodhill is Manager of Policy and Programme Development for Greening Australia, a large environmental NGO. He has a background in agricultural science and the sociology of natural resource management. Prior to working for Greening Australia, Jim lectured in systems agriculture at the University of Western Sydney – Hawkesbury. He played a significant role in furthering the use experiential learning and participatory development methods in both the Faculties, teaching and in natural resource management. He has worked nationally and internationally as a consultant, in particular facilitating participatory planning workshops using a systems learning methodology and has been extensively involved in the development of landcare and catchment management in Australia. He has a particular interest in achieving a better integration of theory and practice in natural resource management, and is working on a PhD in this field. His current role involves him in a broad range of issues related to the role and functioning of an environmental NGO, including advocacy, strategic planning, and performance evaluation.

Contributors

M. Noëlle C. Aarts
Department of Communication
and Innovation Studies
Hollandseweg 1
6706 KN Wageningen
The Netherlands

Abraham Blum
Faculty of Agriculture
The Hebrew University of
Jerusalem
POB 12
Rehovoth 76100
Israel

Jouke Boerma
Roos Vlasmanstraat 8
1391 EH Abcoude
The Netherlands

Andrew Campbell
Environmental Strategies
Department of Environment
GPO Box 787
2601 Australia

Gelia T. Castillo
International Rice Research
Institute
PO Box 933
1099 Manilla
Philippines

John W. Fisk
Food Systems and Rural
Development
W.K. Kellogg Foundation
One Michigan Avenue East
Battle Creek
Michigan 49017-4058, USA

Elske van de Fliert
Jl. Bali 11
Widoro Baru
Ngropoh CC
Yogyakarta 55283
Indonesia

Alexander Gerber
Institut für Sozialwissenschaften
des Agrarbereichs
Universität Hohenheim
7000 Stuttgard 70
Germany

N.A. (Gus) Hamilton
QDPI South Region Australia
PO Box 597
Dalby Queensland 4405
Australia

Oran B. Hesterman
Food Systems and Rural
Development
W.K. Kellog Foundation
One Michigan Avenue East
Battle Creek
Michigan 49017-4058, USA

Volker Hoffmann
Institut für Sozialwissenschaften
des Agrarbereichs
Universität Hohenheim
7000 Stuttgart 70
Germany

Janice Jiggins
Department of Rural
Development Studies
Box 7005
S-750 07 Uppsala
Sweden

Alex Koutsouris
AERD Unit
Development Agency of
Karditsa–ANKASA
5, Artessianou St. and Kolokotroni
43100 Kardista
Greece

Dimitrios Papdopoulos
Agricultural University of Athens
Unit of Agricultural Extension
765 Iera Odos
11855 Athens, Greece

Jules N. Pretty
IIED
3 Endleigh Street
London WC1H 0DD, UK

M.D.C. (Jet) Proost
Department of Communication
and Innovation Studies
Hollandseweg 1
6706 KN Wageningen
The Netherlands

Niels G. Röling
Department of Communication
and Innovation Studies
Hollandseweg 1
6706 KN Wageningen
The Netherlands

Michel Roux
Landwirtschaftliche
Beratungszentrale (LBL)
Eschikon
CH-8315
Lindau/ZH, Switzerland

B.M. (Nadet) Somers
Hoofdweg 3
2333 LH Oostvoorne
The Netherlands

Thomas L. Thorburn
Food Systems and Rural
Development
W.K. Kellog Foundation
One Michigan Avenue East
Battle Creek
Michigan 49017-4058, USA

M. Annemarie E. Wagemakers
Department of Communication
and Innovation Studies
Hollandseweg 1
6706 KN Wageningen
The Netherlands

Mathieu Wagemans
Belerbroeklaan 22
6093 BT Heythuysen
The Netherlands

Willem van Weperen
Winkelseweg 4
7255 PN Hengelo
The Netherlands

Cees M.J. van Woerkum
Department of Communication
and Innovation Studies
Hollandseweg 1
6706 KN Wageningen
The Netherlands

James Woodhill
94 Hopetoun Circuit
Yarralumla, Canberra
ACT 2600
Australia

Abbreviations

ACM	Co-operative union for pesticide supply
AGÖL	The Central Organization of all German Federations of Eco-farming
AKIS	Agricultural knowledge information systems
AKS	Agricultural knowledge system
APSRU	Agricultural Production Systems Research Unit
AVEBE	Co-operative potato processing company
BW	Bacterial Wilt
CAP	Common Agricultural Policy
DAD-model	Decide, announce, defend – model
De Peel EC	De Peel Environmental Co-operative
DLV	Dutch Privatised Agricultural Extension Service
EM	Environmental management
EPA	Environmental Protection Agency
ESU	European Size Unit
ETH	Federal Institute of Technology
ETL	economic threshold level
EU	European Union
EUAD	Environment and Urban Affairs Department of Pakistan
FAO	Food and Agricultural Organization of the United Nations
FAP	Research Station for Plant Cultivation
FAT	Federal Research Station for Farm Management and Agricultural Engineering
FFS	farmer field school
FIBL	Research Institute of Biological Husbandry (Swiss)
FLEW	field level extension worker
FSR	farming systems research
FTF	field training facilities
GATT	Global Agreement on Tariffs and Trade
HYV	high-yielding varieties
IAF	integrated arable farming
ICRE	International Course on Rural Extension
IFOAM	International Federation of Organic Agriculture Movement

IFS	Integrated Farming Systems
IIED	International Institute for Environment and Development
ILEIA	Information Centre for Low External Input and Sustainable Agriculture
IP	integrated production
IPM	Integrated Pest Management
IRRI	International Rice Research Institute
KUD	village unit co-operative
LBL	Inter-cantonal Extension Centre of the SVBL serving the German-speaking areas of Switzerland
LFA	Less Favourable Area
MEKA	scheme of Baden Wurtemburg, which stands for 'Marketeentlastungs und Kulturlandschafsausgleich'
MYCPP	Multi-Year Crop Protection Plan
NAEP	National Agricultural Extension Project
NCS	National Conservation Strategy
NEFYTO	The Netherlands Pesticide Producers' Organization
NFE	non-formal education
NGO	non-governmental organization
NLP	National Landcare Programme
NLTO	Dutch Farmers' Union
NPP	Nature Policy Plan
OFA	Operation Future Association
PAGV	Dutch Experiment Station for Arable Farming and Broad Acre Vegetable
PLAR	Participatory Learning and Action Research
PTD	Participatory Technology Development
QDPI	Queensland Department of Primary Industries
R&D	Research and Development
RAAKS	rapid appraisal of agricultural knowledge systems
REC	regional extension centres
SALT	sloping agricultural land technology
SAPPRAD	South-east Asian Program for Potato Research and Development
SEV	socio-economic advisory services
SIL	Swiss College of Agriculture
SÖL	Foundation of Ecology and Farming
SRVA	Inter-cantonal Extension Centre of the SVBL serving the French and Italian speaking areas of Switzerland
SSM	soft systems methodology
SVBL	Swiss National Union for the Advancement of Agricultural Extension
T&V	Training-and-Visit
ToT	Transfer of Technology

UPWARD	User's Perspective with Agricultural Research and Development
VFSG	Viable Farming Systems Group
VSBLO	Union of the Swiss Associations of Biological Agriculture
WAU	Wageningen Agricultural University
WBP	Werkgroep Behoud de Peel (committee to preserve De Peel)
WKKF	W. K. Kellogg Foundation
WUB	*Wagenings Universiteitsblad* (*Wageningen University Weekly*)

Preface

One really misses a lot by not owning a pear tree. For one, there is the partridge. Without a pear tree, it all becomes hearsay, vicarious experience, but the main thing is that one will never know what a pear really tastes like. One will never have that total experience of eating a really ripe pear, quite apart from the added benefit of doing so in a garden.

We cannot promise you a garden when reading this book, nor can we guarantee partridges, but we can assure the reader that this book is ripe. It has definitely not been picked green to get to the market in time, but immature. In fact, it has been picked over and over again, chapters have been thrown out, new ones included, and editing has been severe. Of course, ripeness does not guarantee quality, it can still be of the wrong variety, there might be a worm in it, or it might be over its prime. That is for the reader and reviewers to decide.

The first idea for this book emerged somewhere in 1993, when a set of papers presented by kindred spirits at a meeting of the European Society for Rural Sociology seemed interesting enough to develop into a bundle of readings. Looking back, one wonders at the light-hearted audacity with which such thoughts are mooted. As in marriage, one enters book-making without too much thought about the implications. One falls in love and that's more or less it, as far as sound decision making goes, but whatever the implications, we, and all the contributors, stuck with it. And here it is, some four years later. It does not bear much resemblance to the original bundle of papers. We feel that much has changed for the better.

The subject area of the book is in rapid, if not turbulent, change. It is surprising how many of our colleagues have been bitten by the same bug. New ideas, perspectives, approaches, and projects arise everywhere. It is truly exciting to witness. For us, it meant that we could ride a crest of rich pickings and incorporate a number of new and exciting developments which emerged during the course of the time we were working on the book. Of course, it also meant that we had to let go of some cherished earlier stepping stones. Especially earlier Dutch chapters, including some of our own, had to be mercilessly cut or amalgamated. We feel this improved the whole considerably. Looking back, we are confident that the rather long gestation period proved to be a boon.

Its history gives us the feeling that this book allows for a moment of consolidation and reflection by the participants in a world-wide network of people with similar concerns. They are co-travellers on an exciting journey, dedicated to learning about social learning as the only way out of the predicament caused by the collective impact of individual choices, informed mainly by self-interest and definitely not by ecological understanding. As such, this book undoubtedly will be eclipsed soon by other and better ones, but we hope it will make its temporary contribution by consolidating the discourse of what we are about, and by helping this along a notch or two.

If we achieve this aim, it is due to a large number of people. In the first place, we must thank the contributors, most of whom have been extremely patient in enduring the ripening process. Some were called upon to make last-minute major changes to update their contributions, and some only joined at the last moment when it turned out that theirs could make an essential contribution.

In the second place, we want to thank the people very warmly who made available their considerable editing skills. These skills are essential for a book such as this one, which brings together people who write in American, Australian and Philippine English, German English (not the worst, we understand), Greek English, and Dunglish, Dutch English, alas a lot of that. We were very fortunate in having editors who are all too aware of the pitfalls, especially of Dunglish, but both did more than just language editing. Janice Jiggins' review led to a revised chapter outline and to rewriting entire chapters. Ann Long had penetrating queries about logic and theoretical consistency. We are very grateful to their major contribution.

We would also like to thank Dr Alan Crowden, the publisher's Editorial Director for Science, Technology and Medicine, for his confidence and support.

Niels Röling and Annemarie Wagemakers
Wageningen, November 1996

Part I: Introduction

1 A new practice: facilitating sustainable agriculture

NIELS G. RÖLING AND M. ANNEMARIE E. WAGEMAKERS

1.1 When Greenland became white

How did our ancestors manage their environment? A green history of the world (Ponting, 1991) and the first conference on how pre-industrial civilizations coped with climate change (Pain, 1994) offer plausible insights.

One case study presented at the conference concerns the medieval Norse settlements in Greenland. Their agriculture prospered during the twelfth century. In 1127, they sent a live polar bear to the Nordic King and received a bishop in return. But by 1500, Greenland had become white, and the only people still living there were the Inuit seal hunters. All that remains of the Norse communities are the ruins of their churches.

A comparative analysis of the strategies of the communities of the two peoples for coping with the Small Ice Age reveals that the Norsemen were more concerned with providing for their bishops and building churches than with changing their way of life or agriculture to accommodate the harsher climate. They persisted in trying to graze their cattle on increasingly poor grassland. The Inuit, however, adapted and developed their hunting and fishing in ways that allowed them to survive through the increasing cold.

Though it seems, at first, that the demise of the Greenlanders was caused by climate change, closer scrutiny reveals that the climate was not the only factor.

> By the time the temperature had plummeted in the 1370s, the productivity of the land was falling because of overgrazing and soil erosion. But while the Inuit adopted appropriate technologies of hunting and fishing, the Norse farmers – constrained by their rigidly ordered society and Christian culture – tried to maintain the way of life they were used to. They failed (Pain, 1994).

From this, and other cases presented during the conference, the following conclusions emerge:

- It is not so much climate change that causes problems, but entrenched modes of adapting to change. 'The ability and willingness of society to respond to changing conditions are the crucial conditions in determining whether it survives' (Pain, 1994).
- Such responsiveness depends on individual and collective choices, which are, of necessity, shaped by the past. This makes us vulnerable to discontinuous events for which there are no historical precedents.

- The development and use of knowledge is our main mechanism for survival in conditions of rapid change. That is, adaptation to changing conditions depends on perceiving and interpreting the signs of impending change, and on the timely development of knowledge, technology and organisation in reaction to those signs. Thus, the adaptative response also demands creativity and inventiveness and a capacity for collective learning and innovation.
- By virtue of their privileged position, the elites who have a formal or social mandate to provide leadership are often shut off from direct or even indirect experience of the signs of change. They have the power to maintain their lifestyles and the way things are when it is no longer prudent to do so.

Are we like the Norsemen on Greenland?

1.2 About this book

The question could imply that this book is about the Apocalypse, but nothing would be further from the truth. This book is about developing appropriate responses to environmental uncertainty and discontinuity. If anything, its authors are possibly too optimistic in their expectation that it is not beyond human society to make the adaptations that now appear necessary. That does not mean, of course, that we are convinced that meaningful and timely change will come about, but we hope that this book will increase the likelihood that it will.

The more specific aims of the book are to 'capture a new practice', i.e. to examine a number of cases of attempts to make farming more sustainable in conditions of uncertainty. Secondly, the contributors tease out the lessons of emerging practice with respect to the kinds of learning, facilitation, supporting institutions, and conducive policy contexts that are required. The book's central questions therefore are: can we learn our way to a more sustainable agriculture? And if so, what does it take?

The case studies suggest that the answer to the first question is 'Yes'!, but also, that it will take a transformation of our epistemology, our technological and organizational practices, our ways of learning, our institutional frameworks and our policies. Such transformations do not come easy, and the question 'Are we Norsemen on Greenland?' remains unanswered now that we have finished the book, but we have become a little clearer about the effort that needs to be made and we feel heartened by that understanding.

1.3 A social science perspective

This book approaches sustainable farming from the point of view of social science. Hence its central focus is the human actor, and not the bio-physical processes occurring in the agro-ecosystem (see Box 1.1). As you will see later, this has important implications for the way we define sustainability. It also means that this book does not deal in detail with the agronomic or other practices that are needed for using natural resources in a more sustainable manner. A great deal of information is available on that issue (e.g., Howard, 1943; National Science Council, 1989; Reijntjes, Haverkort & Waters-Bayer 1992; Pretty, 1995). Jiggins and De Zeeuw (1992) provide pioneering information on participatory technology development (PTD) for sustainable farming, while Pretty's (1995) authoritative book on regenerating agriculture provides a wealth of information on its feasibility and the conditions for transformation.

Box 1.1 The area of discourse of this book

1		2		3		4
farm	⟷	farmer	⟷	facilitator	⟷	etc.

The relationship between farmer and farm (1) is that of a human being dealing with bio-physical resources and processes. Natural science supports the technologies and interventions used by the farmer to make the bio-physical environment yield desired outcomes. These 'technical' aspects do not constitute the area of discourse of this book.

But change in relationship 1 is only possible if the farmer him/herself engages in learning (2). The learning process lies at the heart of this book. As we shall see, the transformation to sustainable agriculture requires a fundamental change in learning processes. These turn out to be very different from the well-established processes of adoption of add-on innovations, in the 'more of the same' fashion, which occur when the farmer tries to improve conventional farm management (Somers, this volume).

Learning can be facilitated (relationship 3). The facilitation of learning is also a core subject of the book. Most of the contributors, including the editors, are engaged in extension and innovation studies. That is, they are interested in fostering voluntary change in behaviour through communication, and in innovation as an outcome of social interaction (e.g. Röling, 1988). Other perspectives on innovation, for example, as a process induced by changes in relative factor prices (Ruttan & Hayami, 1984) are not dealt with, although we recognize that changed relative factor prices might well motivate innovative interaction.

The case studies reaffirm the proposition that learning and facilitation occur in specific institutional frameworks and policy contexts (relationship 4), and that the nature of these frameworks and contexts is of crucial importance for the transformation of farming. Moreover, the cases illustrate how the institutional and policy changes required for scaling-up successes achieved on a pilot scale, cause strife and conflict for which appropriate communication and negotiation strategies and methodologies are needed.

We do not limit ourselves to the field or farm level, but explicitly also take into consideration larger-scale agro-ecosystems, such as water catchments, which need to be managed in their own right in order to allow sustainable management at the field and farm levels. Since this book assumes that the transformation to sustainable farming is social as much as agronomic and ecological (Vartdall, 1995), we shall examine these agro-ecosystems by looking at innovation processes at levels of social aggregation concomitant with the scale of the agro-ecosystem.

The contributors to the book bring along a wide range of experience assembled during widely different life times, from widely different contexts, which allows examination of the transition to sustainable agriculture from widely different perspectives, at different levels and along different dimensions (Box 1.2.).

Box 1.2 Profile of the contributors

Campbell and Wagemans are senior policy makers in Government, while Woodhill plays the same role in an environmental voluntary organization. All three also undertake consultancies in other parts of the world.

Fisk, Hesterman and Thorburn work for the W.K.Kellogg Foundation in various capacities. WKKF is a philanthropic foundation which supports learning in and about agriculture and rural development.

Boerma, Hamilton, Roux, Van de Fliert and Van Weperen play professional roles as consultants, project leaders or implementors.

Aarts, Gerber, Koutsouris, Papadopoulos, Proost, Van de Fliert (who has two jobs) and Wagemakers work for universities as lecturers and researchers, Somers is a researcher in a research institute.

Blum, Castillo, Hoffmann, Pretty, Röling and Van Woerkum are university professors who hold (or held, Castillo is emeritus professor) chairs in departments of extension studies of one kind or another. Jiggins holds a chair in the field of human ecology. Hesterman has worked as a professor in crop and soil science. All undertake consultancies in various parts of the world.

1.4 Sustainable agriculture

Productivity, equity, sustainability and stability have been identified as key goals of agricultural policy (Conway, 1994). But they are not necessarily mutually consistent. Conway speaks of 'trade-offs' among them, especially between the economic and ecological. This book is weighted heavily towards ecological imperatives. Maintaining or enhancing the natural resource base is the precondition for a sustainably productive agriculture.

Box 1.3 outlines the main aspects we shall examine. These aspects relate systemically to one another, in that change in one aspect necessarily affects the others. For example, the learning required for effectively practising integrated pest management can apparently not be achieved by the transfer of technology (ToT) mode of extension. It requires a new approach to facilitation (Matteson, Gallagher & Kenmore, 1992), which in turn has important implications for institutional support (Röling & Van de Fliert, this volume). We come back to the five dimensions in the last chapter, where we examine models of innovation which are suitable for understanding the transformation to sustainable farming.

Box 1.3 Five interlocking dimensions of the transformation to sustainable farming

- agricultural practices, both at the farm and higher system levels;
- learning those practices;
- facilitating that learning;
- institutional frameworks that support such facilitation, comprising markets, science, extension, networks of innovation, etc.;
- conducive policy frameworks, including regulations, subsidies, etc.;

and especially:

- the management of change from conventional to sustainable agriculture along each of the dimensions.

We shall not define sustainability solely in terms of the carrying capacity or other 'hard' characteristics of an agro-ecosystem. We use a social science definition (Box 1.4) that, as our cases show, proves to be eminently practical. We borrowed the definition from the 'Hawkesbury pioneers', a small band of agriculturalists at the University of Western Sidney in New South Wales (Bawden & Packam, 1991; Sriskandarajah, Bawden & Packam, 1989; Ison, 1994; Woodhill, 1993).

Box 1.4 Sustainability defined

Sustainability is an emergent property of a 'soft system' (Woodhill & Röling, this volume). It is the outcome of the collective decision-making that arises from interaction among stakeholders. Stakeholders are identified here as natural resource users and managers. A natural resource can be considered at the field, farm or higher level of aggregation, including watersheds, landscapes, agro-ecological regions, lakes and rivers, and, ultimately, the Earth itself.

The formulation of sustainability in this manner implies that the definition is part of the problem that stakeholders have to resolve (Pretty, 1995 and this volume). That is, securing agreement on what people shall take sustainability to mean for a given environment, is half the job of getting there.

The definition in Box 1.4 incorporates elements which focus on the hard properties of a farm or an agro-ecosystem. Yet we suspect that our seemingly relativist definition will irritate those who want to use scientific definitions to identify the limits beyond which use of natural resources should not go (Korthals, 1994). None the less, our definition, as this book shows, has proven robust because it fits so well the outcomes of research on the interface between people and their environment. It also serves our immediate purpose of illuminating the facilitation of sustainable agriculture.

Our definition is consistent with Wouter de Groot's (1992) 'problem-in-context framework for the analysis, explanation and solution of environmental problems' (Box 1.5). A problem is here defined as an undesirable difference between 'wants' or norms, and 'gets' or impacts. According to De Groot, an environmental problem is an undesirable difference between the environmental impacts of human activity, and environmental norms. Solving environmental problems thus requires changing human activity to fit the norms, and vice versa. The problem-in-context framework provides for an integration of environmental sciences and the derivation of norms for human activity.

Box 1.5 De Groot's 'Problem in Context Model' of environmental problems (simplified adaptation of De Groot, 1992).

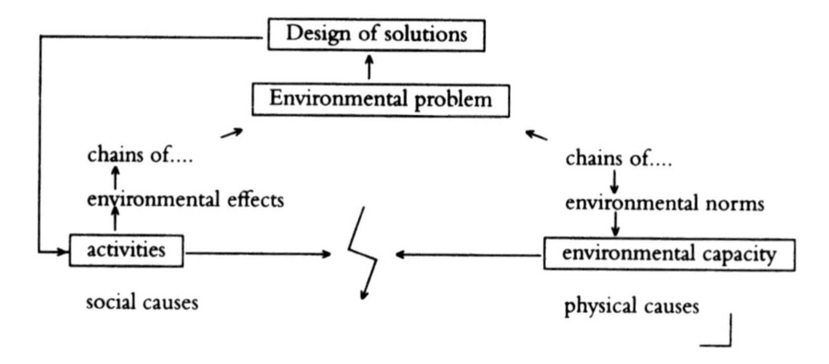

1.5 Is it immoral to be concerned about the health of natural resources?

The term 'sustainable agriculture' implies regenerative practices which optimally use locally available resources and natural processes, such as nutrient recycling; build on bio-diversity; regenerate and develop natural resources; and limit the use of external inputs of agro-chemicals, minerals, and non-renewable energy. Regenerative agriculture requires that, where used, such external inputs are used efficiently so that emissions can be recycled and absorbed (Pretty, 1995), renewable resources are regeared, and non-renewable resource use generates optimal productivity.

Defined in this way, regenerative agriculture, in terms of yield, at present tends to be slightly less productive than high input agriculture in industrial countries, and approximately as productive as such agriculture in 'green revolution' areas. On presently available evidence, state-of-the-art regenerative practices would increase significantly productivity in the rainfed, complex and resource poor areas in developing countries which have so far not benefitted from high external input technologies and are usually heavily degraded (Pretty, 1995).

The question then arises whether, from a global point of view, regenerative agriculture can deliver the output required to meet aggregate demands. The trend line extrapolation of growth in the world's population to more than 8 billion in 2025, an increase of over 2.5 billion in the next 30 years, has led many authors to emphasize the need to double productivity per hectare of available farm land (McCalla, 1994; CGIAR, 1995, Tansey & Worsely, 1995).

The productivity deemed possible under regenerative agriculture, and the doubling of productivity per hectare of available land required by 2025 seem contradictory. Indeed, one expert claims that promoting a form of agriculture which uses no artificial fertilizers or chemical pesticides is immoral because it undermines global food security (quoted from Rabbinge in WUB 17, 11th May, 1995).

It suffices in this introductory chapter to highlight the irreducible uncertainty about what lies ahead over the next two or three decades. On the one hand, while there is at present ample grain available in global food markets to feed the world's present population, many millions continue to go to bed hungry. This has to do with the fact that surplus production is not directly related to the relief of chronic hunger. Many of those who are hungry have neither the means to produce sufficient to feed themselves, nor the money to buy it. The issues of inequity and distribution are likely to remain present into the foreseeable future, at any level of grain production, or population growth, or environmental change.

On the other hand, although present estimates of production potential indicate adequate capacity to meet the food needs in aggregate of up to twice as many people as are alive today, such estimates are by their very nature conditional, and hedged around by assumptions about what might happen in other sectors. For example, estimates of the future adequacy of water supplies to agriculture are dependent on considerations such as demands for water from industrial, recreational and domestic users, the rate at which water loss is controlled, the efficiency of water use, and the rate at which water-winning and water storage technology is introduced.

Further food output scenarios also are strongly affected by assumptions made about the outcome of political negotiations concerning acceptable levels of environmental pollution, degradation and waste, questimates concerning the mobilization of political will to invest in the health of soil amelioration measures, which would open up hitherto unusable areas to cultivation, as well as price relationships, market developments, and changes in consumer preferences.

In addition, the introduction of factory-based production of meat muscle (already foreshadowed in the UK Government's technology foresight programme) could relieve much of the pressure of livestock farming on grasslands, for example, while an acceleration of the already observable shift towards more vegetarian diets in Northern consumer markets (e.g. Kleiner, 1996) would markedly change food demand forecasts. The likelihood of rapid change in global temperature and rainfall patterns adds further uncertainty.

Conversely, continued reliance on high energy input, chemical-dependent, intensive farming as the sole or even the main source of global surplus would appear to be environmentally foolhardy, and increasingly unacceptable politically as the wider consequences to human health, eco-system and valued landscapes become apparent.

1.6 'Making the flip'

Historically, the number of people and their 'wants' have grown, and uses of natural resources to satisfy them have been developed, regardless of the longer-term consequences for the environment (Ponting, 1991). In this book, we consider the alternative, indeed the necessity, of 'making the flip':

- conserving, even enhancing, the natural resource base upon which all agriculture ultimately, and indeed human survival, depends;
- the resultant negotiations involved in the transformation of wants to accommodate the emergent understanding of the natural resource imperative; and
- the kinds of agriculture which then results from such a reversal of attitudes, approaches, and behaviour.

The book illustrates the social energy which is created as people and institutions begin to engage in accommodation of 'wants' to 'effects'. The evidence suggests explicitly that the way to capture the potential for productivity, which is realizable by paying heed to the health of natural resources, is to work with and support the creativity, diversity, and serendipity which emerges from a process of social learning.

In agriculture, the process requires that farmers become experts, instead of 'users', 'receivers', or 'adopters' of other specialists' wisdom and technologies. They must learn to apply general ecological principles to their own locality and time-specific situations. They must be able to manage complex agro-ecosystem systems as businesses in competitive markets. But, as eco-systems do not stop at farm boundaries, local communities and wider consortia of interest groups and resource users also need to engage in learning how to manage landscapes and resources. Societies have to develop, and adjust to, trade-offs among potentially competing interests. We still have a long way to go in that respect. On the whole, a wasteful way of life has so far remained politically non-negotiable.

1.7 The prevalent paradigm for thinking about innovation

The current ways of thinking about the processes of innovation are embedded in a particular *epistemology*, that is in ways or methods for knowing on the basis of knowledge which hitherto has remained largely unchallenged.

The prevalent view of agricultural science is that it deals with 'things', which are as they are, which can be objectively known through research, and about which science can formulate generalizable 'truths'. These objectively verifiable propositions underpin the efforts made to influence agricultural performance. The goals of such intervention are taken to be unambiguous and not of scientific interest. The focus is on the 'best technical means' for achieving any stated goal. Box 1.6 presents typical statements which are informed by this so-called realist–positivist epistemology (Woodhill & Röling, this volume).

Box 1.6 Illustrative realist–positivist statements

- Reality exists independently of the human observer;
- Through scientific research we can build objective, true knowledge (generalizations) about that reality;
- Scientists discover and lay bare the naked truth, lift the veil hiding it, and reveal its secrets;
- The goal of science is to add to the store of human knowledge;
- Scientific research is the source of innovation;
- Technology is applied science;
- Development results from the transfer of the results of science to users;
- Problems can be solved by experts. In fact, we do not have to worry too much about the future. Science will find an answer;
- Social science is not really a science: it has not resulted in any true generalizations and cannot be used to send a man to the moon.

The realist–positivist epistemology is a coherent and internally consistent paradigm which most agricultural professionals have drilled into them, or absorb during their training, whether it is technical or social, and whether it is at an academic, polytechnic, or secondary level institute. But it is an epistemology which is increasingly incompatible with the search for a sustainable society.

Kuhn (1970) has opened our eyes to the notion of coherent and internally consistent paradigms which shift as the number of plausible knowledge claims which are in conflict with them increases. We then move from a period of 'normal science' during which people agree on the fundamental premises and occupy themselves with questions within the paradigm, into a period of 'post-normal' science, during which the paradigm itself is contested.

An increasing number of knowledge claims which are inconsistent with the realist-positivist paradigm (Box 1.7) arise in current debates about agriculture and the environment.

The erosion of realist–positivism as a universally trusted epistemology appears to be associated with other important societal changes. For one, the trust in experts and specialized institutions is waning. A recent survey found, for example, that British respondents do not trust the information about bio-technology which they get from scientists, business corporations, or the Department of Trade and Industry. They prefer to trust organizations such as Greenpeace (Tate, 1995).

Box 1.7 Some claims at odds with the realist–positivist paradigm

- Agriculture has multiple goals which are not mutually compatible. Hence the assumption of unambiguous goals and the focus on 'best technical means' is becoming irrelevant. The need for arbitration among contested goals is becoming one of the key challenges in dealing with our natural resources.
- Decisions about the use of natural resources are less and less a question of expertise or the province of specialist institutions, and more and more determined by negotiation and agreement among stakeholders. The focus shifts from result to process. The problems we are faced with have less to do with instrumental problems, i.e. people–thing problems, and increasingly to do with people–people relationships, i.e. social problems. This has important implications for agricultural science which has so far profiled itself as a bio-physical and technical activity.
- In the conventional paradigm, innovation is seen to originate in science and to be realized through the transfer and adoption of the results of science (the linear model or Transfer of Technology (ToT) model). But it is increasingly clear that, in practice, innovation emerges from interaction among various 'actors', i.e. among people and collectivities as role playing and sense making beings. Each one contributes to the final outcome (e.g. Kline & Rosenberg, 1986; Engel, 1995). Local knowledge, business ingenuity, farmer experimentation and inventiveness are as important as expert knowledge and the role of specialized actors such as scientists and farm advisors.
- Agricultural development is conventionally seen as driven by technological change. However, few would now disagree that price changes, improved institutional support (reducing transaction costs), conducive policy contexts, value shifts, and social organization can be necessary conditions for, and sometimes the stimulus of, innovation.

The erosion of trust also appears to be related to uncertainty about issues for which the stakes are high, such as global food security which was touched upon in Sections 5 and 6. Funtowicz and Ravetz (1990, 1994) argue that we have entered a period of 'post-normal science' in the sense of Kuhn (1970) because 'normal science' cannot deal with conditions of high uncertainty. The development of reliable grounds for knowing must proceed in part along other lines. Self-appointed activists emerge, who become formidably well informed across discipline boundaries about a subject or situation which threatens their values or livelihood. Decision fora include speech-makers and citizens, as well as scientists or specialists, and 'facts' encompass people's values and express cultural meaning. The final arbiters of reliable knowledge are 'extended peer communities' made up of a much wider membership than the conventional narrow professional elites or restricted political circles.

Funtowicz and Ravetz (1994) speak of the 'democratization of science', i.e. a widely shared process of learning and informed public debate about goals, and not just means, seems the only acceptable way to deal with high uncertainty when the stakes are high and the consequences of getting it wrong are potentially catastrophic.

This observation is reminiscent of Habermas' (1984, 1985) argument that society can overcome the momentum of what we have constructed in the past – and thus prevent the Norsemen on Greenland scenario – only by reaching consensus about what action to take next, i.e. not on the basis of controlling things (instrumental rationality), nor on the basis of beating competitors or opponents (strategic rationality), but on the basis of shared learning, collaboration, and the development of consensus about the action to take (communicative rationality).

1.8 Constructionism

Constructionism is the name given to the epistemology which supports the learning processes described in this book. If everyone agrees about the goals, we can afford to worry about the best technical means of securing those goals. If everyone agrees about the facts, we can speak of objective truth. 'Objective' knowledge has therefore, by no means become outdated or unneeded. But, if these conditions do not hold, we have to stretch the positivist epistemology and embrace constructionism. Reality no longer appears as a 'given' but as something actively 'constructed' by people. There are three main strands in constructionist thinking (Knorr Cetina, 1995).

In the first place, reality is said to be socially constructed (Berger & Luckmann, 1967). It is created in the discourse of, and negotiations among, people as social actors. Socially negotiated agreements become experienced as 'objective' truth. Berger and Luckmann concern themselves with the mechanisms by which objective social order emerges from interaction.

Secondly, convincing empirical analysis of 'fact production' in natural science laboratories by, e.g. Knorr Cetina herself (1981); Latour (1987) and Collins (1985, 1992) has focused attention on epistemic practices and human construction in the very area we used to think of as given natural reality (Box 1.8).

Box 1.8 Constructionism and quantum physics

Niels Bohr did not believe in the Newtonian clockwork universe (which was the conventional perspective early this century). 'There is no quantum world. What exists is a quantum physical description. It would be mistaken, therefore, to believe that it is the task of physics to find out what nature is. Physics occupies itself with what we can say about nature' (quoted in NRC/Handelsblad, 18 May 1995). Another quantum physicist, David Bohm, said: 'It is not the task of science to increase the store of knowledge, but to formulate fresh perspectives' (1993). Some researchers and analysts reject constructionism because of its apparent relativism (Röling, 1995). If it is people who construct reality, there must be multiple realities. What the one has constructed can be deconstructed by someone else. Everything can be true.

Thirdly, biological research into the 'observing organism' (Maturana & Varela, 1987) has demonstrated that the environment external to the observer, be it a frog or a person, does not project itself objectively on to the nervous system. Perception is accomplished by the brain, and the brain is an informationally closed system that reconstructs an external environment only from environmental 'triggers', memory, and interaction with itself.

We explicitly reject such an interpretation of constructionism. There is an environment. If an organism loses touch with it, or, as Maturana and Varela (1987) put it, if the structural coupling between organism and environment is broken, the organism cannot survive. But the 'constructions' appropriate for survival are not fixed or self-evident, nor is their interpretation unambiguous. They have selectively evolved, are culturally conditioned, continue to be actively created, or learned experientially on the basis of trial and error, or vicariously on the basis of communication. In this scheme of things, science has a role to play. It is engaged in the active construction of reality. But its impact is not based on the predictive power of its generalizations, however, but on the extent to which it affects other people's reality construction.

Social actor network theory (Callon & Law, 1989; Latour, 1987) takes this a bit further. It claims that the impact of scientific results is based on the extent to which the laboratory conditions, which gave rise to the results, are replicated in society. These conditions are the outcome of the efforts of people who have an interest in creating and maintaining them. That is, 'social actor networks' are necessary to maintain these conditions. Changing the conditions, for example, as a result of trying to reduce the use of chemical inputs in agriculture, invariably leads to resistance as actors' interests are affected (e.g. Aarts & Van Woerkum; Hamilton; Wagemans *et al.*, this volume). The transformation to sustainable agriculture is not politically neutral.

What about the impact of social science? As we have seen earlier, from a positivist point of view, social science has little to contribute, but from a constructionist view, the contributions are assessed in different terms. All forms of knowledge which claim the 'science label', social science included, are seen as special cases of social reality construction. The scientist actively constructs a fresh perspective on reality and this, in turn, affects the way others see it. Giddens (1987) has called this 'the double hermeneutic'. The behaviour of celestial bodies such as the sun and the earth presumably remain the same, whether people think the sun circles around the earth or vice versa, but people and societies do change their behaviour in the light of what science has to say about celestial movements. This also holds for the social sciences. Economics, for example, has been quite effective during the past few decades in making us believe that we are largely driven by maximization of monetary values and that society must be organized to allow the unhindered operation of market mechanisms.

The impact of constructionist thinking becomes abundantly clear if we consider what it implies for the professional field of most of the contributors to this book: extension studies (Box 1.9) (Leeuwis, 1993). Within the realist–positivist epistemology, extension is looked upon as a necessary delivery mechanism of the results of scientific research. We do not need to repeat the criticism of that perspective here (Freire, 1972; Chambers, 1983; Kline & Rosenberg, 1986; Röling, 1988; Long & Van der Ploeg, 1989). Within the constructionist epistemology, extension is a means for socially (re-)constructing agrarian reality through communication and information sharing activities. More generously and truly constructionally, extension can be seen as a societal mechanism for facilitating social learning of appropriate responses to changing circumstance. It is easy to see from this perspective how the transformation of agriculture implies an active social reconstruction of what our natural resources mean to our survival, and how to use them to support our continuing livelihood. It is equally easy to see that the transformation cannot be accomplished only on the basis of positivist science, elite expertise and on transfer of technology to farmers (Campbell, this volume).

Box 1.9 Extension

Although literally hundreds of thousands of people in the world earn their living as professional extensionists, the area of endeavour remains murky and difficult to explain. One reason is the diversity of perspectives with which extension is construed and which are often based on differences in basic assumptions about agricultural development, the role of science, and so forth.

The word 'extension' originally refers to extending scientific education beyond the walls of school or university (Van den Ban & Hawkins, 1996). Often the word 'extension education' has therefore been used. This idea is close to the French concept of 'vulgarization', making accessible scientific or other elevated thoughts to the 'vulgus' or ordinary people. Consistent with this notion is the emphasis on technology transfer from scientific research to farmer users (the ToT model) as the central mandate of extension.

But, other concepts are being used which reflect an entirely different perspective. Thus farm advisory workers, or, in German, 'Beraters', have less of an educational or transfer, and more of a consultant role. Words such as 'mobilizer' or 'facilitator', which try to avoid the implication of external imposition, go even further.

Extension, as a practice, is underpinned by a body of knowledge and accumulated experience which has, at one time, been called 'extension science' (Röling, 1988), but which perhaps can be better labelled as 'extension communication and innovation studies' to reflect a more constructionist perspective on science. Many of the contributors to this book are engaged in such studies (Box 1.2).

But, extension and innovation studies cannot be considered a discipline. Innovation, including the transition to sustainable agriculture, cannot be understood by focusing only on extension communication, but requires taking account of intentionality, culture, power, technology development, institutions, policies, and, of course, epistemology.

1.9 Soft systems thinking

Finally, it is necessary to say by way of introduction that we accept systems thinking as a necessary holistic approach to complex issues such as the sustainability of agro-ecosystems. Such issues cannot be understood by examining only the parts in isolation, nor the whole as a mere aggregation of parts. Tasks such as eco-system management require that the emergent properties of systems as wholes are taken into account (e.g. Hurthubise, 1984). 'System performance must therefore be judged not simply in terms of how each part works separately, but in terms of how the parts fit together and relate to each other, and in terms of how the system relates to its environment and to other systems in that environment' (Dillon, 1976).

Constructionism has deeply affected the thinking about systems which emerged from biology and other 'hard' sciences and was later applied in engineering. Checkland (1981; Checkland & Scholes, 1990) distinguishes between 'hard' and 'soft' systems. The former are treated as if they really exist. Their boundaries and goals are assumed to be given. Analysis and problem solving focus on goal-seeking and the best technical means to reach a goal. Such hard system thinking can be usefully applied to natural systems, such as plants, or designed systems, such as computers.

Soft systems are deliberate social constructs, that is, they exist only to the extent that people agree on their goals, their boundaries, their membership and their usefulness. The crucial assumption is that system goals are not given but contested and that system boundaries are negotiated. The necessary condition for a soft system to exist is agreement among its members on its goals. A soft system may also be defined as a human activity system, e.g. an organization, a task force, or the stakeholders in an agro-ecosystem who have been forced by environmental problems to exert joint agency at the level of social aggregation commensurate with the agro-ecosystem. Thus the agro-ecosystem is a sub-system of a human activity system, and its sustainability is an emergent property of that soft system. In this perspective, hard systems are subsumed by soft systems, just as scientific research is a special approach to the social construction of reality.

A special application of soft systems thinking is represented in what are called agricultural knowledge and information systems (AKIS) (Röling, 1988, Röling, 1990; Röling & Engel, 1991; Engel, 1995). AKIS can be used in a number of ways:

- empirically to discover how social actors in agriculture, such as scientists, advisers, farmers, but also seed suppliers, credit banks, and so on, are linked together in the creation, adaptation, sharing, storage and application of knowledge and information;
- normatively, as a mental construct, to design ideal links and flows;
- analytically to guide interventions to ensure that the actors do, in practice, interact in ways that give rise to desired emergent properties, such as innovation. System boundaries can be drawn widely, to achieve goals such as a competitive and productive, and/or indeed a sustainable agriculture,

or narrowly to achieve goals such as the production of x-litres of milk per cow.

Peter Checkland's real achievement is the development of a soft systems methodology (SSM), which allows a group of actors who are faced with a shared problem to engage in a collective learning process in order to design a human activity system that can help solve the problem through collective action. SSM has been tested extensively over a period of more than 15 years, but especially in corporate environments. At a workshop with Checkland and some of the present contributors, it was noted that sustainability problems are more complex than corporate ones with respect to the social dilemmas which arise between individual and collective interests (Ostrom, 1992; Koelen & Röling, 1994; Maarleveld, 1996) and hence require more preliminary exploration (Leeuwis, 1993). The adaptations to SSM which are required to take into account these dilemmas are discussed especially in Part IV of this book.

1.10 The structure of the book and an overview of contributions

This book originated as a collection of papers presented at a workshop at the 15th Congress of the European Rural Sociological Society in 1993. Few of the original papers have survived. The issues and experiences evolved, new and exciting perspectives and insights emerged, and interesting people presented themselves as contributors. This book thus represents a momentary stock-taking in a rapidly moving field that is generating much theoretical excitement and attracts the attention of those concerned with practical action. The book consists of five parts. This introduction opens Part I. In the next chapter, Pretty draws on the extensive work of the International Institute for Environment and Development (IIED) in London (e.g. Pretty, 1995) to summarize the main lessons learned about 'policies at work'. In his view, policies provide a crucial context which can greatly affect the realization of regenerative agriculture both in positive and negative ways. Woodhill and Röling provide theoretical and philosophical perspectives on social learning and sustainable natural resource use. The metaphor of the platform, as the social counterpart to efforts to scale up the perception of eco-systems that require integral management, was formulated for the first time in an earlier draft of this chapter written in 1993. The chapter is heavily influenced by Australia's experience with Landcare.

Part II focuses on environmental policies and farmers' reactions. Policy making is often seen as a sufficient condition for bringing about societal change, especially fiscal policy and regulatory measures. Some interesting lessons are being learned about the extent to which this is the case, especially with respect to policies which seek to promote sustainable agriculture. The chapters deal with specific country experiences, respectively in Switzerland, Greece and the Netherlands. The Swiss case, by Roux and Blum, shows how policy makers have involved stakeholders in fragile mountain landscapes in participatory norm and goal setting. This has resulted in a unique environmental policy framework and implementation system. The Greek case by Koutsouris and Papadopoulos is based on extensive empirical work that analyses EU-supported efforts to enhance the competitiveness of Greek agriculture within the European Union. What is interesting is that Greek extension policies do not seem to be influenced much by the lessons of countries whose highly competitive agricultures have been developed at huge and unsustainable environmental costs. Van Weperen, Proost and Röling examine the learning experiences of the 35 Dutch farmers engaged in an integrated arable farming innovation project. Their experience contrasts to the way in which arable farmers in general seek to satisfy the requirements of recent environmental laws. The chapter is partly based on extensive empirical research by Proost of the impact of the multi-year crop plan.

Part III looks at the processes of learning a way forward towards a more sustainable agriculture at the farm level, and at the facilitation and institutional support that is required. It begins with a study by Somers of Dutch arable farmers as they learn their way towards new practices. The learning is compared to the classic diffusion of innovations approach (Rogers, 1983). Gerber and Hoffmann describe networks of farmers, facilitators and consumers in Germany, which open fresh perspectives on the institutional supports and linkages which are emerging when new consumer values drive agricultural change. Röling and Van de Fliert describe a unique programme in Indonesia which is introducing integrated pest management in rice. Van de Fliert has written a doctoral dissertation (1993) evaluating the impact of the farmer field school approach employed by the IPM programme. Röling supervised the thesis, and has been involved in the programme as a consultant and a member of a recent World Bank mid-term review. Together, the authors describe the multi-faceted nature of the pioneering attempt to institutionalize experiential learning on a significant social scale. Hamilton's chapter is based on his experience as team leader of a very interesting and successful effort to introduce minimal tillage in Queensland, Australia, which he first elaborated in a doctoral dissertation (1995). The approach followed, i.e. the facilitation of farmers' own processes of discovery and learning, is unique and explicitly uses a constructionist epistemology. He examines with considerable insight the implications for institutional behaviour, research and extension management and for the debate among agricultural professionals about the grounds for knowing. Part III is brought to a close with an overview by Gelia Castillo, the doyenne of research on the Philippine rural and agricultural development experience. She provides an overview of user-responsive, local knowledge orientated, participatory and systems-orientated agricultural research in South Asia. She derives numerous insights into the kinds of agricultural research that can best support the transformation of agriculture.

Part IV takes the discussion to a higher agro-ecosystem level: the platform for natural resource use negotiation. It opens with a chapter by Fisk, Hesterman and Thorburn, which describes the first results of a Kellogg programme to support sustainable agriculture learning communities in the USA. They write: 'Achieving a sustainable agriculture will require integrated farming systems that involve many diverse individuals and institutions in rural communities'. Their chapter is an ideal opening chapter for Part IV. It does not directly address eco-systems at levels above the farm, but emphasizes that change at the farm level also requires community learning. Campbell, Australia's former National Landcare Facilitator (see his book, 1994) describes some of the lessons learned in trying to support this fascinating movement. Wagemans and Boerma deal with the Dutch experience of area-based land use planning, in three cases. The first author has been at the forefront of attempts to facilitate negotiations in one of the cases. The authors draw interesting conclusions about the nature of resource use negotiation. The last chapter of Part IV, by van Woerkum and Aarts, describes farmers' reactions to the Dutch government's efforts to implement its 'Nature Policy Plan' based on policy making according to electoral politics and parliamentary process.

Part V, by Röling and Jiggins, provides the grand finale. They draw together the major lessons learned in a concrete analytical framework. They provide practical guidance on the implications for facilitation methods, university teaching, research funding and inter-agency collaboration. The questions raised in Part I are answered in the light of the case material presented.

1.11 References

Bawden, R.J. & Packam, R. (1991). Systems praxis in the education of the agricultural systems practitioner. Richmond (NSW): University of Western Sidney-Hawkesbury. Paper presented at the 1991 Annual Meeting of the International Society for Systems Sciences. Östersund, Sweden.

Berger, P.L. & Luckman, T. (1967). *The Social Construction of Reality. A Treatise in the Sociology of Knowledge.* Garden City: Doubleday and Middlesex: Anchor Books.

Bohm, D. (1993). 'Last words of a quantum heretic', interview with John Morgan, *New Scientist*, **137** (1862), 27th February, 42.

Callon, M. & Law, J. (1989). On the construction of socio-technical networks: content and context revisited. *Knowledge in Society: Studies in the Sociology of Science Past and Present*, **8**, 57–83. JAI Press.

Campbell, A. (1994). *Landcare. Communities Shaping the Land and the Future.* St Leonards (Australia): Allan and Unwin.

CGIAR (1995). A Vision for CGIAR: Sustainable Agriculture for a food secure world. Ministerial-level meeting, Lucerne, Switzerland, February 9–10, 1995. *Background Documents on Major Issues*, pp. 41–76. Washington: CGIAR Secretariat.

Chambers, R. (1983). *Rural Development: Putting the Last First.* London: Longman.

Checkland, P. (1981). *Systems Thinking, Systems Practice.* Chichester: John Wiley.

Checkland, P. & Scholes, J. (1990). *Soft Systems Methodology in Action.* Chichester: John Wiley.

Collins, H.M. (1985,1992). *Changing Order: Replication and Induction in Scientific Practice.* Chicago: Chicago University Press.

Conway, G.R. (1994). Sustainability in agricultural development: trade-offs between productivity, stability and equitability. *Journal for Farming Systems Research-Extension*, **4** (2), 1–14.

De Groot, W.T. (1992). *Environmental Science Theory: Concepts and Methods in a One-world Problem-oriented Paradigm*. Amsterdam: University of Leyden, published doctoral dissertation. Amsterdam: Elsevier.

Dillon, J.D. (1976). The economics of systems research. *Agricultural Systems*, **1**, 5–22.

Engel, P. (1995). *Facilitating Innovation: An Action-oriented Approach and Participatory Methodology to Improve Innovative Social Practice in Agriculture*. Wageningen: Agricultural University, published doctoral dissertation. Commercial version in press from Royal Tropical Institute, Amsterdam.

Freire, P. (1972). *The Pedagogy of the Oppressed* (transl. M.B. Ramos). Harmondsworth: Penguin.

Funtowicz, S.O. & Ravetz, J.R. (1990). *Global Environmental Issues and the Emergence of Second Order Science*. Luxemburg: Commission for the European Community, DG Telecommunications, Information Industries and Innovation. CD-NA 12803 EN C, Report EUR 12803 EN.

Funtowicz, S.O. & Ravetz, J.R. (1994). The worth of a songbird; ecological economics as a post-normal science. *Ecological Economics*, **10**, 197–207.

Giddens, A. (1987). *Social Theory and Modern Sociology*. Cambridge: Polity Press.

Habermas, J. (1984). *The Theory of Communicative Action. Vol. 1: Reason and the Rationalisation of Society*. Boston: Beacon Press.

Habermas, J. (1985). *The Theory of Communicative Action. Vol. 2: Lifeworld and System. A Critique of Functionalist Reason*. Boston: Beacon Press.

Hamilton, N.A. (1995). *Learning to Learn with Farmers. A Case Study of an Adult Learning Extension Project Conducted in Queensland, 1990–1995*. Wageningen: Agricultural University, published doctoral dissertation.

Howard, Sir A. (1943, 1947). *An Agricultural Testament*. London: Oxford University Press.

Hurthubise, R. (1984). *Managing Information Systems: Concepts and Tools*. Hartford (Conn.): Kumarian Press.

Ison, R. (1994). *Designing Learning Systems: How can Systems Approaches be Applied in the Training of Research Workers and Development Actors?* Synthesis paper for Workshop 6 on Formation and Training of 21 submitted contributions to that subject for the International Symposium on Systems Oriented Research in Agriculture and Rural Development, Montpellier, France, 21–25 November 1994.

Jiggins, J.L.S. & De Zeeuw, H. (1992). Participatory technology development in practice: process and methods. In *Farming for the Future. An introduction to Low-External Input and Sustainable Agriculture*, ed. C. Reijntjes, B. Haverkort & A. Waters-Bayer, pp. 135–62. London: Macmillan and Leusden: ILEIA.

Kleiner, K. (1996). Life, liberty and the pursuit of vegetables. *New Scientist*, **149** (2012), January 13, 5.

Kline, S. & Rosenberg, N. (1986). An overview of innovation. In *The Positive Sum Strategy. Harnessing Technology for Economic Growth*, ed. R. Landau & N. Rosenberg, pp. 275–306. Washington DC: National Academic Press.

Knorr-Cetina, K.(1981). *The Manufacture of Knowledge: An Essay on the Constructivist and Contextual Nature of Science*. Oxford: Pergamon.

Knorr-Cetina, K. (1995). Theoretical Constructionism. On the nesting of knowledge structures into social structures. Paper presented at the Annual Meeting of the American Sociological Association, Washington, August 19–23, 1995 and at the Annual Meeting of the Society for Social Studies of Science, Charlottesville, VA, Plenary on Theoretical Foundations and Achievements in Science Studies. October 17–22, 1995. Submitted to *Sociological Theory*.

Koelen, M. & Röling, N. (1994). Sociale Dilemmas. In *Basisboek Voorlichtingskunde,* ed. N.G. Röling, D. Kuiper & R. Janmaat, pp. 58–74. Meppel: Boom.

Korthals, M. (1994). *Duurzaamheid en Democratie. Sociaal-filosofische Beschouwingen over Milieubeleid, Wetenschap en Technologie.* Meppel: Boom.

Kuhn, T.S. (1970). *The Structure of Scientific Revolutions*. 2nd edn. Chicago: University of Chicago Press.

Latour, B. (1987). *Science in Action*. Cambridge MA: Harvard University Press.

Leeuwis, C. (1993). *Of Computers, Myths and Modelling. The Social Construction of Diversity, Knowledge, Information and Communication Technologies in Dutch Agriculture and Agricultural Extension*. Wageningen: Agricultural University. Wageningse Sociologische Reeks, published doctoral dissertation.

Long, N. & J.D. van der Ploeg, J.D. (1989). Demythologising planned intervention. *Sociologia Ruralis*, **29** (3/4), 226–49.

Maarleveld, M. (1996). Improving Participation and Cooperation at the Local Level: Lessons from Economics and Psychology. Paper presented at the 9th Conference of the International Social Conservation Organization (ISCO), Towards sustainable landuse, Furthering cooperation between people and institutions, August 26–30, 1966, Bonn, Germany.

McCalla, A.F. (1994). Agriculture and Food Needs to 2025: Why We Should be Concerned. Sir John Crawford Memorial Lecture. International Centres Week, October 27, 1994. Washington DC: World Bank, CGIAR Secretariat.

Matteson, P., Gallagher, K.D. & Kenmore, P.E. (1992). Extension and integrated pest management for planthoppers in Asian irrigated rice. In *Ecology and Management of Plant Hoppers*, ed. R.F. Denno & T.J. Perfect, pp. 57. London: Chapman and Hall.

Maturana, H.R. & Varela, F.J. (1987, 1992). *The Tree of Knowledge, the Biological Roots of Human Understanding*. Boston MA: Shambala Publications.

National Science Council (1989). *Alternative Agriculture*. Washington: National Academy Press.

Ostrom, E. (1990, 1991, 1992). *Governing the Commons. The Evolution of Institutions for Collective Action*. New York: Cambridge University Press.

Pain, S. (1994). 'Rigid' cultures caught out by climate change. *New Scientist*, 5 March 1994.

Ponting, C. (1991). *A Green History of the World*. London: Sinclair-Stevenson Ltd.

Pretty, J. (1995). *Regenerating Agriculture. Policies and Practice for Sustainability and Self-Reliance*. p. 20. London: Earthscan.

Reijntjes, C., Haverkort, B. & Waters-Bayer, A. (1992). *Farming for the Future. An Introduction to Low-External Input and Sustainable Agriculture*. London: Macmillan and Leusden: ILEIA.

Rogers, E.M. (1961,1972,1983). *Diffusion of Innovations*. New York: Free Press.

Röling, N. (1988). Extension science. *In Information Systems in Agricultural Development*. Cambridge: CUP.

Röling, N. (1990). The agricultural research–technology transfer interface: a knowledge system perspective. In *Making the Link. Agricultural Research and Technology Transfer in Developing Countries*, ed. D. Kaimowitz, pp. 1–42. Boulder, Co: Westview Press, Special Studies in Agricultural Science and Technology.

Röling, N. (1995). Naar een Interactieve Landbouwwetenschap. Wageningen: Agricultural University. Inaugural Address at the occasion of his installment as Extraordinary Professor of Agricultural Knowledge Systems in Developing Countries.

Röling, N. & Engel, P.G.H. (1991). The development of the concept of agricultural knowledge and information systems (AKIS): implications for extension. In *Agricultural Extension: Worldwide Institutional Evolution and Forces for Change*, ed. W. M. Rivera & D.J. Gustafson, pp. 125–38. Amsterdam: Elsevier Science Publishers.

Ruttan, V. & Hayami, Y. (1984). Toward a theory of induced institutional innovation. *The Journal of Development Studies*, **20**(4), 203–23.

Sriskanadarajah, N., Bawden, R.J. & Packam, R.G. (1989). System Agriculture: A Paradigm for Sustainability. Paper presented at the Ninth Annual Farming Systems Research/Extension Symposium, University of Arkansas, Fayetteville, Arkansas, USA, October 9–11, 1989. *AFSRE Newsletter*, **2**(3),1–5, 1991.

Tansey, G. & Worsley, T. (1995). *The Food System, A Guide*. London: Earthscan.

Tate, J. (1995). Statement as Member of Panel. Wageningen: International Congress, Agrarian Questions, International Agricultural Centre, May 22–24, 1995. Public discussion on: 'The Social Shaping of Bio-science: Public Participation in Debates about Biotechnology', organized by P. Richards, Chair of Joint Wageningen/London Group on Technology and Agrarian Development.

Van den Ban, A.W. & Hawkins, S. (1988, 1996). *Agricultural Extension*. London: Longman.

van de Fliert, E. (1993). *Integrated Pest Management. Farmer Field Schools Generate Sustainable Practices: A Case Study in Central Java Evaluating IPM Training*, WU Papers 93–3. Wageningen: Agricultural University, published doctoral dissertation.

Vartdall, B. (1995). Farmers' approaches to ecological agriculture. *IFOAM Ecology and Farming*, May 1995: pp. 20–2.

Woodhill, J. (1993). Science and the facilitation of social learning: a systems perspective. Paper for the 37th Annual Meeting of the International Society for the Systems Sciences, University of Western Sydney.

2 Supportive policies and practice for scaling up sustainable agriculture

JULES N. PRETTY

2.1 Challenges for agricultural development

As the century draws to a close, agricultural development faces some unprecedented challenges. By the year 2020, the world will probably have 2.5 billion more people than today. Today, even though enough food is produced in aggregate to feed everyone, and world prices have been falling in recent years, some 700–800 million people still do not have access to sufficient food. This includes 180 million children who are underweight and suffering from malnutrition. Cereals stocks are at the lowest levels for 20 years, and neither rice nor wheat yields per ha. have increased in the past 5 years.

Recent models constructed to investigate agricultural production and food security changes over the next quarter to half century all conclude that food production will have to increase substantially (IFPRI, 1995; Crosson & Anderson, 1995; Leach, 1995; CGIAR, 1994; FAO, 1993*a*, 1995).

The views on how to proceed, however, vary hugely. Some are optimistic, even complacent; others are darkly pessimistic. Some indicate that not much needs to change; others argue for fundamental reforms to agricultural and food systems. Some indicate that a significant growth in food production will occur only if new lands are taken into cultivation; others suggest that there are feasible social and technical solutions for increasing yields on existing farmland.

There are five distinct schools of thought for future options in agricultural development (for summaries, see McCalla, 1994; Hazell, 1995; Hewitt & Smith, 1995; Pretty, 1995*a*; IIED, 1996).

2.1.1 Optimists and complacents

The optimists say that, in free market conditions, supply will always meet increasing demand. Recent growth in aggregate food production will continue alongside continuing reductions in the rate of population growth, achieved over the recent decades (Rosegrant & Agcaolli, 1994; Mitchell & Ingco, 1993; FAO, 1993*b*).

As food prices are falling (down 50% in the past decade for most commodities), this indicates that there is no current crunch over demand. Food production is expected to grow for two reasons: (i) the fruits of biotechnology research will soon ripen, so boosting plant and animal productivity; (ii) the area under cultivation will expand, probably by some 20–40% by 2020 (this means an extra 79 million ha in

Sub-Saharan Africa alone). It is also expected that developing countries will substantially increase food imports from industrialized countries (perhaps by as much as five-fold by 2050).

2.1.2 Environmental pessimists

The environmental pessimists suggest that ecological limits to expanded agricultural production are being approached, are soon to be passed, or have already been reached (Harris, 1995; Brown, 1994; CGIAR, 1994; Kendall & Pimentel, 1994; Brown & Kane, 1994; Ehrlich, 1968).

Yield increase of the major cereals has slowed, and will slow more, stop or even fall, particularly because of growing production constraints in the form of resource degradation (soil erosion, land degradation, forest loss, pesticide overuse, fisheries over-exploitation). Dietary shifts, in particular the increasing consumption of livestock products, are an emerging threat, as this results in the consumption by feedstock of an even greater share of cereal products. This lobby does not believe that new technological breakthroughs are likely. Increasing numbers of people are seen as a key problem, so population control has to be a priority policy concern.

2.1.3 Industrialized world to the rescue

This group believes that 'Third World' countries will never be able to feed themselves, for a wide range of ecological, institutional and infrastructural reasons. A food gap will emerge between temperate and tropical countries which will have to be filled by modernized agriculture in the North (Avery, 1995; Wirth, 1995; DowElanco, 1994; Carruthers, 1993; Knutson *et al.*, 1990).

Increased production in large, mechanized operations will allow smaller and more 'marginal' farmers to go out of business, so taking the pressure off natural resources. These can then be conserved in protected areas and wildernesses. These large producers will then be able to trade their food with those who need it, or have it distributed as famine relief or food aid. It is also vigorously argued that any adverse health and environmental consequences of chemically based agricultural systems are minor in comparison with those wrought by the expansion of agriculture into new lands. External inputs (especially pesticides and fertilizers), therefore, are crucial for feeding the world (see Avery, 1995, in particular). The problems of lack of effective demand among displaced small farmers, or lack of employment for the millions who would leave the land in this scenario, tend to be left unaddressed by this group.

2.1.4 New modernists

Another group, who we might call the new modernists, argue that biological yield increases are possible on existing lands, but, further, that this food increase can come only from high-external input farming (Borlaug, 1992, 1994*a,b*;

Sasakawa Global 2000, 1994, 1995; World Bank, 1993; Waggoner, 1994; Paarlberg, 1994; Winrock International, 1994; Crosson & Anderson, 1995).

The target includes the high potential lands already used in industrial agriculture, the existing Green Revolution land in tropical countries, and the 'high-potential' lands that have been missed by the past 30 years of agricultural development. This group argues that farmers use too few fertilizers and pesticides, which are said to be the simplest and most effective and efficient way to improve yields, and so keep the pressure off natural habitats. It is argued that science-based high-input agriculture is more environmentally sustainable than low-input, as low-input agriculture, it is claimed, can only produce low output.

2.1.5 Sustainable intensification

The case is also being made by others for the benefits of *sustainable intensification*, on the grounds that substantial growth is possible in currently unimproved or degraded areas whilst at the same time protecting or even regenerating natural resources (Pretty, 1995*a*,*b*; Hazell, 1995; McCalla, 1994, 1995; Scoones & Thompson, 1994; NAF, 1994; Hewitt & Smith, 1995).

It is argued that the empirical evidence indicates that regenerative and low-input (but not necessarily zero-input) agriculture can be highly productive. The effects are sustainable, provided farmers participate fully in all stages of technology development and extension. The evidence also suggests that agricultural and pastoral land productivity is as much a function of human capacity and ingenuity as it is of biological and physical processes.

2.2 What is and what is not sustainable agriculture

2.2.1 Defining sustainability

A great deal of effort has gone into trying to define sustainability in absolute terms. Since the Brundtland Commission's definition of sustainable development in 1987, there have been at least 80 more definitions constructed, each emphasizing different values, priorities and goals.

Precise and absolute definitions of sustainability, and therefore of sustainable agriculture, are impossible. Sustainability itself is a complex and contested concept. To some, for example, it implies persistence and the capacity of something to continue for a long time. To others, it implies not damaging or degrading natural resources.

In any discussion of sustainability, it is important to clarify what is being sustained, for how long, for whose benefit and at whose cost, over what area and measured by what criteria. Answering these questions is difficult, as it means assessing and trading off values and beliefs.

It is critical therefore, that sustainable agriculture does not prescribe a concretely defined set of technologies, practices or policies. This policy would only serve to restrict the future options of farmers. As conditions change and as knowledge changes, so must farmers and communities be encouraged and allowed to change and adapt. Sustainable agriculture is not a simple model or package to be imposed. It is more a process for learning (Pretty, 1995*b*; Röling, 1994).

2.2.2 Goals for sustainable agriculture

In our view, the basic challenge for sustainable agriculture is to make better use of available physical and human resources. This can be done by minimizing the use of external inputs, by regenerating internal resources more effectively, or by combinations of both. This ensures the efficient and effective use of what is available, and ensures that any changes will persist as dependencies on external systems are kept to a reasonable minimum.

A sustainable agriculture therefore, is any system of food or fibre production that pursues the following farming objectives systematically:

- A thorough incorporation of natural processes such as nutrient cycling, nitrogen fixation, and pest–predator relationships into agricultural production processes, so ensuring profitable and efficient food production;
- A reduction in the use of those external and non-renewable inputs with the greatest potential to damage the environment or harm the health of farmers and consumers, and a more targeted use of the remaining inputs used with a view to minimizing costs;
- The full participation of farmers and rural people in all processes of problem analysis, and technology development, adaptation and extension;
- A more equitable access to productive resources and opportunities, and progress towards more socially just forms of agriculture;
- A greater productive use of local knowledge and practices, including innovative approaches not yet fully understood by scientists or widely adopted by farmers;
- An increase in self-reliance amongst farmers and rural people;
- An improvement in the match between cropping patterns and the productive potential and environmental constraints of climate and landscape to ensure long-term sustainability of current production levels.

Sustainable agriculture seeks the integrated use of a wide range of pest, nutrient, and soil and water management technologies. It aims for an increased diversity of enterprises within farms combined with increased linkages and flows between them. By-products or wastes from one component or enterprise become inputs to another. As natural processes increasingly substitute for external inputs, so the impact on the environment is reduced.

2.2.3 The potential for sustainable intensification

There is emerging evidence that regenerative and resource-conserving technologies and practices can bring both environmental and economic benefits for farmers, communities and nations. The best evidence comes from countries of Africa, Asia and Latin America, where the concern is to increase food production in the areas where farming has been largely untouched by the modern packages of externally supplied technologies. In these lands, farming communities adopting regenerative technologies have substantially improved agricultural yields, often using few or no external inputs (Bunch, 1990, 1993; GTZ, 1992; UNDP, 1992; Krishna, 1994; Shah, 1994; SWCB, 1994; Balbarino & Alcober, 1994; de Freitas, 1994; Pretty, 1995*a*).

The sustainable agriculture programme of IIED has examined the extent and impact of sustainable agriculture in a selected number of countries, and used this empirical evidence to estimate sustainable agriculture's potential contribution to global food production (Pretty, Thompson & Hinchcliffe, 1996). Whilst we were aware that many projects and programmes had improved agricultural yields, these data have never been collated in one place.

In the 20 countries examined, it was found that there are some 1.93 million households farming 4.1 million hectares with sustainable agriculture technologies and practices. Most of these improvements have occurred in the past 10 years (many in the past 2–5 years). The greatest increases following a transition to sustainable agriculture are in rainfed agriculture in the lowest yield countries, where the average new yields for wheat, maize and sorghum–millet are of the order of double the yields of conventional or pre-sustainable agriculture.

These are not the only sites for successful sustainable agriculture, however. In high-input and generally irrigated lands, farmers adopting regenerative technologies have maintained yields whilst substantially reducing their use of inputs (Bagadion & Korten, 1991; Kenmore, 1991; van der Werf & de Jager, 1992; UNDP, 1992; Kamp, Gregory & Chowhan, 1993; Pretty, 1995*a*). And in the industrialised countries, increasing numbers of farmers are demonstrating that it is possible to maintain profitability, even though input use has been cut dramatically (Liebhart *et al.*, 1989; NRC, 1989; Hanson *et al.*, 1990; Faeth, 1993; NAF, 1994; Hewitt & Smith, 1995); and in Europe (El Titi & Landes, 1990; Vereijken, 1990; Jordan, Hutcheon & Glen, 1993; Pretty & Howes, 1993; Reus, Weckseler & Pak, 1994; Somers, this volume).

This empirical evidence is still contested, however. In the USA, some 82% of conventional US farmers believe that low-input agriculture will always be low output (Hewitt & Smith, 1995). Two influential politicians have recently emphasized these beliefs. In 1991, the former Secretary of State, Earl Butz, said

> we can go back to organic agriculture in this country if we must – we once farmed that way 75 years ago. However, before we move in that direction, someone must decide which 50 million of our people will starve. We simply cannot feed, even at subsistence levels, our 250 million Americans without a large production input of chemicals, antibiotics and growth hormones (Byrnes & Liebhart, 1993).

In 1996, Under-Secretary for Agriculture, Eugene Moos, said:

> The prospective increase in world population will double food aid needs in the next decade... and it will be necessary for agricultural producing nations to use biotechnology and hormones to meet growing demand (*Farmers Weekly*, 1996).

Yet recent data show that some 40 000 American farmers in 32 states are using regenerative technologies and have cut their use of external inputs substantially. This includes 2800 farmers in the North Western States, who grow twice as many crops compared with conventional farmers, use 60–70% less fertilizer, pesticide and energy, and yet their yields are roughly comparable. They also spend more money on local goods and services – each farm contributed more than £13 500 to its local economy (NAF, 1994).

Most successes, though, are still localized. This is partly because favourable policy environments are missing. Most policies still actively encourage farming that is dependent on external inputs and technologies. It is these policy frameworks that are one of the principal barriers to a more sustainable agriculture.

2.3 The spread and scaling up of sustainable agriculture

2.3.1 Why we should be concerned with spread

Despite the increasing number of successful sustainable agriculture initiatives in different parts of the world, it is clear that most of these are still only 'islands of success'. There remains a huge challenge to find ways to spread or 'scale up' the processes which have brought about these transitions.

Sustainability ought to mean more than just agricultural activities that are environmentally neutral or positive; it implies the capacity for activities to spread beyond the project in both space and time. A successful project that leads to improvements that neither persist nor spread beyond the project boundary should not be considered sustainable.

When the recent record of development assistance is considered, it is clear that sustainability has been poor. There is a widespread perception amongst both multilaterals and bilateral aid agencies that agricultural development is difficult, that agricultural projects perform badly, and that resources may best be spent in other sectors. Reviews by the World Bank, the EC, Danida and DFID (formerly ODA) have shown that agricultural and natural resource projects performed worse in the 1990s than in the 1970s–1980s and worse than projects in other sectors (World Bank, 1993; Pohl & Mihaljek, 1992; EC, 1994; Danida, 1994; Dyer & Bartholomew, 1995). They are also less likely to maintain achievements beyond the provision of aid inputs. A recent analysis of 95 agricultural project evaluations logged on the DAC–OECD database also shows a disturbing rate of failure, with at least 27% of projects having non-sustainable structures, practices or institutions, and 10% causing significant negative environmental impact (Pretty & Thompson, 1996).

This empirical evidence of completed agricultural development projects suggest four important principles for sustainability and spread:

1. *Imposed technologies do not persist*: if coercion or financial incentives are used to encourage people to adopt sustainable agriculture technologies (such as soil conservation, alley cropping, IPM), then these are not likely to persist.

2. *Imposed institutions do not persist*: if new institutional structures are imposed, such as co-operatives or other groups at local level, or project management units and other institutions at project level, then these rarely persist beyond the project.
3. *Expensive technologies do not persist*: if expensive external inputs, including subsidised inputs, machinery or high technology hardware are introduced with no thought to how they will be paid for, they too will not persist beyond the project.
4. *Sustainability does not equal fossilisation or continuation of a thing or practice forever*: rather it implies an enhanced capacity to adapt in the face of unexpected changes and emerging uncertainties.

2.3.2 The comprehensive technology package

Modern agricultural development has begun with the notion that there are technologies that work, and it is just a matter of inducing or persuading farmers to adopt them. Yet few farmers are able to adopt whole packages of technologies without considerable adjustments in their own practices and livelihood systems, which they might not wish or be able to do.

New technological models may look good at first, and then fade away. For instance, alley cropping, an agroforestry system comprising rows of nitrogen-fixing trees or bushes separated by rows of cereals, has long been the focus of research (Kang, Wilson & Lawson, 1984; Lal, 1989). Many productive and sustainable alley cropping systems, needing few or no external inputs, have been developed. They stop erosion, produce food, fodder and wood, and can be cropped over long periods, but the problem is that very few, if any, farmers have adopted these alley cropping systems as designed. Despite millions of dollars of research expenditure over many years, the systems that have been produced appear suitable only for research stations (Carter, 1995).

There has been success, however, where farmers have been able to take one or two components of alley cropping, and then adapt them to their own farms. In Kenya, for example, farmers planted rows of leguminous trees next to field boundaries, or single rows through their fields in Rwanda. Trees planted in alleys by extension workers soon became dispersed through fields (Kerkhof, 1990).

The prevailing view, however, tends to be that it is farmers who should adapt to the requirements of the technology. Of the Agroforestry Outreach Project in Haiti, it was said that

> Farmer management of hedgerows does not conform to the extension program... Some farmers prune the hedgerows too early, others too late. Some hedges are not yet pruned at two years of age, when they have already reached heights of 4–5 metres. Other hedges are pruned too early, mainly because animals are let in or the tops are cut and carried to animals... Finally, it is very common for farmers to allow some of the trees in the hedgerow to grow to pole size (Bannister & Nair, 1990).

Farmers were clearly adapting the technology to suit their own needs, but were criticized for not carrying out 'correct' management procedures.

In Laos, one project used food-for-work to encourage shifting agriculturalists to settle and adopt contour farming with bench terraces (Fujisaka, 1989). These fields became so infested with weeds, however, that farmers were forced to shift to new lands, and the structures were so unstable in the face of seasonal rains that they led to worsened gully erosion. Farmers then refused to do further work when the incentives were terminated.

What does this mean for sustainable agriculture? How should we proceed so as to ensure farmers are fully involved in developing and adapting regenerative technologies?

2.4 The three steps of sustainability

One option is to see sustainable agriculture as a series of steps along a pathway that never reaches a final goal. The achieved sustainable agriculture is often conceived as requiring a sudden transition, a one-off shift in practices and values. But it is argued here that a more accurate and effective metaphor is that of taking the first small steps along a new pathway.

Step 1: Improved economic and environmental efficiency

This comprises 'precision' farming using existing conventional technologies: including targeted inputs, 'wise-use' of inputs, patch spraying, deep placement and slow-release fertilizers, use of global positioning systems (GPS) and satellite mapping, low volume and minimal dose pesticides, soil testing, weed maps, no-till or non-inversion farming, mechanical and weed harrowing, and pest and disease-resistant crops.

The approach is based on modifying conventional systems in order to reduce consumption of inputs, so that wastes and adverse environmental impacts are substantially reduced. Existing values and rights are not fundamentally challenged, and the goals of farming remain mainly the same.

Step 2: Integration of regenerative technologies

This comprises incorporation of alternative pesticides (biological, bacterial and viral); habitats manipulated to encourage predators; natural enemies released or augmented; animals (e.g. sheep, goats, cattle, pigs, fish) integrated into the system; crops and trees incorporated that fix nitrogen (legumes, green manures, cover crops); and technologies that conserve and collect soil and water.

These regenerative technologies are introduced to make best use of all locally available biological and human resources. Some technologies are dropped (e.g. pesticides) and new system components added. There is more reliance on management skills and knowledge, requiring shifts in attitude and values among farmers.

The capacity of farmers and their communities to promote, experiment with and adapt technologies and processes is not enhanced.

Step 3: Re-design with communities

This comprises resource-conserving technologies that are fitted to time and place, and varied adaptively by farmers. Agriculture is seen as an area-based activity (e.g. in communities, catchments, watersheds, landscapes); and local groups and institutions are strengthened in order to undertake natural resource management and financial resource management.

Attitudes and values are now completely different to those associated with conventional farming. New philosophies emerge to support agriculture. External institutions are reformed as facilitators and enablers of local change. Sustainable agriculture systems become self-regulatory, with the ecological and economic landscape diversified. Local people have self-reliance and cohesion – they have capacity to plan, experiment, problem-solve and interlink with other communities so that solutions spread. Agricultural inputs, practices and outputs are now structured to emphasize local economic regeneration. This step will need supportive policies if it is to be spread over whole nations.

2.5 Enhancing farmers' capacity to innovate

2.5.1 Agricultural recuperation in Honduras and Guatemala

Important evidence of the feasibility of taking Step 3 comes from a variety of soil conservation and agricultural regeneration programmes in Central America (Bunch & López, 1994). The Guinope (1981–1989) and Cantarranas (1987–1991) programmes in Honduras and the San Martin Jilotepeque programme in Guatemala (1972–1979) were collaborative efforts between World Neighbors and other local agencies. All began with a focus on soil conservation in areas where maize yields were very low (400 to 660 kg/ha), and where shifting cultivation, malnutrition, and out migration prevailed. All show the importance of developing resource-conserving practices in partnership with local people.

There were several common elements. All forms of paternalism were avoided, including giving things away, subsidizing farmer activities or inputs, or doing anything for local people. Each started slowly and on a small scale, so that local people could participate meaningfully in planning and implementation. They used technologies such as green manures, cover crops, contour grass strips, in-row tillage, rock bunds and animal manures, that were appropriate to the local area, and which were finely tuned through experimentation by and with farmers. Extension and training was done largely by villager farmers who had already experienced success with the technologies on their own farms.

Each programme substantially improved agricultural yields, increasing output per area of land from some 400–600 kg/ha to 2000–2500 kg/ha. Altogether, improvements have been made in some 121 villages. Over time, soils were not simply conserved but regenerated, with depth increases from 0.1 metres to 0.4 – 1.3 metres not uncommon.

These programmes have helped also to regenerate local economies. Land prices and labour rates are higher inside the project areas than outside. There are local housing booms, and families have moved back from capital cities. There are also benefits to the forests. Farmers say they no longer need to cut the forests, as they have the technologies to farm permanently the same piece of land. Before the programmes, national park authorities sought to keep villagers out of the forests; now there is no such concern since the forests are no longer threatened.

2.5.2 Farmer-induced changes in Central America

There are few published studies that give evidence of impacts years after outside interventions have ended. In 1994, however, staff of the Honduran organisation COSECHA (Associación de Consejeros una Agricultura Sostenible, Ecológica y Humana) returned to three programme areas and used participatory methods with local communities to evaluate subsequent changes (Bunch & López, 1994).

They first divided all 121 villages into three categories, according to where local people felt there had been good, moderate and poor impact. Twelve villages were sampled from these: four from each programme (one of the best, two of the moderate and one of the poor). These villages had some 1000 families (with a range of 30 to 180 per village). The first major finding was that crop yields and adoption of conserving technologies had continued to grow since project termination (Table 2.1).

Surprisingly, though, many of the technologies known to be 'successful' during the project had been superseded by new practices. Had the original technologies been poorly selected? It would appear not, as many that had been dropped by farmers are still successful elsewhere. The explanation would appear to be that changing external and internal circumstances had reduced or eliminated their usefulness. Circumstances included chances in markets, droughts, diseases, insect pests, land tenure, labour availability, and political disruptions.

Altogether, some 80–90 successful innovations were documented in the 12 villages. In one of the sample villages, Pacayas, there had been 16 innovations, including four new crops, two new green manures, two new species of grass for contour barriers in vegetables, chicken pens made of king grass, marigolds for nematode control, use of lablab and velvet bean as cattle and chicken feed, nutrient recycling into fishponds, human wastes in composting latrines, napier grass to stabilize cliffs, and home-made sprinklers for irrigation.

Technologies had been developed, adopted, adapted and dropped. The study concluded that the half-life of a successful technology in these project areas is 6 years. Quite clearly, many of the technologies themselves have not persisted. But, as Bunch and López (1994) put it *'what needs to be made sustainable is the process of innovation itself'*.

Table 2.1. *Changes in adoption of resource-conserving technologies, maize yields, and migration patterns in three programmes in Central America during and after projects*

	At Initiation	At termination[a]	In 1994
No. of farmers with technologies			
Contour grass barriers	1	192	280
Contour drainage ditches	1	253	239
Contour rows	0	100	245
Green manures	0	35	52
Crop rotations	12	209	254
No burning fields or forests	2	160	235
Organic matter as fertilizer	44	195	397
Yields of maize (kg/ha)			
1. San Martin, Guatemala (1972–79)	400	2500	4500
2. Guinope, Honduras (1981–89)	600	2400	2730
3. Cantarranas, Honduras (1987–1991)	660	2000	2050
Migration (number of households)			
1. San Martin			
San Antonio Correjo	65	nd	4
Las Venturas	85	nd	4
2. Guinope: three villages	38	0	(2)[b]
3. Cantarranas: three villages	nd	10	(6)[b]

Notes:
[a] Termination dates were: San Martin 1979; Guinope 1989; Cantarranas 1991.
[b] (2) and (6) refer to negative outmigration, i.e. families returning to their villages.
nd=no data.
Source: Bunch & López, 1994.

2.5.3 Farmer adaptations in India, Australia and Thailand

A similar picture has emerged in Gujarat, where many farmers have developed new technical innovations after support from the Aga Khan Rural Support Programme to undertake simple conservation measures. Farmers have introduced planting of grafted mango trees and bamboo by the embankments, making full use of the residual moisture near the embankment gully traps. They have also introduced cultivation of vegetables, such as brinjal and lady's finger, other leguminous crops, and tobacco in newly created silt traps. This has increased production substantially, particularly in poor rainfall years, as well as diversifying production. Most of these innovations and adaptations have been introduced and sustained with support from the local network of village extensionists (Shah, 1994).

Table 2.2. *The changing phases in the Thai–German Highland Development Project: the case of 113 villages in Nam Lang*

	Period	Phase
1.	1987–1990	Case incentives and free inputs high adoption, but little or no adaptation of technologies *Adoption: withdrawal= 5:1*
2.	1991–92	All incentives stopped; beginning of participatory work adoption rates fell to 25% of phase I withdrawal increased immediately by 3 fold *Adoption: withdrawal= 1:2.2*
3.	1993–94	Participatory village planning; communities fully involved adopters and withdrawers now equal *Adoption: withdrawal= 1:1*
4.	1995–96	Adopters increasing; farmers adapting technologies and diversifying, e.g. pineapple strips, lemon grass, cash crops, soil and water conservation *Adoption: withdrawal= 3:1*

Source: Steve Carson, pers. comm. 1996.

In south Queensland, Australia, extensionists from the Department of Primary Industry, using very simple learning tools that enabled farmers to learn about the impact of rainfall on their soil, have encouraged more than 80% of farmers to adopt conservation technologies. Many of these have gone on to adapt and develop new and different technologies for their own farms (Hamilton, 1995).

Another example comes from Thailand, where the four different phases of the Thai-German Highland Development Project clearly illustrate the importance of genuine participation with local people (TG-HDP, 1995; Steve Carson, pers. comm. 1996). The project has been working with upland communities in Northern Thailand to support the transition towards sustainable agriculture. The resource-conserving technologies developed and adapted for local use include hedgerows on contours, buffer strips, new crop rotations, IPM, crop diversification, and livestock integration.

The approach, however, has changed significantly since the mid-1980s (Table 2.2). In the first phase, cash incentives and free inputs were used to encourage adoption of these technologies. Adoption rates were high, although there was little or no adaptation of the technologies by the farmers. In 1990, all the incentives were stopped and the project adopted a participatory approach. Immediately adoption rates fell and withdrawal from project activities increased three-fold. By 1993–94, participatory village planning had fully involved the communities, and the ratio of adopters to withdrawers was equal. Most recently, the number of farmers using sustainable technologies has grown rapidly. Crucially, they are now actively adapting and innovating new technologies to satisfy their particular needs (Steve Carson, pers. comm. 1996).

2.6 The policy environment

2.6.1 Policy discrimination against sustainable agriculture and farmers' learning

Most, if not all, of the policy measures used to support agriculture currently act as powerful disincentives against sustainability. In the short term, this means that farmers switching from high-input to resource-conserving technologies can rarely do so without incurring some transition costs. In the long term, it means that sustainable agriculture will not spread widely beyond existing localised successes.

The principal problem is that policies simply do not reflect the long-term social and environmental costs of resource use. The external costs of modern farming, such as soil erosion, health damage or polluted ecosystems, are not incorporated into individual decision-making by farmers. In this way, resource-degrading farmers bear neither the costs of damage to the environment or economy, nor those incurred in controlling the polluting or damaging activity.

In principle, it is possible to imagine pricing a measure of the 'clean unpolluted environment', which is presently a free input to farming. If charges were levied in some way, then degraders or polluters would have higher costs, would be forced to pass them on to consumers, and would be forced to switch to more resource-conserving technologies. This notion is contained within the Polluter Pays Principle, a concept used for many years in the non-farm sector (OECD, 1989). However, beyond the notion of encouraging some internalization of costs, it has not proven to be of practical use for policy formulation in agriculture.

In general, farmers are entirely rational within the present policy environment, in continuing to use high-input degrading practices. High prices for particular commodities, such as key cereals, have discouraged mixed farming practices. In the USA, commodity programmes have inhibited the adoption of resource conserving practices by artificially making them less profitable to farmers (Faeth, 1993). In Pennsylvania, the financial returns to continuous maize monoculture and to alternative rotations are about the same. But the continuous maize attracts about twice as much direct support in the form of deficiency payments. Yet, continuous maize farms use much more nitrogen fertilizer, erode more soil and cause three to six times as much damage to off-site resources. Clearly, a transition to the resource conserving rotations would benefit both farmers and the national economy.

In the context of systemic support for high-input agriculture, many countries have sought to 'bolt-on' conservation goals to existing policies. These have tended to rely on conditionality, such as 'cross-compliance', whereby farmers receive support only if they adopt certain types of resource-conserving technologies and practices. Such cross-compliance occurs widely in the South too. The process of agricultural modernization has widely involved encouraging farmers to adopt modern practices through the linkage of credit or other benefits to adoption of modern inputs and practices. If farmers wish to receive one type of support, they must adopt a particular set of technologies and practices. In many cases, heavy coercion has been used to achieve levels of adoption needed or expected.

The problem with all cross-compliance strategies is that they do not necessarily buy the support of farmers and rural people. They can create long-term resentment, lead to perverse outcomes when policies change or the money runs out, and distort the processes of adaptive response to changing circumstances.

2.6.2 Recent policy initiatives in the North

The most common approach to policy reform has been to introduce taxes and input levies on fertilizers and pesticides. Some of this revenue is used to subsidise exports (such as in Finland); to support further the input reduction programme (such as in Sweden, where $3–3.5 million are raised each year); to support research into alternative agriculture (such as in Iowa and Wisconsin), or to return resources to farmers in the form of income support (such as in Norway). There are proposals to introduce similar taxes in Belgium, Denmark, the Netherlands and Switzerland. It is generally felt, however, that the levels have been set too low to affect consumption significantly.

But other factors may be important, including revised advice from agricultural ministries, growing public concern over high rates of chemical application, and general changes in cropping practices. Other policy measures such as new regulations, training programmes, provision of alternative control measures and reduced price support, have also contributed to substantial reductions in input use in recent years. In Sweden, pesticide consumption was cut by half between 1985 and 1990; in Austria there has been a decline in consumption of fertilizers, especially potassium; and in Denmark, there has been a fall in consumption of pesticides by 18% between 1986 and 1991 (Beaumont, 1993; Pretty, 1995*b*).

Levies have been used to limit pollution in the livestock sector in the Netherlands. Levy payments are used to penalize those farmers producing more livestock waste than their land can absorb. The Netherlands has also introduced a levy on manufactured feed to help pay for research and advisory services dealing with pollution from livestock waste.

Several countries have now set ambitious national targets for the reduction of input use. Sweden aims to reduce nitrogen consumption by 20% by the year 2000, The Netherlands is seeking a cut pesticide use of 50% by the year 2000 as part of its 'multi-year plan for crop protection'. The cost of this reduction programme has been estimated at $1.3 billion, most of which will be raised by levies on sales. Denmark is aiming for a 50% cut in its pesticide use by 1997, a plan which relies on advice, research and training, with no taxes or levies, and Canada is aiming for a 50% reduction in pesticide use by 2000 (Quebec) and 2002 (Ontario). In the USA, the Clinton administration announced in 1993 a programme to reduce pesticide use whilst promoting sustainable agriculture. The aim is to see IPM programmes on 75% of the total area of farmland by the year 2000.

Supplemented by other policy measures, such as new regulations, training programmes, provision of alternative control measures and reduced price support, there have been some substantial reductions in input use in recent years. In Sweden, pesticide consumption was cut by 60% between 1985 and 1993 (from 4500 to 1500 tonnes a. i.); in Austria there has been a decline in consumption of fertilizers, especially potassium; in Denmark, there has been a fall in consumption of pesticides by 39% between 1985 and 1993 (from 7000 to 4300 tonnes a. i.); and in the Netherlands, there has been a 43% fall between 1985 and 1993 (from 20 000 to 11 300 tonnes a. i.) (Matteson, 1995; Beaumont, 1993).

The alternative to penalizing farmers is to encourage them to adopt alternative low or non-polluting or non-degrading technologies by acting on subsidies, grants, credit or low-interest loans. These could be in the form of direct subsidies for low input systems or the removal of subsidies and other interventions that currently work against alternative systems. Acting on either would have the effect of removing distortions and making the low input options more attractive.

Such initiatives are much rarer, and tend to be fragmented, with support to farmers offered mainly for maintaining landscapes or habitats. An exception is the MEKA project in Baden Würtemberg, Germany, where the principle is to pay farmers not to damage the environment. The scheme is voluntary and open to all farmers, who are able to choose the aspects of the scheme with which they wish to comply. They then receive payments on a points system. Each point brings them DM20 per hectare. Already 55 000 farmers have signed up, with the result that 43 000 hectares are now farmed with no fertilizers or pesticides; 207 000 ha have grass undersown in cereals to prevent autumn soil erosion; and on 25 000 ha rare breeds of animals are kept.

2.6.3 Recent policy initiatives in the South

Possibly the best example of an integrated National Conservation Strategy in the South comes from Pakistan (EUAD, 1992). It covers an impressive range and depth of issues relating to sustainable development, as well as prescribing a wide range of institutional and societal processes. These include administrative matters, communication, education, legislation, grassroots organisation, and research and development roles.

It also sets out indicators against which national and local progress can be measured. What is important is that the NCS recognizes that it is not the plan or document that is important. Rather, it is enhanced communication and linkages between institutions and actors, the increased mutual understanding, and the development of a national network of people directly committed to continuing the process. These are all valuable resources for sustainable development.

The Group Farming initiative of the Kerala State government in India is another good example of how coordinated action within the agricultural sector can have a significant impact on farming practice (Sherief, 1991). Land reform in the 1970s led to the formation of a new class of small farmers, with some 70% owning less than two hectares, but as the costs of inputs, pest and disease control spiralled, so the area under rice fell from 810 000 to 570 000 ha between 1975 and 1988. Small farmers had been unable to adopt the whole technological package recommended.

In 1989, the group farming for rice programme was launched. Local committees comprising all rice farmers were formed, and given the task of charting a detailed farming plan. Group activities, such as for water management and labour operations, are jointly agreed. Costs are reduced through community rice nurseries; fertilizer application on the basis of soil testing; the introduction of IPM and minimum use of pesticides; and the formation of plant protection squads. The average cost reduction to farmers has been Rs 1000[1] per hectare, and rice yields have improved by 500 kg/ha. As Sherief (1991) put it, *'instead of pampering the cultivators with subsidies, stress was given to self-reliance and timely action for all agricultural operations'*.

In China, agricultural policy is encouraging farmers to grow green manures in the rice fields (Yixian, 1991). During the 1980s, the continuous monocropping of rice was widely observed to have caused soil fertility and pest problems. The Agricultural Ministry set up multiplication bases for green manures, which are expected to produce 5.5m kg seed each year. In some regions, farmers who sold green manure seed to state-run farm co-operatives receive fertilizers at lower prices. Green manures and plant residues are now used on 68% of the 22 million hectares of rice fields.

Similar successes have been observed in Indonesia where, during the late 1980s, pesticide subsidies were cut from 85% of actual cost of the pesticide to zero by January 1989, and farmer field schools established for IPM (Kenmore, 1991; Matteson, Gallagher & Kenmore, 1993; FAO, 1994; Röling & van der Fliert, this volume). The country saves some $130–160 million each year, domestic pesticide production fell by nearly 60% between 1985 and 1990. Rice yields have continued to improve, despite the cut in chemical pesticide use. This raises a crucial policy issue when it comes to cutting inputs. Farmers who are dependent on external inputs do need support to make the transition to a more sustainable agriculture, but this programme demonstrates that, if support is in the form of increasing farmers' capacity to learn and act on their own farms, the cost savings can be high.

The FAO is supporting similar rice-IPM programmes in another eight countries in South and South East Asia (FAO, 1994). Together, these programmes are training hundreds of thousands of farmers, have saved many millions of dollars in pesticides, and generated significant benefits in terms of human and ecosystem health.

2.6.4 Policies that work for sustainability and learning

Policy reform in agriculture is under way in many countries, with some new initiatives supportive of a more sustainable agriculture. Most of these have focused on input reduction strategies, driven by concerns over foreign exchange expenditure or environmental damage. Only a few as yet represent coherent plans and processes that clearly demonstrate the value of embracing integrated produc-

[1] 1 US Dollar = Rs 35.7 (25 September 1996).

Table 2.3. *Policies that work for sustainable agriculture*

Policies
Policy 1: Declare a national policy for sustainable agriculture
Encouraging resource-conserving technologies and practices
Policy 2: Establish a national strategy for IPM
Policy 3: Prioritize research into sustainable agriculture
Policy 4: Grant farmers appropriate property rights
Policy 5: Promote farmer-to-farmer exchanges
Policy 6: Offer direct transitionary support to farmers
Policy 7: Direct subsidies and grants towards sustainable technologies
Policy 8: Link support payments to resource conserving practices
Policy 9: Set appropriate prices (penalise polluters) with taxes and levies
Policy 10: Provide better information for consumers and the public
Policy 11: Adopt natural resource accounting
Policy 12: Establish appropriate standards and licensing for pesticides
Supporting local groups for community action
Policy 13: Encourage the formation of local groups
Policy 14: Foster rural partnerships
Policy 15: Support for farmers' training and farmer field schools
Policy 16: Provide incentives for on-farm employment
Policy 17: Assign local responsibilities for landscape conservation
Policy 18: Permit groups to have access to credit
Reforming external institutions and professional approaches
Policy 19: Encourage the formal adoption of participatory methods and processes
Policy 20: Support information systems to link research, extension and farmers
Policy 21: Rethink the project culture
Policy 22: Strengthen the capacities of NGOs to scale up
Policy 23: Foster stronger NGO–government partnerships
Policy 24: Reform teaching and training establishments
Policy 25: Develop capacity in planning for conflict resolution and mediation

Source: Pretty, 1995*a*.

tion and environment policy goals. None the less, it is clear that many policy reforms are leading to changes in the sustainability of agriculture.

There is much that governments can do within existing resources to encourage and nurture the transition from modernized systems towards more sustainable alternatives (Table 2.3). The first action that governments can take is to declare a national policy for sustainable agriculture. This would help to raise the profile of these processes and needs, as well as giving explicit value to alternative societal goals. It would also establish the necessary framework within which the more specific actions listed in Table 2.3 can fit and be supported.

New policies must be enabling, creating the conditions for development based more on locally available resources and local skills and knowledge. Policy makers will have to find ways of establishing dialogues and alliances with other actors, and farmers' own analyses could be facilitated and their organized needs articulated. Dialogue and interaction would give rapid feedback, allowing policies to be adapted iteratively. Agricultural policies could then focus on enabling people and professionals to learn together so as to make the most of available social and biological resources.

It is important to be clear about just how policies should be trying to address the issues of sustainability and learning. Precise and absolute definitions of sustainability, and therefore of sustainable agriculture, are impossible. Sustainable agriculture should not, therefore, be seen as a set of practices to be fixed in time and space. It implies the capacity to adapt and change as external and internal conditions change. Yet, there is a danger that policy, as it has tended to do in the past, will prescribe the practices that farmers should use rather than create the enabling conditions for locally generated and adapted technologies.

Throughout the world, environmental policy has tended to take the view that rural people are mismanagers of natural resources. The history of soil and water conservation, rangeland management, protected area management, irrigation development, and modern crop dissemination shows a common pattern: technical prescriptions are derived from controlled and uniform conditions, supported by limited cases of success, and then applied widely with little or no regard for diverse local needs and conditions (Pretty & Shah, 1994; Benhke & Scoones, 1992; Pimbert & Pretty, 1995). Differences in local environments and livelihoods then often make the technologies unworkable and unacceptable. When they are rejected locally, policies shift to seeking success through the manipulation of social, economic and ecological conditions, and eventually through outright enforcement.

For sustainable agriculture to spread widely, policy formulation must not repeat these mistakes. Policies will have to arise in a new way. They must be enabling, creating the conditions for sustainable development based on locally available resources and local skills and knowledge. Achieving this will be difficult. In practice, policy is the net result of the actions of different interest groups pulling in both complementary and opposing directions. It is not just the normative expression of government will. Effective policy will have to recognize this, and seek to bring together a range of actors and institutions for creative interaction and joint learning.

2.7 References

Attah-Krah, A.N. & Francis, P.A. (1987). The role of on-farm trials in the evaluation of composite technologies: the case of alley farming in Southern Nigeria. *Agricultural Systems*, 23, 133–52.

Avery, D. (1995). *Saving the Planet with Pesticides and Plastic.* Indianapolis: The Hudson Institute.

Bagadion, B.U. & Korten, F.F. (1991). Developing irrigators' organisations; a learning process approach. In *Putting People First.* ed. M. M. Cernea, 2nd edn. Oxford: Oxford University Press.

Balbarino, E.A. & Alcober, D.L. (1994). Participatory watershed management in Leyte, Philippines: experience and impacts after three years. Paper for IIED/ActionAid Conference, New Horizons: The Social, Economic and Environmental Impacts of Participatory Watershed Development. Bangalore, India: November 28 to December 2.

Bannister, M.E. & Nair, P.K.R. (1990). Alley cropping as a sustainable agricultural technology for the hillsides of Haiti: experience of an agroforestry outreach project. *American Journal of Alternative Agriculture* **5**, (2), 51–9.

Beaumont, P. (1993). *Pesticides, Policies and People*. London: The Pesticides Trust.

Benhke, R. & Scoones, I. (1992). *Rethinking Range Ecology: Implications for Rangeland Management in Africa*. Drylands Programme Issues Paper No. 33, London: IIED.

Borlaug, N. (1992). Small-scale agriculture in Africa: the myths and realities. *Feeding the Future (Newsletter of the Sasakawa Africa Association)*, **4**, 2.

Borlaug, N. (1994*a*). Agricultural research for sustainable development. Testimony before US House of Representatives Committee on Agriculture, March 1, 1994.

Borlaug, N.E. (1994*b*). Chemical fertilizer 'essential'. Letter to *International Agricultural Development* (Nov.–Dec.), p. 23.

Brown, L. (1994). The world food prospect: entering a new era. In *Assisting Sustainable Food Production: Apathy or Action?* Arlington VA: Winrock International.

Brown, L.R. & Kane, H. (1994). *Full House: Reassessing the Earth's Population Carrying Capacity.* New York: W.W. Norton and Co.

Bunch, R. (1990). Low input soil restoration in Honduras: the Cantarranas farmer-to-farmer extension programme. *Sustainable Agriculture Programme Gatekeeper Series SA23*. London: IIED.

Bunch, R. (1993). EPAGRI's work in the State of Santa Catarina, Brazil: major new possibilities for resource-poor farmers. COSECHA, Tegucigalpa, Honduras.

Bunch, R. & López, G.V. (1994). Soil recuperation in Central America: measuring the impact four and forty years after intervention. Paper for IIED Conference New Horizons:The Social, Economic and Environmental Impacts of Participatory Watershed Development, November, Bangalore, India.

Byrnes, K. & Liebhardt, W. (1993). Sustainable agriculture gets its due. *Forum for Applied Research and Public Policy* (Fall), 68–82.

Carruthers, I. (1993). Going, going, gone! Tropical agriculture as we knew it. *Tropical Agriculture Association Newsletter,* **13** (3), 1–5.

Carter, J. (1995). Alley cropping: have resource poor farmers benefited? *ODI Natural Resource Perspectives No 3*, London.

CGIAR (1994). *Sustainable Agriculture for a Food Secure World: A Vision for International Agricultural Research*. Expert Panel of the CGIAR, Washington DC and SAREC, Stockholm.

Crosson, P. & Anderson, J.R. (1995). Achieving a sustainable agricultural system in Sub-Saharan Africa. *Building Block for Africa Paper No. 2*, AFTES, Washington DC: The World Bank.

DANIDA (Danish International Development Agency) (1994). *Agricultural Sector Evaluation. Lessons Learned.* Copenhagen: Ministry of Foreign Affairs.

De Freitas, H.V. (1994). EPAGRI in Santa Catarina, Brazil: the micro-catchment approach. Paper for IIED New Horizons conference, Bangalore, India. November 1994. London: IIED.

DowElanco (1994). *The Bottom Line.* Indianapolis: DowElanco.

Dyer, N. & Bartholomew, A. (1995). Project completion reports: evaluation synthesis study. *Evaluation Report Ev583.* London: ODA.

EC (1994). *Evaluation des Projets de Developpement Rural Finances Durant les Conventions de Lomé I, II, et III.* European Commission, Brussels.

Ehrlich, P. (1968). *The Population Bomb.* New York: Ballantine.

El Titi, A. & Landes, H. (1990). Integrated farming system of Lautenbach: A practical contribution toward sustainable agriculture. In *Sustainable Agricultural Systems.* ed. C.A. Edwards, R. Lal, P. Madden, R.H. Miller, & G. House, Ankeny: Soil and Water Conservation Society.

EUAD (1992). *The Pakistan National Conservation Strategy.* Environment and Urban Affairs Division, Government of Pakistan, Islamabad.

Faeth, P. (ed.) (1993). *Agricultural Policy and Sustainability: Case Studies from India, Chile, the Philippines and the United States.* Washington DC: WRI.

FAO (1993*a*). *The State of Food and Agriculture.* Rome: FAO.

FAO (1993*b*). *Strategies for Sustainable Agriculture and Rural Development (SARD): The Role of Agriculture, Forestry and Fisheries.* Rome: FAO.

FAO (1994). Intercountry programme for the development and application of integrated pest control in rice in South and South-East Asia, Phase I and II. *Project Findings and Recommendations. Terminal Report,* Rome: FAO.

FAO (1995). *World Agriculture: Towards 2010.* Rome: FAO.

Farmers Weekly (1996). Report of Oxford Farming Conference. January, 1996.

Fujisaka, S. (1989). The need to build on farmer practice and knowledge: reminders from selected upland conservation projects and policies. *Agroforestry Systems,* **9,** 141–53.

GTZ (1992). The spark has jumped the gap. Deutsche Gessellschaft für Technische Zusammenarbeit (GTZ), Eschborn.

Hamilton, N.A. (1995). Learning to learn with farmers. PhD thesis. Wageningen: Wageningen Agricultural University, the Netherlands.

Hanson, J.C., Johnson, D.M., Peters, S.E., & Janke R.R. (1990). The profitability of sustainable agriculture on a representative grain farm in the mid-Atlantic region, 1981–1989. *Northeastern Journal of Agriculture and Resource Economics,* **19**, (2), 90–8.

Harris, J.M. (1995). World agriculture: regional sustainability and ecological limits. Discussion Paper No 1. Center for Agriculture, Food and Environment, Tufts University, MA.

Hazell, P. (1995). Managing Agricultural Intensification. IFPRI 2020 Brief 11. IFPRI, Washington DC.

Hewitt, T.I. & Smith, K.R. (1995). *Intensive Agriculture and Environmental Quality: Examining the Newest Agricultural Myth.* Greenbelt MD: Henry Wallace Institute for Alternative Agriculture.

IIED (1996). Sustainable agriculture and food security in East and Southern

Africa. *A report for Swedish International Development Authority, Stockholm.* London: IIED,

IFPRI (1995). *A 2020 Vision for Food, Agriculture and the Environment.* Washington DC: IFPRI.

Jordan, V.W.L., Hutcheon, J.H. & Glen, D.M. (1993). *Studies in Technology Transfer of Integrated Farming Systems. Considerations and Principles for Development.* Bristol: AFRC Institute of Arable Crops Research, Long Ashton Research Station.

Kamp, K., Gregory, R. & Chowhan, G. (1993). Fish cutting pesticide use. *ILEIA Newsletter* 2/93, 22–3.

Kang, B.T., Wilson, G.F. & Lawson, T.L. (1984). *Alley Cropping: A Stable Alternative to Shifting Agriculture.* Ibadan: IITA.

Kendall, H.W. and Pimentel, D. (1994). Constraints on the expansion of the global food supply. *Ambio*, **23**, 198–205.

Kenmore, P. (1991). *How Rice Farmers Clean up the Environment, Conserve Biodiversity, Raise More Food, Make Higher Profits. Indonesia's IPM – A Model for Asia.* Manila, Philippines: FAO.

Kerkhof, P. (1990). *Agroforestry in Africa. A Survey of Project Experience.* London: Panos Institute.

Knutson, R.D., Taylor, J.B., Penson, J.B. & Smith, E.G. (1990). *Economic Impacts of Reduced Chemical Use.* Texas A&M University.

Krishna, A. (1994). Large-scale government programmes: watershed development in Rajasthan, India. Paper for IIED New Horizons conference, Bangalore, India. November 1994. London: IED.

Lal, R. (1989). Agroforestry systems and soil surface management of a Tropical Alfisol. I: Soil moisture and crop yields. *Agroforestry Systems*, **8**, 7–29.

Leach, G. (1995). Global land and food in the 21st century. *Polestar Series Report No 5*, Stockholm Environment Institute, Stockholm.

Liebhardt, W., Andrews, R.W., Culik, M.N., Harwood, R.R., Janke, R.R., Radke, J.K. & Kand Rieger-Schwartz, S.L. (1989). Crop production during conversion from conventional to low-input methods. *Agronomy Journal*, **81**(2), 150–9.

McCalla, A. (1994). *Agriculture and Food Needs to 2025: Why We Should be Concerned.* Sir John Crawford Memorial Lecture, October 27. CGIAR Secretariat, Washington DC: World Bank.

McCalla, A. (1995). Towards a strategic vision for the rural/agricultural/natural resource sector activities of the World Bank. World Bank 15th Annual Agricultural Symposium, January 5–6th, Washington DC.

Matteson, P. (1995). The 50% pesticide cuts in Europe: a glimpse of our future? *American Entomologist*, **41**(4), 210–20.

Matteson, P., Gallagher, K.D. & Kenmore, P.E, (1992). Extension and integrated pest management for planthoppers in Asian irrigated rice. In *Ecology and Management of Plant Hoppers*, ed. R.F. Denno & T.J. Perfect, pp. 57. London: Chapman & Hall.

Mitchell, D. O. & Ingco, M.D. (1993). The world food outlook. International Economics Dept. World Bank, Washington DC.

NAF (1994). *A Better Row to Hoe: The Economic, Environmental and Social Impact of Sustainable Agriculture.* St Paul, Minnesota: Northwest Area Foundation.

NRC (1989). *Alternative Agriculture.* National Research Council. Washington DC: National Academy Press.

OECD (1989). *Agricultural and Environmental Policies.* Paris: OECD.

Paarlberg, R.L. (1994). Sustainable farming: a political geography. IFPRI 2020 Brief 4. IFPRI, Washington DC.

Pimbert, M. & Pretty, J.N. (1995). Parks, people and professionals: putting 'participation' into protected area management. UNRISD Discussion Paper DP 57, United Nations Research Institute for Social Development and WWF International, Geneva.

Pohl, G. & Mihaljek, D. (1992). Project evaluation and uncertainty in practice: a statistical analysis of rate-of-return divergences of 1015 World Bank projects. *World Bank Economic Review*, **6**(2), 255–77.

Pretty, J.N. (1995*a*). *Regenerating Agriculture: Policies and Practice for Sustainability and Self-Reliance.* London: Earthscan Publications; Washington DC: National Academy Press; Bangalore: Action Aid.

Pretty, J.N. (1995*b*). Participatory learning for sustainable agriculture. *World Development*, **23**(8), 1247–63.

Pretty, J.N. & Howes, R. (1993). *Sustainable Agriculture in Britain: Recent Achievements and New Policy Challenges.* IIED Research Series Vol 3, No 1. London: IIED.

Pretty, J.N. & Shah, P. (1994). *Soil and Water Conservation in the 20th Century: A History of Coercion and Control.* Rural History Centre Papers No. 1, Reading: University of Reading.

Pretty, J.N. & Thompson, J. (1996). *Sustainable Agriculture at the Overseas Development Administration.* Report for NRPAD, ODA, London.

Pretty, J.N., Thompson, J. & Hinchcliffe, F. (1996). *Sustainable Agriculture: Impacts on Food Production and Challenges for Food Security.* Gatekeeper Series SA60, IIED, London.

Reus, J.A.W.A., Weckseler, H.J. & Pak, G.A (1994). *Towards A Future EC Pesticide Policy.* Utrecht: Centre for Agriculture and Environment (CLM).

Röling, N. (1994). Platforms for decision making about ecosystems. In *The Future of the Land.* ed. L. Fresco. Chichester: John Wiley.

Rosegrant, M.W. & Agcaolli, M. (1994). *Global and Regional Food Demand, Supply and Trade Prospects to 2010.* Washington DC: IFPRI.

Sasakawa Global 2000 (1994). *Developing African Agriculture: New Initiatives for Institutional Cooperation.* Tokyo: SAA.

Sasakawa Global 2000 (1995). *Feeding the Future.* Newsletter of the Sasakawa Africa Association, Tokyo.

Scoones, I. & Thompson, J. (1994). *Beyond Farmer First: Rural People's Knowledge, Agricultural Research and Extension Practice.* London: IT Publications.

Shah, P. (1994). Village-managed extension systems in India: implications for policy and practice. In *Beyond Farmer First: Rural People's Knowledge, Agricultural Research and Extension Practice.* ed. I. Scoones & J. Thompson. London: IT Publications.

Sherief, A.K. (1991). Kerala, India: group farming. *AERDD Bulletin* (University of Reading), **32**, 14–17.

SWCB (1994). *The Impact of the Catchment Approach to Soil and Water*

Conservation: A Study of Six Catchments in Western, Rift Valley and Central Provinces, Kenya. Nairobi: Ministry of Agriculture.

TG-HDP (1995). *Thai-German Highland Development Project Annual Report.* Thailand: Chiang Mai.

UNDP (1992). *The Benefits of Diversity. An Incentive Toward Sustainable Agriculture.* New York: United Nations Development Program.

van der Werf, E. & de Jager, A. (1992). *Ecological Agriculture in South India: An Agro-Economic Comparison and Study of Transition.* The Hague: Landbouw-Economisch Institut; and Leusden: ETC-Foundation.

Vereijken, P. (1990). Research on integrated arable farming and organic mixed farming in the Netherlands. In *Sustainable Agricultural Systems.* ed. C.A. Edwards, R. Lal, P. Madden, R.H. Miller & G. House. Soil and Water Conservation Society, Ankeny.

Waggoner, P.E. (1994). *How Much Land Can Ten Billion People Spare for Nature?* Council for Agricultural Science and technology. Task Force Report 121.

Winrock International (1994). *Assisting Sustainable Food Production: Apathy or Action?* Arlington VA: Winrock International.

Wirth, T.E. (1995). US policy, food security and developing countries. Undersecretary of State for Global Affairs, presentation to Committee on Agricultural Sustainability for Developing Countries, Washington DC.

World Bank (1993). *Agricultural Sector Review.* Washington DC: Agriculture and Natural Resources Department.

Yixian, G. (1991). Improving China's rice cropping systems. *Shell Agriculture,* **10,** 28–30.

3 The second wing of the eagle: the human dimension in learning our way to more sustainable futures

JAMES WOODHILL AND NIELS G. RÖLING[1]

Science consists not in the accumulation of knowledge, but in the creation of fresh modes of perception (David Bohm, 1993).

3.1 Introduction

With increasing frequency, society is presented with graphic media images of the 'environmental crisis': disappearing forests, destruction of the ozone layer, polluted rivers and oceans and the extinction of species. Meanwhile, a scanning of advertisements for environmental management positions reveals an almost exclusive demand for professionals from the biophysical sciences. One could be forgiven for gaining the impression that 'the problem' is to do with the 'environment' and not with people.

Why has society's response to the perceived threats of environmental decay been so predominantly 'outward'? This is the paradox at the heart of this chapter. We believe there is a very real 'environmental crisis', but, the crisis is not 'out there' in the 'environment', it is within. It is a crisis that needs to be understood in terms of competing values, beliefs, perceptions and political positions. It has to do with our 'way of life' and how we understand, explain and create our existence.

Environmental management (EM), we argue, has been perceived primarily as a technical task. The major investments have gone into understanding the biophysical dimensions of change in the 'natural environment' and the quest for technological 'solutions'. Societal response has centred on ecological consequences rather than on the social causes. This is remarkable as, despite the 'outward portrayal' of the crisis, most people would readily agree that the causal agents are indeed people.

This chapter argues that environmental management is in need of 'fresh modes of perception'. The profession and the disciplines that inform it must, we believe, deal with not only the *consequences* of human activity, but *human activity* itself. This is not to deny the physical dimension of environmental issues. What we are arguing for is balance and interconnection between the biophysical and social dimensions. Hence, the 'two wings' metaphor. However, this is not simple. To take the task seriously challenges traditional and often unquestioned beliefs about science, the 'natural world', knowledge and political power.

[1] The authors wish to acknowledge the helpful comments of Janice Jiggins (who also suggested the title) and Cees Leeuwis, and the valuable editorial work of Virginia Woodhill and Felicity Woodhill.

A crisis, by definition, is a time of danger and great difficulty, a time for decisions. It is a turning point when the self-understanding and self-definition of individuals is in a state of flux and life situations *as a whole* are in question (Oxford Dictionary, 1992; Brunkhorst, 1993). There are two aspects to the 'environmental crisis within' that we wish to explore. The first has to do with values and beliefs. Environmental degradation is a threat because it confronts our system of beliefs and values and shows up internal contradictions. For example, we can not endlessly exploit natural resources for material wealth and at the same time live in an environment that is healthy, aesthetically pleasing, and that will provide these same qualities for future generations.

The second aspect has to do with beliefs about knowledge. As we struggle to deal with these contradictions in values and their political ramifications, we are forced to question our assumptions about knowledge and science. This is not a disguised attack on science. It is an attack on ways of thinking, approaches to problem solving and political institutions founded on the ideological view that empirical science is the epitome of human reason and the primary route to truth and human understanding. Contemporary philosophy, social science and physics have shown many earlier beliefs about science and its place in human affairs to be demonstrably false. However, despite this emerging realisation, the professional practice and organisational culture of many institutions, and notably those with EM responsibilities, remain dominated by patterns of what Miller (1985) refers to as 'technological thinking'.

We are not alone in reflecting on this connection between the practicalities of EM, philosophy and science. Indeed, there is an emerging discourse, which is challenging past approaches and seeking out a new framework for action. It is crucial that the biophysical sciences monitor the environmental consequences of human activity. Without this contribution, we might not know what's 'hitting' us. It is the effort of many natural scientists that alerts us to the unsustainable nature of modern human activity. There is also no question of the need for further scientific understanding and technological innovation in the quest for 'sustainable development'. However, technical 'tools' are not an answer in themselves. They need to be directed by sound policies and integrated with broad-ranging social, political and economic transformation.

The implication of this is the need for more creative, forward thinking and socially engaging processes of change. We argue the need to focus on integrating the creative capacities of people, whether they be land users, lay people, natural scientists, social scientists, policy makers or politicians. The processes of social change, cultural transformation and institutional development necessary to achieve this we refer to as *social learning.* To address environmental issues effectively from an integration of biophysical and social dimensions requires understanding, decision making and action at and across all levels of society. New *platforms* and *processes* for facilitating social learning are required. It is to this task that the social sciences need to make a more relevant contribution and to which biophysical science needs to be more open.

The chapter has it origins in the our experience of the difficulties, tensions and contradictions in the field of EM (Röling, 1993*a*,*b*,*c*; Woodhill, 1992; 1993), in particular those situations where the need for an integrated approach and community participation is recognized and espoused, yet where the practice does not match the rhetoric. To explore these tensions and difficulties, we have found it necessary to cross disciplines and a diversity of theoretical frameworks. This makes a difficult task for both author and reader. It is made more so by what is too often an enormous divide between academic pursuits and professional practice and hence the disconnection of theory and practice. Moving into a truly transdisciplinary mode is both an intellectually and personally challenging process. There is an inevitable struggle with unknown language and concepts that may call into question firmly held professional and personal convictions. Taking a 'holistic', 'integrated' or 'systemic' approach requires much more than simply forcing together perspective's from what remain fragmented disciplines. New conceptual frameworks and languages that are shared across disciplines are required. It is down such a path we have gone in full recognition of the pitfalls, not the least being those who judge such endeavours not in terms of the broader challenge but from the safety of their own disciplinary perspective.

In this chapter the philosophical and pragmatic perspectives are merged. We begin by outlining the complex human dimensions of the 'environmental crisis' and then set a broader context by raising instrumental reason and the expert culture as two features of modernity that have shaped the practice of EM. The management dilemmas of Yellowstone National Park provide a context for conceptualizing EM as 'social ecology' and stressing the importance of systemic approaches to inquiry. Such a perspective raises questions about the assumptions (epistemological and ontological) that underpin positivist approaches to science and EM. We discuss these in relation to the theme of 'philosophy for action'. This leads back to the implications for social learning and the need for a new research agenda for EM; an agenda which we define and outline.

3.2 Social dimensions of the environmental crisis

Why is the environmental crisis so challenging, so difficult, so multidimensional? The following features begin to answer this question.

Interdependence

Increasingly, local communities are interconnected through a global world order. The activities of individuals, local communities and nations have become highly interdependent. What one actor does has consequences which effects others. Consequently, no one is immune. The consequences of environmental decay cross all borders.

Social imbeddedness

The causes and potential strategies for dealing with the environmental crisis are intimately connected with all aspects of our social and cultural life.

Environmental issues cannot be hived off as problems to be resolved in isolation, or as 'technical' problems which can be safely left to scientists.

Complexity

We are faced with the implications of a world which is a network of multitudinous interacting pieces, often too complicated to understand, yet organized and functioning as a whole. Things relate in unknown ways, leading to the emergence of structures and outcomes which can often not be predicted. The complexity with which humans have to deal has escalated as forms of knowledge expand and as social and environmental influences become global.

Uncertainty

High levels of uncertainty exist in two domains. First, in the prediction of the rate, scale and consequences of environmental change; and secondly, in the appropriateness and timeliness of the response of individuals, communities and governments. Because of the increasing sophistication of life for humans and the power of our inventions, uncertainty is amplified.

Skewed criteria for development

Economic optimization and material wealth have been widely seen as the key factor in 'development'. Growing evidence suggests that the dominance of this belief is incompatible with ecologically sustainable development. There is growing realization that 'quality of life' is not simply a function of pecuniary wealth and unlimited control over nature. It involves the freedom to access clean air and water, spiritual well-being and hope.

Individual engagement

In a democratic world, there is little scope for policies that do not reflect public opinion. The engagement of individuals in understanding the environmental crisis and in developing a willingness to promote or at least accept substantial change is essential.

Local action, global coordination

Because of the complex and uncertain nature of our times, there is a need for both local action and broad-scale coordination. The old adage of 'think globally, act locally' is insufficient. It is also necessary to act globally while thinking of both global and local consequences.

Paradigm dilemmas

The environmental crisis challenges many assumptions that underpin modern industrialized culture. Reflection on the processes of science and human understanding has shown how, for periods of time, particular sets of assumptions and theoretical frameworks hold sway. Over time, contradictions emerge which challenge the assumptions of the dominant paradigm and new paradigms emerge.

The environmental crisis is inseparable from what Guba (1990) refers to as the paradigm dialogue: a contemporary struggle to articulate the theoretical grounds for action in the face of the complexity created by philosophical recognition that there are no absolute foundations on which to base the validity of human knowledge and understanding.

These above dimensions fit and need to be understood within a historical context and the patterns of modern Western (global) culture. This is a difficult philosophical and sociological task, yet one that can no longer be shirked by the theory and practice of EM.

3.3 Features of modernity and the tradition of environmental management

Modernity is defined by Giddens (1990: 1) as the 'modes of social life or organisation which emerged in Europe from about the seventeenth century onwards and which subsequently became more or less worldwide in their influence'. These modes of 'social life' include industrialization, capitalism, consumerism, bureaucratic systems of administration, the dominance of the nation state, and globalization. A full discussion of these is not our intention.[2] Rather, we wish to highlight two features of modernity: *instrumental reason* and the *expert culture* which have been dominant in shaping the theory and practice of EM in contemporary society.

3.3.1 Instrumental reasoning

Instrumental reasoning is a broadly used term that refers to 'goal-directed, feedback-controlled interventions in the world of existing states of affairs' (Habermas, 1984: 11–12). It relates to a means-ends rationality based on using instrumental tools to manipulate objects or events according to an assumed knowledge about cause and effect relationships. The dominance of instrumental reasoning in Western culture is linked to the perceived power of science and technology to provide 'the most important things' for human development. For our purposes here, we shall characterize instrumental reason in the following way:

- dealing with complexity by reducing it to constituent parts (reductionism);
- assuming that all information relevant to a problematic situation can, and should be, accumulated prior to making decisions or taking action;
- adopting a linear approach to problem solving and valuing the 'logic' of working in a step-by-step fashion: identifying the problem; gathering information; making decisions, designing solution strategies; and implementing action;

[2] An analysis of current thought on such issues can be found in the works of Pepper (1984), Carely & Christie (1992), Eckersley (1992), Giddens (1990).

- assuming linear cause-and-effect relationships between phenomena in both the natural and social worlds;
- focusing on the achievement of specific and quantifiable technical or material outcomes;
- placing total reliance on the utility of science and technology in problematic situations;
- assuming that knowledge (scientific truth) can be separated from human values and political power, and paying little regard to ethical or moral implications of its use;
- attempting to deal with all aspects of social life according to the 'rules' of instrumental reason, for example trying to place all aspects of societal governance in economic terms.

There is no problem with instrumental reason itself. For example, if we wish to build a house or a road, or develop 'environmentally friendly technologies', we need to use such 'logical' and linear thought. However, problems arise when:

- complex multifaceted problems are reduced to a form in which they can be tackled by instrumental reason alone;
- it is assumed that all social phenomena are, in essence, little different from natural phenomena and can be dealt with in a purely instrumental way; or
- those aspects of life not amenable to instrumental reason are ignored or devalued.

It is a reduction of human reason to instrumental reason alone that leads Habermas to talk of the neglect of the moral–practical and aesthetic–expressive dimensions of everyday social life and the consequent breakdown or colonization of the 'lifeworld' (White, 1988: 138). An exclusive focus on instrumental reason, which is what most environmental management professionals are highly trained in, is not conducive to a being able to deal appropriately with the complex multifaceted issues that characterize the environmental crisis.

3.3.2 The expert culture

Instrumental reasoning is embodied in various 'modern' expert institutions, such as the health, education, transport and EM systems. To a large extent they operate through bureaucracies which rely on highly trained and specialized experts who develop policy, make decisions and implement programmes on behalf of society. Habermas (1984: 18–159), following Weber, discusses this phenomena in terms of the 'rationalization' of society, while Giddens (1990: 21–29) talks of 'disembedding mechanisms' and the increasing reliance on 'abstract systems' in everyday life. In relation to EM, a critique of the expert culture can be summarized as follows.

- The broader question of the overall direction of social change and the implicit values that underlie policy are dealt with in very limited and uncritical ways;

- the values, beliefs, and knowledge of lay people in society are devalued relative to expert knowledge. It is generally assumed that the 'experts' are right and others wrong, ignorant or ill-informed;
- an intervention mentality develops. It assumes the right and ability of experts to intervene, based on, and legitimated by, criteria of absolute and objective truth; and
- citizens, expecting problems to be solved by specialized agencies, are disempowered from taking action, and disengage politically.

The trap of increasing specialization associated with the expert culture is well put by Giddens (1990):

> Expert problem-solving endeavours tend very often to be measured by their capacity to define issues with increasing clarity or precision (qualities that in turn have the effect of producing further specialisation). However, the more a given problem is placed precisely in focus, the more surrounding areas of knowledge become blurred for the individuals concerned, and the less likely they are to be able to foresee the consequences of their contributions beyond the particular sphere of their application.

Bureaucracy, as we know it, owes its existence to these instrumental and expert orientations. 'In its pure form it maximises calculability, impersonality, ethical neutrality and instrumentalism by operating strictly according to formal rules'. One of the political ramifications of bureaucracy is that 'ethically guided politics are displaced by bureaucratic politics which translate questions of values into problems appropriate to administrative procedures' (Pakulski, 1991: 162–3).

3.4 The case for social learning

How the environmental crisis will unfold is by no means clear. Without new modes of political activity and social interaction, we are fearful of the following scenarios:

1 Paralysis

An overwhelming level of complexity, hopelessness, unresolved conflicts of interest or beliefs in 'delivery' or 'doom' according to external forces lead to paralysis. The issues are viewed in such a way that any effective action is believed impossible. Such a scenario is likely to be associated with various forms of escapism (drink, drugs, cults and religious or political fanaticism).

2 Environmental totalitarianism

The threats of environmental degradation will lead those with power and resources, both economic and military, to protect privileged access to environmental resources and/or coerce the wider populace to adopt particular strategies.

'The US way of life is not negotiable' cried the former US president, George Bush, at UNCED.

3 Naïve faith in the market-driven technological fix

When environmental problems become serious enough they will automatically be remedied by the existing economic system and new economically 'induced' technologies will emerge to help. Thus, there is no need for deliberate environmental strategies. The 'invisible hand of the market' will function, along with technological ingenuity, to mitigate environmental disaster.

In relation to the third scenario, we should make clear that we are not arguing against the use of market-driven environmentally friendly technologies, particularly if the real environmental costs can be accounted for. However, this alone is not adequate. There is little evidence to suggest that the current economic system, in which environmental issues and the depletion of natural capital are seen as 'externalities', will be 'self-correcting'. Indeed, quite the opposite seems likely, that the economic system left to its own devices will exploit natural capital and overload ecosystems with waste to the point where ecosystems collapse, inducing massive and rapid change. The capacity of human societies to survive such environmental change is uncertain, and the potential for human suffering and misery on an unprecedented scale all too real.

The popularity and growth of environmentalism makes it clear that many people care deeply about the environmental situation. They are actively seeking new ways in which to live and conserve the natural resources on which all life ultimately depends. How, as a society, might we work towards the development of alternatives to the scenarios outlined above? It is here we turn to the concept of *social learning*.

Social learning is an approach and a philosophy which focuses on participatory processes of social change. It is a concept which has recently entered the discourse on issues of the environment and development (Korten & Klauss, 1984; Milbrath, 1989; Weale, 1992). It encompasses a positive belief in the potential for social transformation based on:

- critical self-reflection;
- the development of participatory multi-layered democratic processes;
- the reflexive capabilities of human individuals and societies; and
- the capacity for social movements to change political and economic frameworks for the better.

Social learning, as we envisage it, must addresses issues of *social structure* (Giddens, 1984). For instance, it must not be blind to the influences of money and power. One can, of course, fatalistically accept that economic power and the relationship between the modern nation state and the forces of capitalism will conspire to prevent all social learning towards more sustainable futures. Alternatively, one can recognize that actors have considerable room to manoeuvre in realizing their 'projects' even in the most oppressive situations (Long, 1984; Long & Long, 1992; Giddens, 1984: 16). Such an actor-orientated perspective allows for social learning as a potentially powerful force for change. Such change emerges as actors 'change their minds' through interaction and dialogue with others. Social learning pays particular attention to how learning processes can be facilitated and enhanced through appropriate institutional and policy contexts.

Social learning should be thought of as a society-wide process. It is not an exclusive or elite task for 'scientists', 'experts' or 'intellectuals'. The importance of such community learning and the role of 'the self-taught activist' in relation to environmental issues has been articulated by Funtowicz and Ravetz (1990).

A central concern of this chapter is the lack of a cultural disposition, institutional framework and political will to work towards engaging wider society in social learning.

3.4.1 Critical questions

Social learning demands *critical thinking*. This involves being able to question the assumptions that underlie our actions, values, and claims to knowledge (Brookfield, 1987; Flood, 1990). The improvements to which we are orientated should be ethically defensible. Bawden (1995) calls such a way of being – one that integrates theory, practice and ethics into a holistic approach to learning – *praxis*.

Social learning is action orientated. It is intended to help improve the quality and wisdom of the decisions we take when faced with complexity, uncertainty, conflict and paradox. However, the ways of thinking and acting that flow from an appreciation of praxis do not sit easily with the modes of instrumental, linear and narrowly scientific 'problem solving' that have permeated much decision making in the modern world.

Consider, for example, environmental issues such as global warming or the pollution of a river system. Is it possible to gain all the scientific 'facts' about the situation? Certainly science can give us a lot of useful information. However, as anybody who has dealt with such situations knows, there are always gaps, unproven hypotheses, insufficient data, and difficulties in long-term prediction. The more variables introduced into the situation, the greater the difficulty of comprehensive scientific understanding. In complex and uncertain situations, different interpretations, hypotheses, understandings and perceptions are inevitable, yet the continuous inescapable unfolding of social life means that, irrespectively, decisions are made and action taken. As Giddens (1984) points out, there are always unacknowledged conditions and unintended consequences of action. Hence, human judgement and political activity become things not to be simply made 'scientific', but rather are more encompassing dimensions of the human condition fundamental to our ability to 'go on' in the face of uncertainty and our partial understanding of the worlds we inhabit. Science can certainly contribute greatly to our judgement and the wisdom of our political stances, but it cannot and should not subsume these dimensions.

This theme leads to the following type of questions:

- What ethical grounds justify the claims of different groups for the use of natural resources in particular ways?
- On what grounds is humankind's relationship with nature to be understood?
- What differing values do stakeholders have and what potential conflicts to these differences give rise to?
- How is knowledge created and used by different stakeholders to support particular interests?
- How has political history and past government policy influenced the type, dissemination and validity of scientific information available for EM?
- What social structures are constraining or enabling of the development of holistic and integrated approaches to EM.

These are open-ended philosophically and politically difficult questions. Yet they cannot be ignored as technically focused approaches to EM have tended to do. This demands thinking more deeply about what is involved in EM.

3.5 Environmental management as social ecology

3.5.1 The case of Yellowstone National Park

Yellowstone National Park provides a good example of the diversity and interconnected nature of issues that have to be confronted in EM (based on Keiter & Boyce, 1991). The name 'Yellowstone' calls to mind a wilderness of geysers, mountain streams and forests, where grizzly bears, buffalo and other animals roam free. One of the few places left on earth where modern visitors can partake in a world unspoiled by man. Unfortunately, this image of an unspoiled wilderness and an ecosystem protected from the ravages of industrialization is an illusion.

Outside the Park, tin mines are in operation which provide the State of Wyoming with a large share of its income. The mining threatens the aquifers which feed the geysers. The State Forestry Commission has extensive timber plantations next to the park. The conflict here is in the threat to the plantations posed by forest fires which the Park's administration considers a 'natural' phenomenon and allows to burn unchecked. Cattle ranchers also share a boundary with the Park. They complain that buffalo, which roam widely outside the Park, carry brucellosis.

Inside the Park, strife is rampant between people who want to optimize 'nature' and those who want to optimize the learning experiences for visitors – the Park's original mandate. With 3.5 million visitors a year, man is clearly its most prolific species and the facilities to accommodate it seriously affect the non-human species.

After a severe fire in 1988, it was recognized that the Park could not be considered a self-contained single purpose EM unit. It was recognized that the ecosystem involved covers a much larger area than the Park boundaries. This

Greater Yellowstone Area comprises actors with diverse and conflicting, but inter-dependent claims on the natural resource. This recognition led to the formation of the Greater Yellowstone Co-ordination Committee, a *'platform' for decision making.* The aim of the committee is to take into account the various interests, provide opportunities for conflict resolution, and build the consensus necessary for concerted action. Far from being shaped and managed by natural forces, the future of Yellowstone is determined by the messy business of humans trying to reach compromise about fires, the reintroduction of wolves (exterminated in 1930), the migration routes of animals and the depth of mining operations.

Managing the Park, let alone the Greater Yellowstone Area, is not only a question of bio-physical information and technical intervention. Managing the area is impossible without accommodation between the various human actors who are dependent on the same natural environment but with differing purposes and interests. They are inter-dependent in that each affects the desired outcomes of the others.

This scenario illustrates the complex interactions between humans and the environment. The general pattern of this situation, if not the specifics, occur globally. Three themes emerge and are receiving attention.

1. The need for **'ecosystem management'**, that is recognition that a holistic and integrated approach to the environment is imperative;
2. The need to focus on the *'human dimension'* thus dealing with issues that range from improved decision making processes in EM to transformed attitudes and social norms about the environment. This entails the restructuring of political and economic institutions;
3. Commensurate human *platforms* for learning and decision making: This means creating mechanisms and processes for decision making on a scale appropriate to particular environmental issues or bioregions. This idea of 'creating agency at a higher level' needs to take account that what is good for one local community is not necessarily good for another or society at large, yet without support and understanding at the local level change will be immensely more difficult if not impossible.

3.5.2 Conceptualizing environmental management

Given the 'environmental crisis' we have outlined, the critical issues raised and the pragmatic imperatives that emerge from situations such as Yellowstone, the question remains. How is environmental management to be understood?

We have represented environmental management in terms of the model in Fig. 3.1. The focus is on the *relationship* between social life and the natural world. A holistic perspective is crucial, yet comprehension of the whole also embodies understanding of the parts, particularly the relations among them. The development of ways of thinking, modes of science and strategies for action that attempt to address the problems of the modern world from such a holistic perspective is

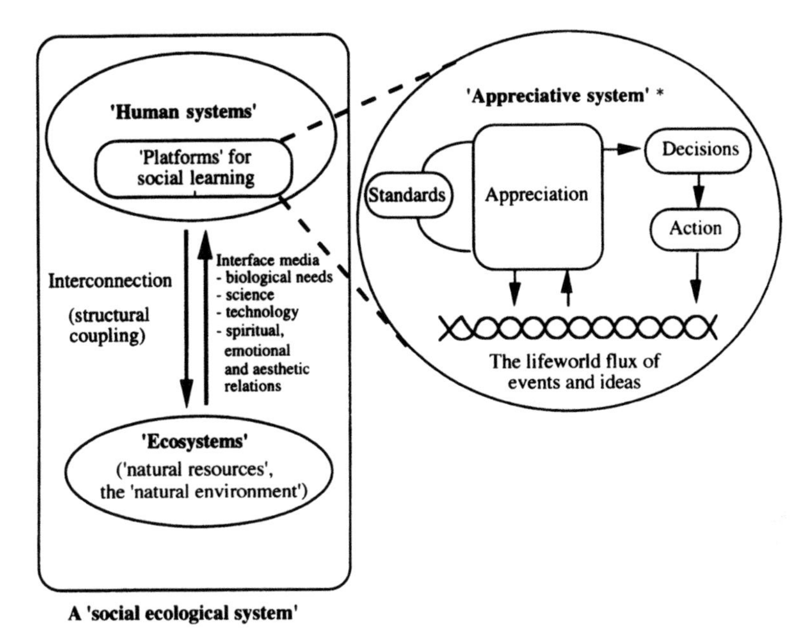

Fig. 3.1. A conceptual model of a 'social ecological system' – elements of a holistic approach to environmental management. (*After Vickers, 1983 and Checkland & Casar, 1986.)

what we refer to as *social ecology*. Our model is not intended to reinforce a human/nature dualism. Nor, however, is it intended to collapse a useful distinction between an organism and the environment or medium in which it lives to some trite undifferentiated 'whole'.

A major contribution to the theory and practice of holistic and integrated problem solving, and hence social ecology is to be found in the *systems sciences* (Checkland, 1981; Flood, 1990; Packham, 1993; Vickers, 1983). Consequently we describe the model of EM in systems terms. A perceived entity is named as a 'system' to reflect emergent properties. Ecosystems or human societies, for example, exhibit properties that can be neither predicted nor understood in terms of their constituent parts (Vickers, 1983: 1–11).

In our model, as in our perspective on systems thinking, 'systems' are not 'real' structures. They are *intellectual constructs* that help us understand the complexity of human experience. We need to remember that "the name is not the thing named" (Bateson, 1988) and resist the temptation to reify 'systems'. An 'ecosystem' is the name we give to our perception of a complex set of relationships and interactions in nature, but an 'ecosystem' as such does not exist. In other words, 'systems' do not exist independently of human processes of inquiry. To make clear that 'human systems', 'ecosystems', and 'social ecological systems' are constructs and not 'real' entities, we place the terms in quotation marks.

From this perspective, EM can be understood as a *'process of systemic learning;* (Bawden, 1995) rather than 'learning about systems'. This distinction reveals a crucial philosophical difference. It is the difference between *constructivist* and *realist/positivist* concepts of knowledge (Guba, 1990: 17–27). This difference leads to different paradigms of inquiry, different understandings of science and different approaches to EM. These issues will be explored in the following section of the chapter.

We have distinguished *'human systems'* and *'ecosystems'* to draw attention to some perceived differences between social and natural phenomena. Our cognitive abilities enable us to perceive, to change, and reflexively monitor our 'relationship' with the *natural milieu.* To do this, we construct understandings and explanations, whether of a spiritual, mythical or scientific nature to guide action and to monitor the consequences of that action (Giddens, 1984: 5–7; Maturana & Varela, 1987: 205–35). It is this ability for highly developed reflexive behaviour, made possible by language, which enables us to even entertain the concept of environmental management.

The following points expand on this model of EM.

1. The focal point for EM is the relationship between 'human systems' and 'ecosystems', in particular the question: is the relationship sustainable or unsustainable?
2. To understand the dynamics of this 'social ecological system', it is vital to pay particular attention to the way humans make sense of their worlds and construct various 'realities' in accordance with their standards and intentions. Following Vickers (1983) and Checkland and Casar (1986), we refer to this process as an *'appreciative system'.*
3. To the notion of 'appreciative systems' we can add the concept of *agency.* Giddens (1984:5–16) defines this as the power to intervene in a course of events to make a difference. As we discussed above, it is crucial to be able to attain agency on a social and geographical scale appropriate to tackling the systemic influences of a particular environmental issue. For example, to address global warming requires agency (co-operation, policy development, decision making, and strategy) at a global level.
4. 'Human systems' and 'ecosystems' are dynamic. Humans (and other species) can only exist while there is a *congruence* with the environment or 'ecosystem' in which they live, and a congruence in the evolution of each. No species is infinitely plastic or adaptable. Changes in the environment can drift in a direction and/or at a rate that makes life for a particular species impossible. Such a drift may well be triggered by changes in that species or its community characteristics. Humans risk triggering changes that may, ultimately, make the natural environment uninhabitable for us and many other species. A valuable way of understanding this is to use the notions of *'structure', 'structural plasticity'* and *'structural coupling'* as developed by Maturana and Varela (1988). These concepts will be developed in the following section.

5. 'Human systems' and 'ecosystems' can be seen as interconnecting or 'structurally coupling' via a range of media. Most directly is a biological connection, breathing, eating, drinking and excreting. Humans are unique in the degree to which they couple with their environment through the use of scientific understanding and technological innovation. Spiritual, aesthetic and emotional dimensions of humans' connection with nature have throughout history been of immense importance. However, these psychosocial dimensions have tended to assume a secondary importance in the face of modern science and technology.

Given these points and the above model we can now define EM as:

> The concurrent manipulation, management, and facilitation of, and/or adaptation to, social ecological change so as to ensure the continuity of human society and other forms of life according to the socially constructed standards of a particular era and society.

Adaptation, facilitation, management and manipulation represent a continuum in the degree of predicability and control of situations that is possible or desirable. Prediction and direct control of 'social ecological systems' is very often beyond our grasp. It is thus necessary to think in terms of a facilitative and adaptive 'co-evolution'. We need to be aware of the limits to our understanding and of the uncertainty and potentially chaotic behaviour of complex 'social ecological systems'.

Recognition that decisions made in the name of EM are inseparable from 'the socially constructed standards of a particular era and society' is fundamental to the idea of social learning. This implies the need for political processes that maximize the potential for all citizens and society as a whole to engage in the construction and review of such standards in an informed, open and critical way. It puts human values and beliefs centre stage. EM can be viewed as processes of *systemic learning* about:

- the dynamics of change in 'human systems',
- the dynamics of change in 'ecosystems',
- the biophysical and social consequences of change as triggered reciprocally in one 'system' by the other through their interconnection,
- the design of technological and political 'systems' that can affect ethically defensible change in 'human systems' and/or 'ecosystems',
- meta-level reflection on the premises, assumptions, values, ethical standards and institutional orders that underpin and make possible such learning activities.

EM has tended to focus primarily on the dynamics of change in 'ecoystems'. Furthermore, its theoretical frameworks and methodological approaches, rooted as they are in the tradition of naturalistic inquiry, are inappropriate or limiting when it comes to learning about these other dimensions.

This framework encourages a number of further considerations. First, what is the potential for contribution and complementarity between aesthetic, spiritual, intuitive, artistic and scientific understanding? Secondly, how can we know the limits or boundaries of our understanding? Thirdly, *who* in society understands *what*? Hence, *who* should be involved in processes of ethical deliberation, or technological and political system design? (Uhlrich, 1983, 1987).

The perspective on EM outlined above is based on a constructivist epistemology, the foundations of which we can now discuss.

3.6 Philosophy for action

Politics, when understood broadly in terms of governance and public life, is central to the model of EM we are espousing. Given that politics involves the exercise of power and, as the saying goes, 'knowledge is power' we are drawn inextricably to philosophy: the study of knowledge. Philosophical thought thus becomes a practical necessity to the task of EM and not some luxury 'academic' pursuit as it is often perceived. While as a society we have little difficulty in mastering the concepts and language (or jargon) of technological change, the same cannot be said for our ability to master the concepts and language required to think deeply about the social ecological predicaments we find ourselves in. This is potentially catastrophic if, as Maturana and Varela (1987) argue, '. . . at the core of all the troubles we face today is our very ignorance of knowing'.

In this section we shall draw on the work of Maturana and Varela to introduce a distinction between *realist/positivist* and *constructivist* assumptions about knowledge and conceptions of science. Instrumental reason, the expert culture and technological thinking are based on assumptions rooted in the realist/positivist tradition. Social learning, while not outrightly rejecting the utility of positivist methodologies, is predicated on a constructivist position. The heart of this difference is captured by Maturana's (undated) notion that 'science deals with the explanation and understanding of human experience (human life) and not with the explanation and understanding of some external reality'. This may not sound radical but it is.

Western thought has been marked by a profound dualism. That of mind and matter, subject and object. This gives rise to the representational model of knowing. In this model, the individual subject gains knowledge about an external world of objects as they are represented in the mind through the senses. The 'reality' that we 'see' out there is taken for granted. The empirical methods of science build on this model by introducing experimentation, and rigour in observation. Along with this dualism in Western thought has been a drive to find the unshakeable foundations for knowledge. Positivist science, which is based on the representational model, was for a time, and for many still is, that unshakeable foundation, the source of 'truth'.

Philosophically it is necessary to distinguish between ontology – what is believed to exist, and epistemology – questions about how knowledge can be acquired and validated. Positivist science is based on a realist ontology, 'the belief that there exists a "reality" *out there*, driven by immutable laws of nature' (Guba, 1990: 19). Science, positivism holds, enables these laws to be uncovered through an epistemology based on objective, experimental and empirical techniques.

Positivist science, as an absolute foundation for knowledge and as an exemplar for human reason, falls down in a number of ways:

1. The human mind is not blank when an observation is made. What is observed and how it is interpreted is influenced by our existing theoretical frameworks, a whole history of ideas. Hence, what science discovers and its day-to-day practice is far from the 'objective' ideal.
2. Positivist science has been unable to provide explanations of social life in anyway comparable to the explanations of natural phenomena.
3. The model of empirical, quantitative, experimental and reductionist methods, along with the objective impassionate researcher, works poorly, if at all, in providing useful knowledge for the resolution of complex 'human' problems.
4. The ideology of positivist science as being 'truthful' and 'objective' is often used to deceive and manipulate by providing a false mask of validity for the arguments of sectional interests.

The constructivist alternative to positivism is based on recognizing the primary importance of language. Humans are reflexive knowledgable beings because of language. Consciousness and 'reality' arise from language and not vice versa. This shift places the emphasis for understanding knowledge not on the subject–object relationship but on the relationship between human subjects. What we experience as 'reality' and hence knowledge is to a very large extent constructed by social processes. Thus, different groups in different eras can have different but equally valid (though not necessarily equally desirable) 'realities' 'brought forth' through language (Maturana & Varela, 1987: 241–5).

We should make clear at this point that a constructivist position does not equate to solipsism or some individualistic relativism where moral and ethical considerations have no place, as critics of constructivism are wont to argue.

We shall explore this notion of constructivism in relation to Maturana and Varela's theory of autopoesis. However, very similar conclusions are reached by hermeneutic philosophy (Bernstein, 1983) which has greatly influenced modern philosophy, social theory and psychology.

3.6.1 Maturana and Varela's theory of autopoesis and cognition

Cognition, Maturana and Varela argue, is a phenomenon of living, self-reproducing (autopoetic) beings. Any being, or entity, has a particular structure, 'a set of components and relations that make it real'. For example, as humans, we have bones, organs and a nervous system that all interconnect to give a particular 'organization' that we identify as being human. Our notion of structure is often limited to the tangibly physical, but in Maturana and Varela's terms, structure encompasses the neurological 'wiring' of the brain which enables perception, and gives rise to the seemingly intangible tangle of human thought and personality.

This structure has significant implications for the interactions between 'human systems' and 'ecosystems'. Any entity (thing or living being) is 'structurally determined', that is, it can only behave in a way that its structure will enable it to. Birds can fly because of their structure, humans cannot fly because of theirs. If we take this a step further, the way we perceive that which is external to us is also determined by our structure, in both the functioning of our sense organs and in the way *information is processed by our brain. Any living entity is an informationally closed system*, which means that changes in the entity can only be *triggered*. External stimuli cannot *direct* change. Thus a process of 'structural change' in a person's thinking can be triggered but not directed. The nature of the change will be determined by the pre-existing structure of the person's ideas and theories of the world which have been learnt during life and form their cultural heritage.

The implication of this is that it makes no sense to talk of an external 'reality'. This is different from saying nothing exists externally to the human mind. Of course there is an environment or medium in which we exist. However, the reality we percieve is determined not by what is external to us but rather by our own physical and cognitive structures. Because we are informationally closed systems we can only ever talk of our experience.

Different entities and parts thereof have varying degrees of *structural plasticity*. Structural plasticity is the potential for structural change without disrupting the organization that defines a particular entity. A rock has very limited structural plasticity. The tissues of the human body which grow, repair themselves and continually reproduce themselves are an example of greater structural plasticity. The human nervous system is the most structurally plastic entity we know of. This plasticity gives rise to the potential for human cognition, reflexive behaviour and complex social phenomena.

Given this definition of structure, we can now turn to the question of interaction between entities and between entities and their environments. Maturana and Varela call these interactions *structural coupling* and define it as 'a history of recurrent interactions leading to the structural congruence between two (or more) systems' (Maturana & Varela, 1987).

This understanding about the biology of cognition requires us to think of knowing and science in a very different way to the positivist tradition. As Maturana and Varela point out, knowing and acting are inseparable. Knowing is not simply the accumulation of 'objective' knowledge about a perceived external world, rather, it is the *effective action of an organism in its environment*. The validity of knowledge and the processes of inquiry depend on whether the actions that flow from that knowledge are judged to be effective by an observer (who can be oneself or others). In relation to EM, a significant criterion is whether particular actions are perceived as leading to a decay of the structural congruence between humans and ecosystems. The inevitable recursiveness is impossible to escape. Past actions give rise to presuppositions and theories. These are used to perceive consequences, take action and

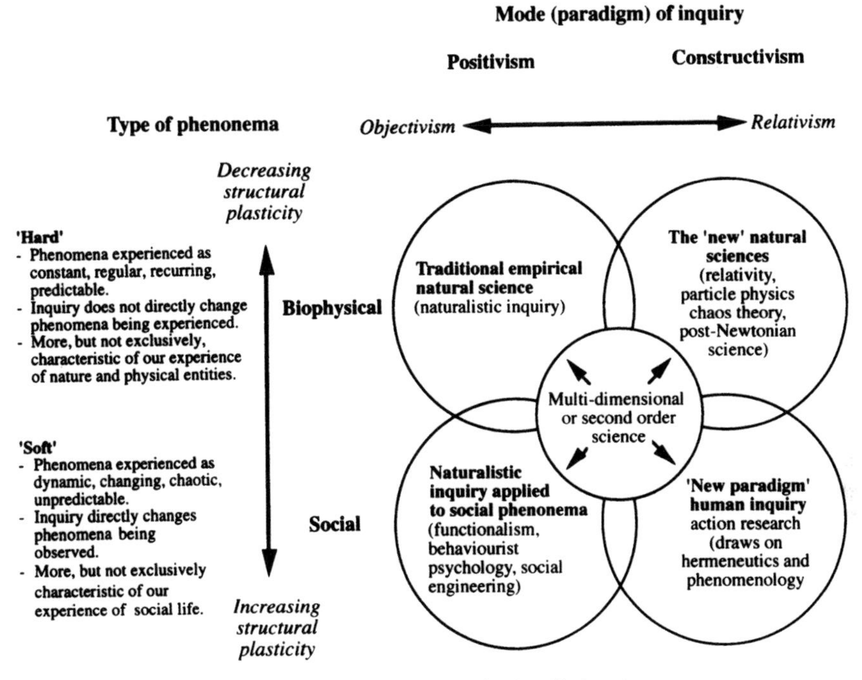

Fig. 3.2. Epistemology and dimensions of scientific inquiry.

judge the effectiveness of those actions as new circumstances unfold and in turn new presuppositions and theories emerge.

3.6.2 Phenomena, ways of knowing and science

This constructivist perspective enables us to examine the distinctions between 'hard' and 'soft' science, 'natural' and 'social' science, that are often made with little clarity of meaning. In this section we shall explore the distinctions and introduce a model that helps to explain them in relation to different dimensions of scientific inquiry. From our perspective, it is crucial that science be understood in an encompassing way; liberated from its exclusive association with positivist and empiricist methods of inquiry. In Fig. 3.2 we illustrate four 'ideal type' characterizations of different modes of science.

Mainstream thought has associated the notion of 'hard' with biophysical science and the notion of 'soft' with social science, implying that the former has tangible validity and the latter is somewhat questionable. These are neither accurate nor useful associations and the negative effects of the division are compounded by the derogatory connotation of the term 'soft'. Using Maturana and Varela's concept of structural plasticity we wish to clarify the notions of 'hard' and 'soft' and the connections between biophysical and social science. To do this, we associate 'hard' and 'soft' with the way human beings experience particular phenomena.

We define 'hard' phenomena as those which humans experience as being relatively constant, unchanging and predictable. Such phenomena have limited structural plasticity and we are able to consider them as obeying 'laws of nature'. 'Soft' phenomena are those that have a high degree of structural plasticity, and which we are unable to explain easily or at all according to laws of nature and linear cause–effect relationships. Human social phenomena are the most striking examples of 'soft' phenomena. Certainly biophysical science tends to deal more with 'hard' phenomena and social science more with 'soft' phenomena. However, this is far from an exclusive association with a growing and fascinating area of overlap.

Dramatic increases in the structural plasticity of social systems result from the fact that the explanations we generate about our experience of social life, whether by lay people or 'social scientists', re-enter and change it. Giddens (1990) refers to this as the 'double hermeneutic' and points out the implications for social inquiry, in particular that it is not possible to arrive at immutable 'laws of society' in any way comparable to the 'laws of nature'. This is not a 'failure' of the social sciences but rather a feature of the difference of the phenomena concerned. Our knowledge or theories about natural phenomena do not change them in any direct way. For example, whether we see the earth as being flat or round does not change the earth itself (though it may well change our experience of it). In contrast, a new theory about organizational development such as Total Quality Management, for example, has a very direct impact on the social phenomena involved.

The central region of Fig. 3.2 represents the notion of a 'space' for the development of multi-dimensional or 'second-order' science (Funtowicz & Ravetz, 1990) and the integration of science with other forms of inquiry and understanding. The underlying epistemological and ontological assumptions of this 'space' must, in our view, lie toward the constructivist position. However, this does not negate the validity of using positivist modes of inquiry or methodology to generate empirical knowledge. This 'space' is one of methodological pluralism where validity arises from the critical awareness of the assumptions being made in developing and using a particular methodology. Ultimately, the validity of the inquiry centres on, returning to Maturana and Varela's words, 'effective as action in a domain of existence'.

3.7 Social learning, a research agenda

3.7.1 The concept

Social learning[3] is a framework for thinking about the knowledge processes that underlie societal adaptation and innovation. As the rates of change, risks of environmental decay, and complexities of globalization escalate, the demands on societies' capabilities to evolve become immense.

[3] Social learning, as we are using the term, should not be confused with the narrow definition found in education textbooks.

A social learning perspective focuses on social actors at all levels as 'circumstance appreciators', who can learn to adapt on the basis of discourse and legitimation of political action. Meaningful interaction and communication between individuals is central to social learning. Through such interaction appropriate responses and ethical social practice can be sought. This is tied to the common theme seen by Bernstein (1983) in the work of contemporary philosophers (Gadamer, Habermas, Rorty & Arendt). Bernstein writes,

> In all of them we have felt a current that keeps drawing us to the central themes of dialogue, conversation, undistorted communication, communal judgement, and the type of rational wooing that can take place when individuals confront each other as equals and participants. . . . they draw us toward the goal of cultivating the types of dialogical communities in which phronesis, judgement and practical discourse become concretely embodied in our every day practices.

They are concerned with new models of public life characterized by widespread participation and engagement. We, as 'social ecologists' see a task for providing and enticing theoretical perspectives on social learning. An important focus is the facilitation of learning about learning by all social actors (Bawden, 1995).

In essence, we need to take stock at all levels. The French call it a *'prise de conscience'*. The result, hopefully, will be that greater priority is given to the required learning and relativization of present realities in education, public service broadcasting, information campaigns, political platforms, religious preaching, business strategic planning and other social structures. 'Prise de conscience' gives the social sciences the task and fundamental contribution of providing the material for self-reflection. It involves criticizing and deconstructing the realities of yesteryear, feeding societal discourse with alternatives, providing theory and tools for collective social action, and creating an action-orientated framework for the practitioner, activist and politician. Admittedly, we have a lot more to learn before such a role can be effectively performed and we are all too aware of the weakness of our present language when entering the public sphere.

We remain sceptically optimistic about the potential for positive change given that,

- people have been able to reconstruct their realities to adapt to circumstance quickly if crisis is upon them;
- people's aspirations and intentions are subjective and socially determined. Valiance, aestheticism, family life, simplicity, creativity, gardening, and many other operationalizations of intentionality have been known to give as much satisfaction as owning a BMW;
- a greater degree of common sense and wisdom that might be expected to prevail if people have the opportunity to communicate in more open ways (Habermas' 'communicative rationality');
- considerable 'fat' where it is most needed: industrial society has the resources to adjust the economic system if political will and public opinion require it.

- according to our experience, coming to grips with the fundamentally constructivist nature of human knowledge is liberating, energizing and empowering.

3.7.2 An agenda for research

It has not been our intention to elaborate a comprehensive theory of social learning. What we shall do is use our perspective on social learning to raise some issues for research. By 'research' we mean active and deliberate construction of realities which might be useful in the public sphere of EM.

1 Social ecological systems

A first and crucial area for research is the 'social ecological system' (Fig. 3.1). Many issues suggest themselves under this heading.

One very interesting aspect is *the role of the biophysical sciences in providing feedback about the state of the environment.* How is this feedback used and at what level? Where does it arrive? Who are the actors that use it? How do political decision makers and public opinion interact on this matter? How do special interest groups for whom the feedback is threatening deform or weaken it? How can such feedback be more optimally handled?

A second issue which merits attention is the nature of the *levels of social aggregation and concomitant eco-systems.* In the biophysical sciences, levels gain importance as the systems perspective becomes accepted as an inter-disciplinary framework for dealing with complexity. We have spoken of the levels of 'human systems' or platforms for decision making about EM. We could be more specific and distinguish among:

- individuals/households;
- local organizations;
- regional agencies and institutions;
- state or national political parties, industries, media, and institutions; and
- international agencies and coalitions.

Obviously, there are different relationships between 'ecosystems' and 'human systems' at different levels (scales). An area of research which merits attention is this interface. What are the tasks to be accomplished at each level? What are the relationships between levels? What are the weaknesses of each level and how can we balance the 'bottom up' and 'top down' approaches?

We have singled out as a key issue the need to *create agency at higher levels of social aggregation.* Studying this process of moving up the level involves the dilemma of choosing between apparent autonomy and self-identity on the one hand, and integration, co-operation and solidarity on the other. Or, in Habermas' terms (1984, 328–37), it means the move from strategic (manipulative, self-seeking) to communicative rationality in our dealing with others. Obviously, different rules and ethics apply.

2 Learning as a social process

It is a common experience that staying out of one's country for several years means missing out on its lived history. Somehow, 'history' is a development of the public sphere, in terms of language, concepts, criteria and behaviours which are considered acceptable. This public sphere, whether at community, regional, national or global level, seems crucial for developing more sustainable coupling between human systems and eco-systems. It is the source of social pressure, of ethics, ideology, of public opinion and political legitimation, of criteria for self-actualization and achievement, and of shared philosophical perspectives.

This public sphere is an important area of research. At present, the views of politicians and policy makers in the public sphere and the ways of influencing them seem based on instrumental or, at best, strategic thinking. The fact that politicians and policy makers have a dismal view of the public sphere means they do not take seriously processes which actually occur and which could be used to great advantage in moving towards more sustainable futures. Research needs to look at how *perceptions* of environmental crises and other competing concerns of individuals affect public support for environmental measures.

There is also a need to explore the *learning paths* that people go through as they move from a consumeristic to a co-operative perspective on the use of the environment. What is the regressive role of commercial advertising and why do people discount it? What are the models or theories people hold of society, the environment, social change and the future? If social learning towards more sustainable futures is to be actively facilitated, it is essential to understand how people make sense of the present complexity and uncertainty which they face.

3 Power and special interests

In our discussion of 'modernity', we have pointed to the role of specialized institutions, and to the 'capitalization' of science to occupy niches for employment and economic gain which are detrimental to human survival. A societal shift to more sustainable forms of coupling with 'eco-systems' means a major realignment of the economic system. This in turn means struggle and conflict, but especially mobilization of human effort.

To reach a situation where such mobilization is possible, major power struggles may well be necessary. A key issue will be 'objective truth'. Since so much is at stake, the specialized interests will fight on the grounds that the truth of certain claims has not been established. In other words, the battle will be fought within the positivist paradigm. This is extremely dangerous. Since absolute criteria do not exist, and there is no external referent or external court of appeal regarding the interpretation of empirical data, and truth depends, in the end, on agreement, accepting the positivist paradigm as the arena for struggle leads to disempowerment of the one factor that counts: political legitimation of measures by the public sphere. *Studies of the strategies and tactics of special interest groups, the sources of their disproportionate political power, and their use of positivist tenets are a crucial area for further research.*

4 Methodology

All that we have said here implies pragmatic changes in *what we do* in the name of EM. On the one hand, we are confronted with very real and very immediate 'problems'. On the other hand, in addressing these problems there is no escaping the larger philosophical questions. Neither frenetic action nor endless philosophical speculation is helpful. New strategies for informed action need to emerge from attention to methodology: the study of how to act effectively given a particular problematic context. Methodology can be the meeting ground for the philosophical and the pragmatic.

Methodologies have been developed for constructing 'rich pictures' to deal with multiple perspectives, for jointly appreciating problems, for establishing joint mission and for collectively learning to realize the mission (e.g. Checkland & Scholes, 1990; van Beek, 1991; Engel & Salmon, 1993). The contours of a praxis for facilitating such platforms are beginning to emerge (Woodhill, Wilson & Mckenzie, 1992; Campbell, 1992; Röling, 1993*b*). However, these efforts are only scratching the surface.

There is an enormous need, opportunity and potential for research on the methodological dimension of social learning, particularly in regard to EM. This is the case for EM projects aimed at managing large-scale 'social ecological systems' where the ideals of an integrated and public participation approach have been espoused. Unfortunately, in many of these cases recognition of the need for, and articulation of, the ideals of integration and public participation are not matched by the development of appropriate methodology.

Such methodological development needs to be truly transdisciplinary. How the barriers and fragmentation between disciplines can be overcome to achieve this is, in itself, an important area of research.

5 Facilitation

Finally, we recognize the issue of actively enhancing social learning in ways that lead to a sustainable future. We are no longer in the realm of the transfer and utilization (adoption and diffusion) of scientific knowledge. That approach to change fits squarely in the positivist tradition (Röling, 1993*a*). What we are after is the facilitation of learning through making things visible, helping people to reconstruct realities through experimentation, discourse, observation and meaningful experience. Although some initial study of facilitation has been done (e.g. Woodhill *et al.*, 1992; Röling & van de Fliert, 1993), it has focussed on rather low levels of social aggregation. *Very little is known about these processes at higher levels of aggregation.* Such knowledge is important for activists, politicians and policy makers. Of special interest is a better understanding of the role of *education, media and communication technology.*

6 Summing up

Social learning for more sustainable futures emerges as an area of inter-disciplinary integration for social philosophy, the philosophy of science, adult education and communication, psychology, sociology and political science. The coupling between 'human systems' and 'ecosystems' requires a heuristic perspective.

The research agenda we have introduced goes way beyond the traditions of academic research. It requires a participative and co-operative approach to research such that new understandings, insights and approaches emerge not only in academia, but more importantly within the communities and organizations that are directly confronted with the need to improve their praxis of EM. This, in turn, demands structural change in the way research is funded and the type and culture of institutions that have research mandates. It also requires an openness from EM professionals and community members to be more critically reflective about what they do and why they do it.

3.8 Conclusion

Our aim is to get the eagle flying again. At present it is skirting disaster, madly beating the one wing it knows how to use. We have approached the task by creating a framework for discourse, the coupling between 'human systems' and 'ecosystems', which gives the eagle a chance to use both wings. Just as flying is the emergent property of the eagle as a majestic whole, so sustainability emerges from adapting the structural coupling between 'human systems' and 'ecosystems', at all levels, to the perceived circumstances. We have called this social learning. It seems time to use this wing and emerge from the destructive downward spiral.

3.9 References

Bateson, G. (1988). *Mind and Nature – A Necessary Unity*. p. 205. New York: Bantam.

Bawden, R.J. (1995). On the systems dimensions of farming systems research. *Journal of Farming Systems Research and Extension*, **5**(2), 1–19.

Berstein, R.J. (1983). *Beyond Objectivism and Relativism*. p. 31, 223. Oxford: Blackwell.

Bohm, D. (1993). In 'Last words of a quantum heretic', interview with John Morgan, *New Scientist*, **137**, 1962, 27 February, 42.

Brookfield, S.D. (1987). *Developing Critical Thinkers – Challenging Adults to Explore Alternative Ways of Thinking and Acting*. pp. 11–14. Milton Keynes: Open University Press.

Brunkhorst, H. (1993). In *Blackwell Dictionary of Twentieth-Century Social Thought*. ed. W. Outhwaite & T.B. Bottomore. Oxford: Blackwell.

Campbell, A. (1992). Taking the long view in tough times: landcare in Australia. *Third Annual Report of the National Landcare Facilitator.* Canberra: National Soil Conservation Program.

Carley, M. & Christie, I. (1992). *Managing Sustainable Development*. London: Earthscan.

Checkland, P. (1981). *Systems Thinking, Systems Practice*. Chichester: John Wiley.

Checkland, P.B. & Casar, A. (1986). Vickers' concept of an appreciative system: A Systemic Account. *Journal of Applied Systems Analysis*, **13**, 3–17.

Checkland, P. & Scholes, J. (1990). *Soft Systems Methodology In Action*. Chichester: John Wiley.

Eckersley, R. (1992). *Environmentalism and Political Theory: Toward an Ecocentric Approach*. London: UCL.

Engel P. & Salomon, M. (1993). RAAKS manual: rapid appraisal of agricultural knowledge systems. Wageningen (the Netherlands): Agricultural University, Department of Extension Science. Unpublished manuscript.

Flood, R. (1990). *Liberating Systems Theory*. pp. 3–4. New York: Plenum.

Funtowicz, S.O. & Ravetz, J.R. (1990). *Global Environmental Issues and the Emergence of Second Order Science*. Luxembourg: Commission of the European Communities.

Giddens, A. (1984). *The Constitution of Society*. pp. 162–221. Cambridge: Polity.

Giddens, A. (1990). *The Consequences of Modernity*. pp. 31–45. Cambridge: Polity.

Giddens, A. (1992). *Modernity and Self-Identity: Self and Society in the Late Modern Age*. p. 5. Cambridge: Polity.

Guba, E.G. (ed.) (1990). *The Paradigm Dialogue*. London: Sage.

Habermas, J. (1984). *The Theory of Communicative Action. Vol. 1: Reason and the Rationalisation of Society*. Boston: Beacon Press.

Habermas, J. (1987). *The Theory of Communicative Action. Vol. 2: Lifeworld and System. A Critique of Functionalist Reason*. Boston: Beacon Press.

Jiggins, J.L.S. (in press). *Women and Sustainable Development*. Washington (DC): Island Press.

Keiter, R.B. & Boyce, M.S. (1991). *The Greater Yellowstone Ecosystem: Redefining America's Wilderness Heritage*. Boston: Yale University Press.

Korten, D. C. & Klauss, R. (1984). *People Centred Development – Contributions Toward Theory and Planning Frameworks*. Connecticut: Kumarian.

Long, N. (1984). Creating space for change: a perspective on the sociology of development. *Sociologia Ruralis*, **24**, 168–84.

Long, N. & Long, A. (eds.) (1992). *Battlefields of Knowledge: The Interlocking of Theory and Practice in Research and Development*. London: Routledge.

Maturana, H.R. (undated). *Science and Daily Life: The Ontology of Scientific Explanations*. Chile: University of Chile.

Maturana, H.R. & Varela, F.J. (1987). *The Tree of Knowledge – The Biological roots of Human Understanding*. p. 248. Boston: New Science Library.

Maturana, H.R. & Varela, F.J. (1988). Reality: the search for objectivity, or the quest for a compelling argument. *Irish Journal of Psychology*, **64**, 5.

Milbrath, L.W. (1989). *Envisioning a Sustainable Society: Learning Our Way Out*. New York: State University of New York Press.

Miller A. (1985). Technological thinking. *Environmental Management*, **9**(3), 179–90.

Oxford University Press (1992). *The Pocket Oxford Dictionary*. Oxford.

Packham, R. (1993). Ethical management of science as a system. *Proceedings of the Thirty Seventh Annual Meeting of the International Society for the Systems Sciences*, University of Western Sydney – Hawkesbury, 5 July (1993). Sydney: International Society for the Systems Sciences.

Pepper, D. (1984). *The Roots of Modern Environmentalism*. London: Routledge.
Pakulski, J. (1991). *Social Movements: The Politics of Moral Protest*. Melbourne: Longman Cheshire.
Röling, N. (1993*a*). Agricultural knowledge and environmental regulation in the Netherlands. *Sociologia Ruralis*, **XXXIII**, 261–80.
Röling, N. (1993*b*). Facilitating sustainable agriculture: turning policy models upside down. Invited paper for: 'Beyond Farmers First: Rural People's Knowledge, Agricultural Research and Extension Practice', Workshop at the Institute of Development Studies, University of Sussex, Brighton, UK, October 27–29 (1992), in a collaboration between IDS and the International Institute for Environment and Development (IIED), London.
Röling, N. (1993*c*). Platforms for decision making about eco-systems. Keynote address at the 75–year Anniversary Conference of the Wageningen Agricultural University, 'Future of the Land: Mobilising and Integrating Knowledge for Land Use Options'. August 22–25 (1993) in Wageningen, the Netherlands. The paper will be published in the proceedings of the conference. Chichester: John Wiley, Spring 1994.
Röling, N. & Van de Fliert (1994). Transforming extension for sustainable agriculture: the case of Integrated Pest Management in rice in Indonesia. *Agriculture and Human Values*, **2**(2,3), 96–108.
Ulrich, W. (1983). *Critical Heuristics of Social Planning: A New Approach to Practical Philosophy*. Switzerland: Haupt.
Ulrich, W. (1987). Critical heuristics of social systems design. *European J. Operations Research*, **31**, 276–83.
Van Beek, P. (1991). Using a workshop to create a rich picture: defusing the ponded pastures conflict in central Queensland. Paper for the Workshop on Managing Complex Issues in Uncertain Environments: Systems Methodologies in Agriculture. Brisbane: Queensland Department of Primary Industries, 26–27 August.
Vickers, G. (1983). *Human Systems Are Different*. London: Harper & Row.
Weale, A. (1992). *The New Politics of Pollution*. Manchester: Manchester University Press.
White, S.K. (1988). *The Recent Work of Jürgen Habermas: Reason, Justice and Modernity*. Cambridge: Cambridge University Press.
Woodhill, J. (1992). *Landcare in NSW Taking the Next Step – Final Report and Recommendations for the Development of Landcare in NSW from the 1991 Landcare Review*. Faculty of Agriculture and Rural Development, University of Western Sydney – Hawkesbury.
Woodhill, J. (1993). Science and the Facilitation of Social Learning – A Systems Perspective. Paper presented to the Thirty-Seventh Annual Meeting of the International Society for the Systems Sciences. 5–7 July 1993, University of Western Sydney – Hawkesbury.
Woodhill J., Wilson A. & McKenzie, J. (1992). Land conservation and social change – extension to community development a necessary shift in thinking. Paper presented at the 7th International Soil Conservation Conference, Sydney, 27 September 1992.

Part II: Environmental policies and farmers' reactions

4 Developing standards for sustainable farming in Switzerland

MICHEL ROUX AND ABRAHAM BLUM

4.1 Introduction

This chapter is a case study of an approach to the preservation of natural values in rural landscapes. It is markedly different from the Dutch case described by van Woerkum and Aarts and by Wagemans and Boerma (this volume). Whereas the Dutch have used Draconian measures to create space for nature by replacing or modifying agricultural activity, an approach which has resulted in failure and conflict, the Swiss have introduced direct payments to farmers for preserving the diversity and multi-functionality of the Swiss landscape. Procedures have been established for close co-operation between farmers and Government in determining ecological standards for agriculture, requiring in turn a number of institutional changes in the agricultural knowledge system. The chapter illustrates the inter-relations among policy, institutional support, and voluntary change in farmers' behaviour.

The first part of the chapter describes the Swiss agricultural knowledge system, without which it would not be possible to understand the institutional implications of the policy changes. The second part lays out the policy changes and the concomitant institutional support necessary for implementation.

4.2 The Swiss agricultural knowledge system[1]

In order to understand better how ecological standards for farming are established through various institutions and applied by farmers, it is helpful first to present a map of the Swiss agricultural knowledge system (AKS) (Blum, 1992) and point out its main characteristics (Fig. 4.1).

The unique federal structure of Switzerland, in which 26 cantons (or half-cantons), based mainly on language groupings, form a Confederation, explains much of what is specific to Swiss agricultural education. Agricultural policy and research are the responsibility of the Confederation. However, agricultural education and extension are in the hands of the cantons, which receive considerable financial assistance from the Confederation for this purpose.

Most of the seven federal research institutes have a country-wide

[1] The description of the Swiss Agricultural Knowledge System is based on a detailed analysis available in German (Blum, 1992) and in English in a paper presented in August 1993 at the Congress of the European Society for Rural Sociology, the Netherlands (Blum, 1993).

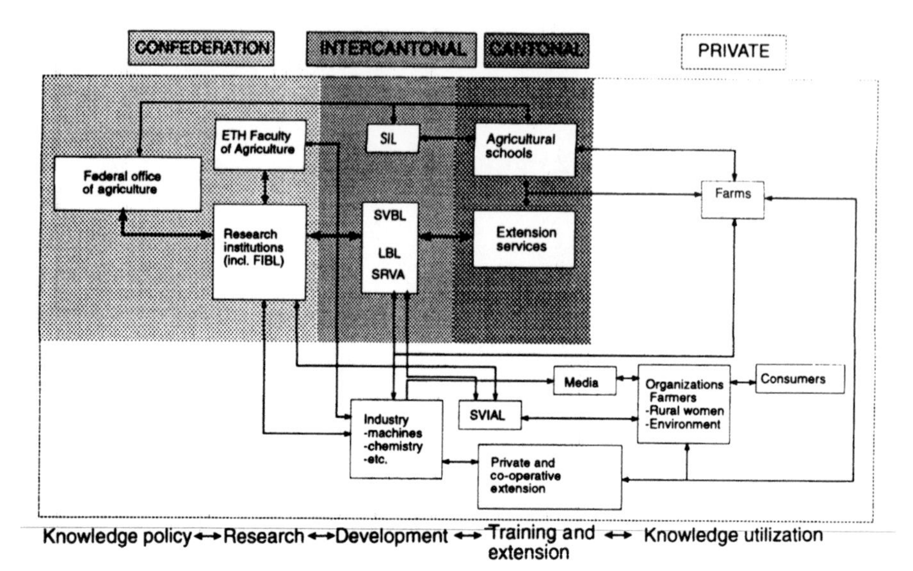

Fig. 4.1. The Swiss agricultural knowledge system (simplified).

responsibility, each for a particular sector. They typically each have sub-stations in the different agro-ecological zones of the country. Mainly for historical reasons, research on arable and horticultural crops is conducted at two parallel institutes, one in Francophone Switzerland, the other in the German speaking part. The Research Institute of Biological Husbandry (FIBL) is unique in its status as a private institute, subsidized by the Confederation, but needing to look elsewhere for other sources of income. The Agricultural Faculty of the Federal Institute of Technology (ETH Zürich) is under the Department of the Interior and its faculty devote their full time to teaching and research.

The formal training of farmers is more developed in Switzerland than in many other countries. Farming is placed on an equal footing with other professions through a system of vocational training which begins with apprenticeship coupled to periods of compulsory vocational school training. Training for increased proficiency conducted at one of the 38 agricultural schools leads to the Master examinations and access to higher education. This stepwise progression through a sequence of formal examinations enables the majority of farmers to access research results independently of agricultural advisors or extension workers.

The cantonal extension services employ in the region of 560 advisers, although only around 210 have extension as their main job. The majority serve also as teachers in agricultural schools, which forces them, in turn, to remain in close contact with practising farmers. But teachers who are also advisers are so overworked that farmers may have to wait for weeks until an adviser can visit their farm. Cantonal advisers also carry out statutory functions, as they do in those German Länder where advisers are state employees. The canton of Vaud is an exception.

Extension there is in the hands of a farmers' organization, and statutory functions are strictly separated, even though the state finances a part of the costs.

In many countries, group extension was only introduced when individual extension services became too expensive. Not so in Switzerland. Group extension was introduced in the mountains there in 1958, when animal producers first received federal subsidies providing they attended group extension meetings. Up to the present, group extension in German-speaking Switzerland is more common in the mountain cantons than in the plains, where it has been in existence from the late 1950s in only one canton – that of Thurgau, and is under the control of a farmers' organization (as in Vaud). But, as shown in Section 4.5, group extension is now becoming important in the plains because of the ecological farming programmes offered to farmers since 1993.

For some tasks, especially for higher education and training, the cantons work together in inter-cantonal organizations. The institutes that have hitherto trained agricultural engineers, and are presently being transformed into non-university agricultural polytechnics (*Fachhochschule*) to comply with the norms of the European Union, provide an example. Another example are the extension centres. The Swiss (national) Union for the Advancement of Agricultural Extension (SVBL) has two such extension centres: LBL, serving the German, and SRVA serving the French and Italian-speaking areas of Switzerland. These organizations, with their large decision-making bodies, consisting of representatives of all cantons, are more complicated to manage than federal institutions. However, they enable the smaller cantons to offer training and education (for which they are responsible), when each of them could not afford to do it alone.

It is expected, as in other European countries, that the number of farms in Switzerland will diminish in the coming years. This will have repercussions on the number of cantonal advisers in the different cantonal institutions, in spite of a growing number of tasks, which will require even more inter-cantonal co-operation. In the case of specialized branches of agriculture and of agricultural schools (which already compete for students), the need for greater co-operation will be especially marked.

As in other countries, agribusiness is deeply involved in agricultural development and employs its own farm advisors. The farm press is also influential. It was consolidated in Francophone Switzerland, and a similar process is now occurring in the German speaking part, where smaller cantonal monthlies compete with two major weeklies. The monthly magazines with the largest distribution are those of the co-operatives and the large powerful agricultural unions. The SVIAL, an organization for agricultural and food diplomas is especially active in providing courses and the development of teaching materials.

Blum carried out comparative studies of the effectiveness of agricultural knowledge systems in Israel (1989), the Netherlands (1991a) and Switzerland (1992, 1993), using eight characteristics of effective knowledge systems recognized by Rogers, Eveland and Bean (1976) for the US AKS, and an additional eight identified in the course of the study. When the Swiss AKS was analysed (Blum, 1992, 1993), it was observed that, notwithstanding its complexity in terms of its federal structure and language divisions, the Swiss AKS shares a number of the characteristics of an effective knowledge system. Of particular concern here is the intensive formal and informal interaction among the key players in the system. Three factors explain the intensity of co-operation among partners.

1. The pluralistic political system in Switzerland makes intensive use of the '*Vernehmlassung*' process – inviting interested parties to air their views on a proposed project or law, before a decision is taken. The idea is to take account of legitimate counter-arguments in order to come to a compromise with opposing groups, rather than enforcing a majority decision top-down;
2. The system is relatively small, 'everyone knows everyone else' as a result of joint periods of study, co-operation in the many agricultural organizations, or through other shared activities;
3. The two inter-cantonal extension centres LBL and SRVA create an important bridge between research stations and the cantonal extension services, especially in the form of written materials and courses.

Some of the weak points of the Swiss system are:

1. Research is burdened with statutory tasks. Scientists at the federal research institutions have to devote 40–50% of their time to such work (e.g. pesticide control and feed mixtures; soil tests);
2. Advisors are motivated and able, but cannot visit farms as much as needed because they too have to fulfil statutory duties;
3. Farmers and advisers seldom participate in field trials;
4. The agricultural press is well developed but has difficulties in surviving because of its fragmentation.

While advisory work is often combined with formal teaching, the direct involvement of extension in research seldom occurs. One reason is that researchers are employed by the Confederation and advisers by cantons, which can make it difficult for an adviser to take part in a research project. There is also a widely held view that researchers should research and advisers advise. This view is often underpinned by the argument that each was trained at a different institution and for different tasks.

4.3 The ecological orientation of agricultural policy

Sustainability in relation to agriculture has two facets: one economic, the other ecological. In the case of Switzerland, the two facets have become intertwined. Farmers used to be a heavily subsidized minority, facing GATT restrictions and the European integration process. Swiss agricultural policy has had to react and redefine its goals and measures under the pressure of fundamental changes in foreign trade policy. Guaranteeing the nation's food supply became a less important policy goal. However, the Swiss population and economy are still interested in farming as

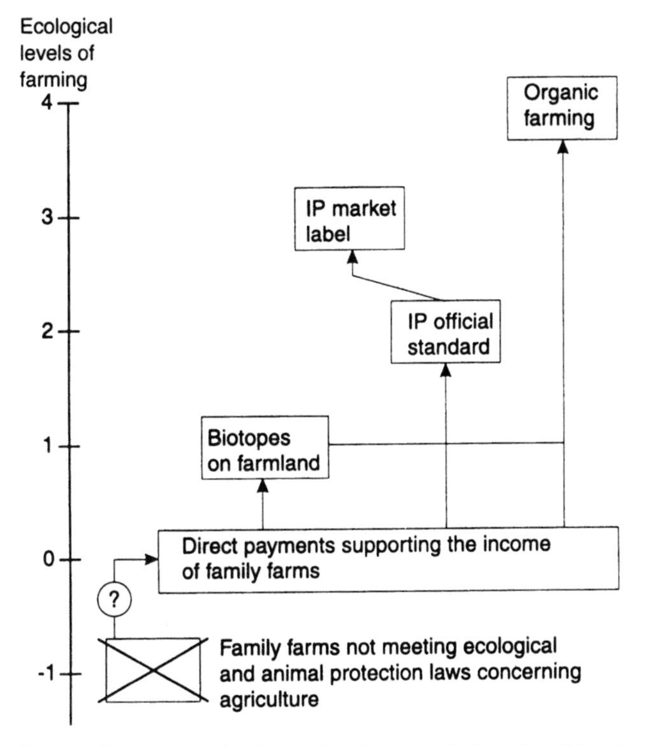

Fig. 4.2. Programmes for increasing the ecological quality of farming. IP=integrated production.

long as it also contributes to environmental protection and to a well-cared for countryside, which should cater for tourist activities.

The new policy, which takes account of GATT and European Union norms, could be implemented only through new statutory measures. In October 1992, Article 31a was added to the Federal Agricultural Law, providing the basis for direct government payments to family farms. Publicly owned farms, very large private farms or farms not conforming to ecological and animal protection laws concerning agriculture, cannot apply for these direct payments (see Fig. 4.2). The direct income support of family farms is supposed to assure a multifunctional agriculture, even if product prices continue to fall and production is being extensified as a result of the measures. It is questionable whether all farmers can afford the investment needed to meet the new legal norms. Those that cannot may be forced to give up their farms.

Article 31b defines additional government support that farmers can receive, over and above the level defined by Article 31a, for special ecological performance (Schweizer Bundesrat, 1993). That is, all family farms can participate on a voluntary basis in programmes which promote sustainable farming. Three levels of compensation are offered in line with three standards of ecological performance:

1. 'Biotopes on farmland', such as extensive cultivation of grassland, litter, high stem fruit trees and hedges. In Fig. 4.2, this programme is located at level one, because the ecological standard here covers only a part of farming activities. More ambitious is:
2. 'Integrated production' (IP) with reduced chemical input, defined as a farming system in which all activities have to meet ecological standards above the basic norms established by the laws concerning the environment. Cantons can define IP on a level even above federal requirements.

   In general, the standards for receiving a recognized IP market label are higher than official IP requirements. The labelling is controlled by large retail organizations and farmers' co-operatives. In some cases, the labels are restricted to one sector only.
3. The highest ecological level is 'organic farming', which is now recognized by the government as a distinct farming system.

In addition to these three ecological standards, a programme called 'pasture farming' has also been introduced on the basis of Article 31b. It provides payments for livestock husbandry which meet higher standards than the norms of the animal protection law.

The new payments are based on the following strategy. Farmers are expected to move to environmentally compatible activities on their own, based on Article 31a. The additional financial incentives under Article 31b are intended to encourage further sustainable agriculture. In this context, extension is seen as an important instrument to facilitate the learning necessary for change in the designed direction. The strategy is in accordance with the conclusions of an empirical study of the awareness of Swiss farmers of environmental problems and of their willingness to act in an environmentally responsible way (Roux, 1988).

In the first year of implementation of Article 31b, in 1993, farmers already showed a great interest in ecological standards. About 11 000 farm owners agreed to manage about 18 000 ha of meadows in an extensive way. Some 1.5 million high stem fruit trees are being protected from being cut down over the next 6 years. More than 9000 farms proved that they were able to meet the requirements of integrated production. Another 1500 farms met the organic farming standard, and about 3 500 farms took part in the programme of pasture farming. Taking into account that a farm holder can participate in more than one programme, the Federal Office of Agriculture assumes that already in the first year more than 20% of the 75 000 Swiss farm owners officially joined at least one of the programmes (Widmer, 1994). In the following sections we shall explain this quite successful start.

4.4 Developing and testing ecological standards in a network of pilot farms

A significant difficulty for promoting sustainable agriculture hitherto has been the absence of standards which could be accepted by both farmers and authorities in the different regions of Switzerland. These standards had to be effective, practical and controllable under all ecological and agricultural conditions. The two inter-cantonal extension centres in Lausanne and Lindau proposed to test the existing standards on a sample of farms throughout the country. By the end of 1990, the Federal Office of Agriculture had asked a project management group, with members from the federal office, the federal research institutions and the two extension centres, to build up a pilot farm network for the following purposes:

- field testing the requirements proposed by experts to meet ecological farming standards;
- exploring the economic and ecological effects of these requirements;
- developing an administrative system based on contracts with farmers so as to ensure voluntary adherence to the requirements.

Over 200 farms participated in the pilot farm network, including 125 farmers in the German-speaking part, coordinated by the Lindau Extension Centre. They did so in collaboration with the cantonal extension services. In Francophone Switzerland, 60 farms were involved and were assisted directly by the Lausanne Extension Centre. Another 20 farms affiliated to organizations for organic farming took part in the network in collaboration with the Research Institute for Biological Husbandry.

The experts who took part in the pilot project participated in the testing of 32 requirements, although they found it hard to agree on the methodology for measuring them (Nationale Projektgruppe IP-Pilotbetriebe, 1993*a*,*b*). In 1991, only 13 farms (6%), and in 1992, not more than ten farms (5%) were able to meet all the requirements. The testing proved that 28 requirements were effective, practical and controllable, even though most of them had to be modified. Figure 4.3. shows the pilot farms in the context of the Swiss AKS.

By April 1993, the Federal Office of Agriculture was able to publish federal standards and rules for participating in ecological farming programmes (*Bundesamt für Landwirtschaft*, 1993). As far as the integrated production of fruit, vegetables, grapes and herbs is concerned, the already existing standards of the national grower organizations were accepted as legal norms for a transition period until 1996. No accepted standards at federal level existed for integrated crop and forage production and for livestock husbandry. The Federal Government therefore based the minimum requirements on the results achieved in the pilot farm network.

In order to receive the payment for integrated production, present conditions require that:

- at least 5% of farmland has to be conserved as biotope;
- nitrogenous and phosphatic fertilizers have to be well balanced according to the fertilizer requirement of the farm;
- chemical plant protection measures have to be reduced to defined risk levels;

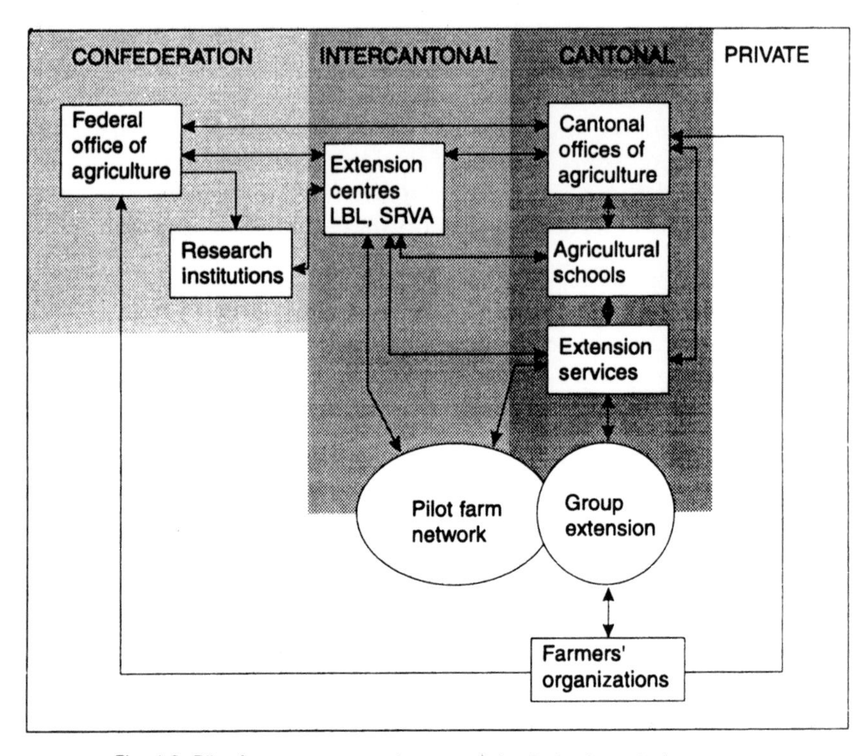

Fig. 4.3. Pilot farms as a new element of the Swiss knowledge system.

- livestock husbandry has to meet defined standards of 'animal friendly' production;
- records have to be kept on all technical aspects of farming activities which are ecologically relevant;
- participation in specific extension activities on integrated production is compulsory.

The cantonal authorities, in co-operation with the cantonal organizations for integrated production and pasture farming, adapted the federal standards to the different conditions in various parts of Switzerland. In 1993, the Federal Office of Agriculture accepted 13 cantonal guidelines concerning integrated production and pasture farming that are, in some cases, slightly higher than the minimum federal norms. Additional costs resulting from more restrictive requirements will have to be covered by the cantons as soon as the legal basis for this has been established.

In some cantons, more ambitious ecological requirements were defined, not only for ecological reasons but also to motivate farmers to go for the highest standards in order to compete in the eco-label markets. With respect to organic farming, the already existing directions of the Union of the Swiss Associations of Biological Agriculture (VSBLO) were accepted as legal norms. These also fulfil the requirements of integrated production, as mentioned above, and the regulations of the European Union concerning organic farming (see Gerber and Hoffmann, this volume).

The use of the pilot farm network has led to some remarkable results:

1. On the basis of the empirical data, collected from over 200 farms, those responsible for research, extension and government, and representatives of agricultural organizations, have been able to agree on standards for integrated production within a short period of time. Environmental organizations that were not involved in the expert group of the pilot farm network have, on the whole, been able to accept the standards and now support this new element of the Swiss agricultural knowledge system.
2. In the beginning, the concepts of integrated production were quite different in the French- and German-speaking parts of Switzerland with regard to contents and methods. In the Francophone cantons, IP was a synonym for 'good agricultural practice', whereas in the German-speaking cantons, IP already stood for lower input and higher ecological standards in livestock husbandry. The steady communication between representatives of both cultural areas of the country on the basis of the results of the pilot farm network, led to the common conviction that 'integrated farming' had to be more than 'good agricultural practice' in order to deserve recognition as special ecological performance.
3. At the same time, the empirical results helped to define which norms should be defined as national and which would be wiser left to the cantons to decide. A few cantons used this margin to further define more ambitious requirements.
4. For the first time a big effort was also made to deal with technical and environmental questions in terms of management requirements. On the basis of record keeping and planning methods, the farming system as a whole had to be taken into consideration.
5. A good empirical database also now existed on the actual technical and ecological standards of farming. Hitherto, only the norms on how ecological farming should be done had been available, with little knowledge about actual farming performance. Comparison of the two databases led to new questions for research and revealed the specific needs for extension support.
6. Their participation in the pilot farm network made it easier for the cantonal authorities, extension services and farmer organizations to prepare in time for the implementation of the new federal programmes (which explains also the good start made in 1993). Many of these organizations say that this was perhaps the most important aspect of the pilot farm network, because they would never have been able to start so smoothly without that opportunity to discuss all kinds of different problems in advance.

Since the programmes based on Article 31b will become increasingly important and encouraged by these first results, the Federal Office of Agriculture decided to continue to work with the pilot farm network. In 1993, about 55 million francs, about 45 million US$ (at time of writing) were used to finance the programmes; the Federal Government expects costs to rise to a level of 470 million francs by 1997. Indicators for evaluating the programmes have yet to be tested, and more attention will have to be paid in the future to the following objectives:

- reduction of energy input and pollutant emissions;
- increased biodiversity on farmland;
- avoiding long-term soil damage;
- more animal-friendly standards of livestock husbandry.

Three criticisms have been raised within the Swiss agricultural knowledge system. The first concerns the role of the federal research institutions. While the Research Station for Farm Management and Agricultural Engineering (FAT) is involved closely in the pilot farm network, the Research Station for Plant Cultivation (FAP) has found it more difficult to participate. A dynamic process of developing sustainable farm management systems under the guidance of national project management, in which not only scientists were represented, takes a great deal of flexibility and an adaptive research methodology.

Secondly, in order to reach ecological goals, it is not enough to optimize farm management systems at the level of the individual farm only. There is also a need for management of agro-ecosystems above the farm level (Röling, 1994). Case studies of different extension projects emphasize this need, but also point out that many difficulties have to be overcome if adequate platforms for decision making about agro-ecosystems above the farm level are to be established (Roux, 1996).

Thirdly, it should be noted that Article 31b was prepared largely outside government institutions. The results justify this approach. It is not only in the field of agriculture that new policies are increasingly formulated as a result of the learning processes in so-called 'policy networks', which consist of relevant private groups and non-governmental institutions (see also van Woerkum and Aarts, this volume). If new policies are prepared in policy networks, then the question of who participates and how the results agreed upon are communicated to government and administrative institutions become key questions (Kissling-Näf & Knoepfel, 1993).

4.5 Facilitating ecological farming by extension associations: the case of the Canton of Berne

The canton of Berne covers a large part of Switzerland, from the Alps over the Central Plains to the Jura. Many types of farming are represented in these areas, which are agro-ecologically very different. Seven agricultural training and extension centres are spread throughout the different areas. From 1988 to 1990, a workgroup of extension officials were given the task of clarifying the objectives and activities of extension in the canton in the light of the changing political and economic conditions facing Swiss agriculture (*Landwirtschaftsdirektion des Kantons Bern, 1990*).

The workgroup emphasized the increasing importance of ecological issues in extension work and of the main functions of a public extension service. The latter was seen to be the facilitation of farming development to systems meeting high ecological standards. However, the extension service was unable to provide the staff and funds needed for the required intensification. A new type of extension association was therefore formed to support the new policy trends, consisting of farmers interested in integrated production and ecological pasture farming. The purpose of these associations is to organize and promote ecological extension groups, and to function as pressure groups in favour of ecological farming systems.

In May 1991, the Cantonal Office of Agriculture accepted the extension associations. However, it was not able to make more resources available to them. By August 1993, six extension associations, with about 2,500 members, had been established in the plains. In the alpine area, no associations were created because group extension was already well established there.

To date, the associations have been financially supported by the extension service, although the cantonal authorities are pushing for more private financing of association activities. The office of such an extension association is located at the area extension centre and its staff is covering a lot of work. In addition, each extension association typically manages to hire a couple of master farmers as part-time advisers, thus increasing the total capacity of the extension staff. This is an interesting development, because, in the main, farmers had up to this point not been actively involved in the organization of extension activities.

In the beginning, the main task of extension associations was to pass on to farmers information about the ecological orientation of agrarian policy and the specific standards to be met if farmers wanted to participate in ecological farming programmes. A second task was to help farmers to decide whether they should apply for these programmes or not. This included establishing whether a particular farm had a realistic chance of meeting the official standards. The costs of the written information on the programmes and for group extension are covered by membership subscriptions. Farmers pay at a subsidised rate for individual services, such as the calculation of the nutrient balance, or other individual technical services.

In future, the main activity of the extension association will be to support the compulsory extension groups for farmers participating in the programme of integrated production. In these groups, farmers compare their ecological and economic decisions and farming activities in order to optimize their farming systems. These extension groups also have the potential to function as a platform for decision making about agro-ecosystems above farm level, if all farmers in a given area participate.

The six extension associations together form the cantonal organization for integrated production and pasture farming. In 1993, they were already able to give feedback to the executive authorities on the applicability and effectiveness of the requirements. Thus, the extension associations help to improve the quality of the knowledge system. However, neither the extension service nor the extension associations are involved in statutory control functions, which are the tasks of a special control organization.

4.6 Conclusions

The relationship between agricultural extension and financial assistance to farmers in Switzerland has gone around the track one and a half times. When agricultural extension was introduced to the alpine region in the 1950s, attendance at extension meetings was a condition for obtaining a subsidy. Farmers did not like such paternalistic treatment; to be told what was good for them. In the end, the obligation to attend was dropped, but in the meantime farmers had become aware of the value of extension as a free service. When policy started to promote ecological approaches to farming, policy measures proved to be too weak to persuade the majority of farmers to respond. Thus, payments to farmers were introduced to stimulate them to follow the rules laid down by the ecological agriculture programme. At the same time, compulsory extension was brought back 'to assure that farmers knew how to do the right thing.' Apparently, the authorities (and the law-making bodies) were not convinced that they could rely on the farmers' own willingness to adopt extension advise. However, when the government could no longer fully subsidize ecological extension, farmers organized themselves into associations and proved ready to share extension costs, as the Bernese example shows.

The creation of programmes to support farm performance at different levels of ecological achievement, and the readiness to adapt standards to regional needs, is typical of the pluralistic character of confederate Switzerland. However, the consensus reached at present covers only the environmental laws, and contains only direct payments available under Article 31a of the Federal Agricultural Law. However, there is no endorsement of the higher level ecological farming system standards by the majority of farmers. None the less, many law makers, consumers and farmers appear ready to support and join the ecological programmes at a higher level, once they are convinced of their environmental value and/or their economic justification.

An important element in the success of the programmes so far has been the consensus reached on what constitutes realistic, acceptable criteria for the definition of standards for the different levels of ecological farming. This has been achieved through an analytical approach and on the basis of the empirical data collected in the relatively large pilot farm networks. A normative approach, based on environmental ideals as proposed by scientists and political pressure groups, would not have succeeded. There is still a need for further regionalization in order to improve the farming systems in all agro-ecological conditions. At present, the main weakness seems to be the lack of management, above farm level, in which all those who influence the agro-ecological system can participate collectively.

4.7 References

Blum, A. (1989). *The Agricultural Knowledge System in Israel*. Rehovot: The Hebrew University of Jerusalem, Faculty of Agriculture.

Blum, A. (1991*a*). What can be learned from a comparison of two agricultural knowledge systems? The case of the Netherlands and Israel. *Agriculture, Ecosystems and Environment*, **33**, 325–39.

Blum, A. (1991*b*). The agricultural knowledge transformation cycle. *Quarterly Journal of International Agriculture*, **30**, 321–33.

Blum, A. (1992). *Das landwirtschaftliche Wissenssystem Schweiz*. Lindau: Landwirtschaftliche Beratungszentrale.

Blum, A. (1993). A confederate agricultural knowledge system: The special case of Switzerland. Paper presented at the Congress of the European Society for Rural Sociology, Wageningen, August 1993. The Netherlands.

Bundesamt für Landwirtschaft (1993). Weisungen zur Verordnung über Beiträge für besondere ökologische Leistungen in der Landwirtschaft (Weisungen Mindestanforderungen) vom 26. April 1993, überarbeitete Fassung vom 28 Februar 1994. Bern.

Kissling-Näf, I. & Knoepfel, P. (eds.) (1993). *Politikorientierte Lernprozesse, Konzeptuelle Überlegungen zu Lernprozessen in Verwaltungen*. Lausanne: IDHEAP, Cahiers de l'IDHEAP No. 120, November 1993.

Landwirtschaftsdirektion des Kantons Bern (1990). *Landwirtschaftliches Beratungskonzept für den Kanton Bern*. Bern.

Nationale Projektgruppe IP-Pilotbetriebe (1993*a*). *Resultate des IP-Pilotbetriebsnetzes 1992*. Tänikon: Eidg. Forschungsanstalt für Agrarwirtschaft und Landtechnik (FAT).

Nationale Projektgruppe IP-Pilotbetriebe (1993*b*). *Auswertung der gesamtbetrieblichen Anforderungen an die Integrierte Produktion 1992*. Tänikon: FAT.

Rogers, E.M., Eveland, J.D. & Bean A.S. (1976). *Extending the Agricultural Extension Model*. Stanford, California: Stanford University, Institute for Communication Research.

Röling, N. (1994). Platforms for decision making about eco-systems. In *Future of the Land: Mobilising and Integrating Knowledge for Land Use Options*, ed. L.O. Fresco *et al.*, pp. 386–93. Chichester: John Wiley.

Roux, M. (1988). *Umweltrelevantes Handeln von Landwirten*; Bericht 11 des Nationalen Forschungsprogramms 'Nutzung des Bodens in der Schweiz'. Liebefeld-Bern.

Roux, M. (1997). Lernprozesse föndern für eine nachhaltige Landwirtschaft in Kulturlandschaften der Schweiz. Dissertation ETH Nr. 12028. Lindau: Landwirtschaftliche Beratungszentrale.

Schweizer Bundesrat (1993). *Verordnung über Beiträge für besondere ökologische Leistungen in der Landwirtschaft* vom 26. April 1993 und Änderungen vom 26. Januar 1994. Bern: EDMZ.

Widmer, C. (1994). Rege Beteiligung am IP-Programm. *Agrarforschung*, **1**(2), 76–8.

5 Extension functions and farmers' attitudes in Greece

ALEX KOUTSOURIS AND DIMITRIOS PAPADOPOULOS

5.1 Introduction

Greece provides an interesting and contrasting case to the other chapters. Through its Common Agricultural Policy (CAP), the European Union provides large amount of subsidies to Less Favourable Areas (LFAs), of which Greece has many.

The selective bureaucratic mechanisms and inadequate services to farmers in the LFAs have implications from the point of view of a reformed CAP. The need to preserve the social fabric, the rural economy and the environment in the LFAs asks for a re-examination of the context and prospects of the farmers living there. The dynamic farmers in the LFAs need to be considered as a key population which, if properly served by extension, could develop into a valuable ally in implementing policy. The custom followed so far, of providing allowances to the LFA population through the CAP's compensatory or environmental regulations is limited in the absence of a long-term development vision. Under the policy followed at present, it is certain that the most dynamic segments of the population, especially the youth, will sooner or later leave the LFA villages (Daoutopoulos, Theofilou & Ananikas, 1989; Panagiotou *et al.*, 1990).

Reversal of this situation requires that research be undertaken in the broader context of sustainability (environmental, economic and social) which will allow the relevant CAP regulations and the proposed additional measures to be utilised in the LFAs. Unfortunately, under the current status and mode of operation of the extension service, this opportunity is largely ignored and the regulations are not utilised to their potential. Hence, there is a need to reconsider the role and function of the Greek extension service. Its policy with respect to the dynamic farmers of the LFAs needs to be changed. They need to become a special target clientele. Research (technical and socio-economic) could assist by defining the specific opportunities of particular agro-climatic zones and local farming systems in these LFA areas.

The findings of the present study allow two main conclusions. The first concerns the way the extension service conceives the situation of farmers, and the second the way the service functions. With respect to the former, one can conclude that an approach based on Farming Systems Research (Hildebrand, 1982; Farrington & Martin, 1987; Collinson, 1981; Chambers & Jiggins, 1986; Chambers, 1987; Röling & Engel, 1991) is needed to understand farmers' conditions and needs. Differences of altitude, corresponding agro-climatic zones and resource availability, together with the nature of Greek agriculture, especially in the LFAs, with their diversity of crops and different types of animal husbandry call for abandoning normative 'scientific' stereotypes of the way farmers are perceived and their needs understood.

Small-scale, low-resource farming systems are complex environments where agricultural production is risk-prone and heavily dependent on climatic factors, particularly on the availability of water. However, risk can be reduced and production increased by exploiting the diversity of such farming systems, as well as by introducing elements that can create additional opportunities (Lightfoot & Noble, 1993). Moreover, local knowledge has to be an essential component in understanding these complex systems. Farmers are the most appropriate people to develop new management practices and to experiment with changes in their farming systems (Chambers, 1993; Etling & Smith, 1994). Of course, 'not all types of local knowledge are in 'harmony' with local eco-systems' (Murdoch & Clark, 1993) and a populist romanticizing of farmers' knowledge must be avoided (e.g. Scoones & Thompson, 1993).

The issue of sustainability requires the development of new conceptual frameworks and modes of operation. A livelihood is said to be environmentally sustainable when it maintains or enhances the local assets on which that livelihood depends. At the local level, the question is whether livelihood activities maintain, regenerate and enhance, or deplete and degrade the local natural resource base. Furthermore, a livelihood is socially sustainable when it can not only gain, but maintain an adequate and decent livelihood and provide for future generations (Chambers & Conway, 1992). Sustainability may involve deliberate policy measures, new research agendas, extensive knowledge of locality-specific systems, appreciation of indigenous knowledge, experimentation with and training of farmers, highly skilled facilitators, etc. (Röling, 1992, 1994).

The development needs of the different farmer categories in Greece call for a clear conceptualization of the roles of extension workers. The ideotypes of the 'change agent' and the 'advisor' cannot be found as such in everyday reality, nor can one maintain in any *a priori* manner which of these roles is the most appropriate. Social, political and development aspects are heavily involved in determining the nature of any service. However, it seems that what most people in the LFAs are in need of are change agents who can cater for a 'mix' of essentials required to get desirable developments off the ground. Meanwhile, the 'dynamic' and 'intermediate' farmers of the well-endowed plain areas appear to be served more appropriately by some type of advisory service.

A further conclusion regards the re-organization of the training division of the service. To date, only farmers using Regulation 797/85 (now R. 2328/91, the main modernization scheme requiring eligible farmers to allocate a certain minimum number of labour units to agriculture each year, as well as to introduce innovations to stabilize or increase the productivity of labour in the medium term) have some form of training of limited duration (150–300 hours), with no evaluation of the results obtained. The lack of a 'Green Certificate', and new prospects for agro-forestry, agro-tourism, alternative crops, organic farming and environment preservation (Koutsouris, 1994), certainly calls for big efforts in terms of training, especially in non-prosperous areas where the continuity of farming is questionable (Bennet, 1990).

For the Greek extension service, these conclusions imply a 'revolution', given its present policy, structure and mode of operation.

5.2 The Greek extension service

The Greek extension service, finally established in 1950 with assistance from the Marshall Plan, was the first systematic attempt by the state to implement an integrated system of occupational training for farmers. The basic aim was the re-organization of the agricultural sector, which had been largely destroyed during World War II and the subsequent Civil War. During the 1950s and the first half of the 1960s, the service was effective in achieving its targets. This period is thought of as the golden age of extension in Greece. Extensionists became 'change agents' in the classic sense of the word, integrating faith in their mission, enthusiasm for their job, even if the circumstances were difficult, the trust of farmers (since extension was working towards the satisfaction of their felt needs), and an integration with the rural population which was aided by a missionary attitude and a popular image. The problems with which extensionists were confronted were relatively easy to solve by means of the existing technical knowledge and the introduction of new/improved inputs (seeds, fertilizers, chemicals, machinery, etc.).

After the mid-1960s, the nature of the agricultural sector changed. The extension service proved unable to tackle emergent issues such as structural problems, which could not be solved by an increase of production *per se*. In addition, the service changed its perception of its clients from peasants to farmers. That is, the perceived role of the extension worker changed from change agent to advisor (Archibald, 1968). Farmers were supposed to be able to deal with most of their problems without assistance, unlike peasants who were seen to need assistance initiated by the extension worker. Extension was to provide mainly technical advice, usually upon request.

At roughly the same time, the extension service was called upon to fulfil new administrative tasks. Extension workers were required to spend the larger part of their time on paper work. As a consequence, extensionists gradually changed into office workers. The vacuum thus created in the field was filled by agronomists who either worked for, or established, private commercial companies promoting various kinds of inputs. By and by these agronomists became antagonistic to the official extensionists. For its part, the extension service did not co-operate with the private agents, given the network to which it belonged (Schwartz, 1994).

After Greece's entry into the European Community in 1981, the burden of administering the implementation of the CAP regulations and subsidies was automatically assigned to the extension service. The CAP requires Greek farmers to become competitive entrepreneurs, first within the European Union, but, in the light of the GATT negotiations, also worldwide. This development implies a change of the extension role from the advisory model back to the change agent, be it in a different manner than during 'the golden age'. Instead of adopting an extensive social and educational role, extensionists now have to segment farmers into categories according to criteria established by the European Commission, and detect their deficiencies. These deficiencies are thought of as obstacles to modernization and in need of remediation.

Meanwhile, no major reorganization of the extension service has been undertaken. Hence the mainly bureaucratic role of extension workers is maintained, but with a vengeance; extensionists have been given further duties to control the implementation of the CAP regulations (Panagiotou & Louloudis, 1984; Ministry of Agriculture, 1989). They are severely restricted in their capacity to provide advice and are able to assist information seeking individuals only in a rather fragmented, inadequate and inefficient manner.

Given this situation, the present study attempts to analyse the interface between farmers and the extension service in terms of the mutual perceptions of different categories of farmers and extension workers, and in terms of the intervention policies of the service.

5.3 Methodological approach

The study draws partly on the quantitative data collected for an investigation of the modernization of Greek agriculture.[1] In addition, data were collected through an ethnographic case-study which aimed to clarify some of the findings of the quantitative study.

The Prefecture of Phthiotis was selected as the area of study because it can be considered representative, at the national level of, on the one hand, the existing diversity among natural environments, and of production systems on the other (Panagiotou, 1974).

5.3.1 The survey

The quantitative study was based on a survey of a random sample of farmers. The sample was derived by first purposefully selecting rural communities considered to be representative of the Prefecture, and then sampling farmers within those communities. The rural community can be considered as the basic unit of agricultural development, constituting an outstanding social, geographic and to a large extent ecological entity, with quite specific structural characteristics and production systems.

[1] The study, 'The Modernisation of Greek Agriculture and the Greek Farmer' was funded by Greek agriculture 1989 to 1991 and managed by A. Panagiotou and L. Kazakopoulos.

The 187 communities in the Prefecture were divided into six categories on the basis of their production system (Panagiotou, 1986). Using these categories, 30 communities were designated as representative of the Prefecture. Subsequently, in each of these 30 communities, the farms belonging to permanent residents were ranked on the basis of acreage. A sample was then drawn using stratified random sampling procedures. The survey was divided into two stages. A total of 738 farmers were selected for the first stage of the survey. The questionnaire administered to them mostly concerned information on 'objective data' (farm and family profile, contacts with the service, knowledge and use of Regulations, etc.). In the second stage, an equivalent sample of 413 farms were selected to be interviewed using a questionnaire concerning 'subjective' socio-psychological issues.

The data used for this chapter comprise community characteristics (secondary data) and data from the two questionnaires. The latter can be categorised into three broad groups:

- variables referring to the household (demography, succession), the farm head (age, general and agricultural education, occupation, etc.) and the farm (owned, hired, cultivated and irrigated land),
- variables referring to communication aspects (personal contacts with rural services and extensionists) and variables referring to the knowledge of schemes related to farm modernization as well as to changes on the farm level (introduction of new crops and/or cultivars, and introduction of new techniques), and
- variables referring to farmers' perceptions of extension interventions (adequacy of the information provided, of the role of extensionists, of the informal occupational training they receive, etc.).

The multi-variate statistical analysis of the data, to identify and analyse relationships among groups of variables, employed factor analysis (Labrousse, 1983; Lagarde, 1983; Volle, 1985), and cluster analysis as a supplementary method to allow a better understanding of the results provided by the factor analysis.

5.3.2 Ethnographic study

The ethnographic study took place in Phthiotis in 1991–92 and aimed to clarify the social processes at the interface between the Extension service and the farmers (Long, 1989). Open-ended interviews with the staff of the Directorate of Agriculture in Phthiotis and participatory observation techniques were used to understand the attitudes of extensionists towards different categories of farmers, and the negotiation processes among extension staff and farmers, on an ordinary, day-to-day basis, both in the Directorate of Agriculture and in local extension offices. Discourse analysis (Potter & Wetherell, 1987) was employed to understand the attitudes of extensionists towards farmers as clients of a bureaucratic organisation, as well as farmers' responses to the Service's mode of intervention.

5.4 Results of the quantitative study

5.4.1 The Prefecture of Phthiotis

The statistical analysis of the survey data yielded two main dimensions among the variables included in the multi-variate factor analysis. The first dimension can be called the 'dynamism' of farm households: the extent to which the households are modernized, entrepreneurial and can be considered to have high potential for further development. The scores on this dimension allow a distinction to be made between dynamic and less dynamic farm households.

Three basic farm household profiles emerge. First are households characterized by a high total ESU[2] ratings (10 ESU and above), large total area under cultivation (10 ha and above) and area owned (8 ha and above), and large total area under irrigation (6 ha and above). These households also have farm operators with a high level of education (at least third-year secondary school), who are relatively young (below 44 years) and have some form of agricultural training. These households also have a good demographic structure, while the contribution of crop production to the total ESU is considerably higher than that of animal production.

The second category of household shows figures for the above mentioned variables in the medium range. The third category comprises farms whose operators are elderly, mostly pensioners who farm to a limited extent, with low educational levels (incomplete primary schooling), small freehold farms (3 ha or less) with limited rented land and no or negligible irrigation, practically no machinery and low ESU ratings (3.5 or less). Animal production makes a higher contribution to farm income than in the two previous profiles.

The outcome of the factor analysis allows the conclusion that the differentiation among the three farm profiles is quite pronounced, particularly between the two extremes. These results were supported by the cluster analysis which confirmed the plausibility of the three household profiles. However, we should allow for the possibility that they are, perhaps, an artefact of the stratified sampling method used.

The second dimension which emerged from the factor analysis comprised variables mainly referring to the characteristics of the communities. These variables are not altogether independent from those of the first dimension. For example, if we take the community variables of altitude and population size, households in mountainous areas, as well as those in smaller communities seem to be less 'dynamic', while those living in the plains' communities and in highly populated communities tend to be 'dynamic'.

Altitude is one of the main components of the second dimension. Knowledge from secondary data and our own experience suggest that differences

[2] The European Size Unit (ESU) was, for the period under consideration, based on a unit of 1200 ECU. That is, the economic size of a holding is obtained by dividing the total standard gross margin of the holding by 1200 ECU (1 ECU = 1.26 US Dollars at 1 October 1996).

in socio-economic indicators, as well as in environmental factors and production systems, are related to altitude. This led us to factor analyse the three categories of communities (plain, semi-mountainous and mountainous) separately.

5.4.2 The plains communities

Five dimensions emerged which together explained 50% of the variance. We focus on the first two. The first dimension again expresses the 'dynamism' of farming households, allowing the three household profiles. Variables describing characteristics of the communities again load on the second dimension, but with them are variables describing the contribution of crop and animal production to the total ESU of the farm and the age of the farm operator.

Further analysis revealed that households with the 'intermediate' profile are characterized by negligible changes in cultivars and cultivation practices, but do show change in cultivated crops and best knowledge and use of programmes and schemes. The farm heads are aged between 45 and 64 years, have a medium general education but no agricultural training, and tend to hold non-agricultural jobs in addition to their farm work. They do not actively seek information, do not maintain close contact with rural services, think of extensionists as bureaucrats and are not satisfied with them.

By contrast, the 'dynamic' households, characterized by a good demographic structure, youth and education level (both in terms of general education and agricultural training), and the largest (mostly irrigated) farms, and which are exclusively occupied with farming, appear to have restricted knowledge and use of programmes but, at the same time, do introduce changes in term of new varieties and practices. These dynamic households maintain communication with all agricultural agencies (including extension). The farm operators actively seek information, visit research stations and agricultural exhibitions and are satisfied by the services provided by extensionists in terms of advice and help. They tend to think of extensionists as active individuals, not heavily involved in bureaucratic tasks, who play a rather significant role.

The numbers of 'less dynamic' households are negligible in the plains and marginal in terms of the analysis.

5.4.3 The semi-mountainous communities

Again, five main dimensions emerged, which together accounted for 47% of the variance. Attention focused on the first two dimensions, with the first again expressing 'dynamism', and the second comprising variables describing the characteristics of the communities as well as the contribution of crop and animal production to the total ESU of the farms.

The households with the intermediate profile, that is, households with middle-aged operators, medium general educational achievement, no agricultural training and medium sized farms, are characterized by minor, in fact negligible, changes in the system and practices of cultivation, and have negligible contacts with rural services.

By contrast, the 'dynamic' households have made changes both in the system of cultivation (introduction of new crops and varieties) as well as in practices. These households maintain extensive communication with all agricultural agencies and have the best knowledge of programmes and schemes.

5.4.4 The mountainous communities

Five dimensions account for 50% of the variance. The first dimension, 'dynamism', again allows three household profiles to be identified, with the most 'dynamic' one being clearly distinct from the remaining two which can be taken as a single profile. The variables referring to the contribution of crop and animal production to the total ESU of the households and those describing the characteristics of the communities load on the second dimension.

The analysis allows us to observe that the households with the 'dynamic' profile are characterized by changes both in the system of cultivation (introduction of new crops and varieties) as well as in practices, concurrent with visits to research stations and agricultural exhibitions. Within the intermediate and less 'dynamic' households, characterized by medium to poor demographic household structure and medium to small farms as well as by a combination of advanced age, lack of (agricultural) education, there appear to have been negligible changes in systems of cultivation and farming practices. At the same time, the less 'dynamic' farmers seem to be satisfied with the services provided by extensionists, despite the fact that they do not actively look for information, have no contact and do not favour training. They generally tend to think of extensionists of as being active individuals, not heavily involved in bureaucratic tasks and holding a significant role.

The 'dynamic' farmers tend to know and use programmes and schemes, have contacts with veterinary staff and extensionists, and are willing to attend short courses on technical and economic issues if provided. They also tend to seek actively for information, but they are not satisfied with the extension service and think of extensionists as bureaucrats whose role in local development is not significant.

5.5 The ethnographic study

In the perception of the extension service, dynamic farmers can be found only in the plains areas. While extensionists recognize that both dynamic and non-dynamic farmers can be found in the plains, only less dynamic farmers are found in the mountainous and, to a larger degree, in the semi-mountainous areas (the last two being defined as LFAs). The justification of this perception is that there are no prospects for agricultural development in LFAs. Hence dynamic farmers in those areas are exceptional.

The perceived nature of an entrepreneurial farmer, according to the 'productionist' model, became apparent through the analysis of both the extensionists' discourse and the Service's documents. The farming enterprise which is considered 'competitive' excludes all forms of alternative practices mainly suitable to the LFAs. According to extension discourse and practice, therefore, a farmer must not only display dynamic personal characteristics, but also live in the proper (i.e. plains and/or irrigated) area, in order for his farm business to be considered viable and competitive.

Thus, the service still operates according to the typical 'progressive farmer' strategy (Röling,1988), deliberately directing its bureaucratic services to the more resource-rich farmers in the plains areas, and virtually ignoring the other categories as policy targets. Local extensionists systematically develop practical relationships with some of the farmers in the plains so that they can delegate development tasks to them. On their part, resource-rich farmers construct personal social networks with extensionists at the local and regional level so as to gain access to the Service's material and information resources which are particularly valuable to them. Furthermore, the existing CAP regulations (e.g. R.797/85) enable their participation in development schemes, while these same regulations are an obstacle to the participation of the other farmers.

In conclusion, the effect of these social processes is that only a minority of farmers (the resource-rich farmers in the plains areas) have regular access to the Service's resources. The mode of the Service's intervention, the existing 'progressive' farmer-orientated regulations and the deficient and unstable institutional framework which determines the relationships between the Service and its clients, form an organizational, political and institutional complex which favours the 'progressive' farmers, leaving the rest at the margins of the Service's intervention policies and practices.

5.6 Combining survey and ethnographic case-study

This section explores the interventions made by the extension service and farmers' responses to them. The exploration is made on the basis of a synthesis of survey data and the ethnographic case-study in the Prefecture of Phthiotis.

As we saw, the quantitative analysis allowed us to identify profiles of households which differed on a number of indicators such as acreage, irrigated land, age and education of the farm owner, etc. The so-called 'dynamic' households displayed the highest values of such variables. Their farms tend to be located in the plains, the more prosperous areas in the Prefecture. On the other hand, the so-called 'less dynamic' households, demonstrating the 'less favourable' characteristics tend to be found in the LFAs of the Prefecture.

It should be noted, however, that the findings from the aggregation according to altitude zones do not differentiate between the areas when it comes to variables pertaining to the number of innovations adopted and attitudes towards extension. Therefore, separate analyses were run for each altitude zone. The results obtained justified this analysis since the three types of households were found to play different roles in each zone.

In the plains areas, two distinct categories of households were identified, namely the 'dynamic' and 'intermediate' ones, while 'less dynamic' households were hardly present. The first were found to have made changes, to keep extensive communication networks with all the agricultural services, to actively seek information and to express their appreciation for the services provided by extensionists. The 'intermediate' households keep pace with modernization schemes but have neither extended contacts with services nor actively looked for information, which is reflected in their negative attitudes towards extension.

In the semi-mountainous areas, further analysis revealed that the 'dynamic' households display extensive changes, well-established networks and good knowledge of modernization opportunities, but these farmers are dismissive of extensionists' contribution. As far as the other two profiles are concerned, the corresponding indicators reflect absence of change and contacts, and a negative attitude towards agricultural training on the part of 'less dynamic' households.

In all, differences can be observed among households similar in terms of their background characteristics, but which live in different altitude zones. The dynamic farmers in the plains appear to have different perceptions of extensionists to those in the LFAs. The latter, though informed about modernization issues, are not satisfied with extensionists and depict them as bureaucrats who do not substantially contribute to development.

Intermediately dynamic households also differ among themselves. The ones located in the plain areas are the most informed about, and most active in applying for modernization schemes, but do not have contacts with services and display negative feelings about extensionists. In contrast, the ones located in the LFAs, while not engaging in changes and contacts, seem to be more or less satisfied by the extensionist services provided. Finally, the less dynamic households, mostly located in the LFAs, are not involved in any kind of modernization scheme or communication network yet appear to appreciate extensionists but not agricultural training.

These rather baffling findings must be discussed in the light of the function of the extension service and the CAP regulations in Greece.

5.7 Discussion

The more or less homogeneous development ideology within which the Greek extension service operates is manifested in the uni-dimensional development discourse deployed around the topic of agricultural competitiveness in the framework of productivist agriculture. As our ethnographic study clearly indicates, the target group is confined to the category of the 'dynamic' farmers located in the plains areas. The information and training needs of other categories of farmers are largely ignored. The application of R. 797/85 (now R. 2328/91) has had a major influence on thinking within the extension service. As mentioned earlier, this regulation is the main modernization scheme in use.

For the LFAs, the provision of other, compensatory, allowances has been the main policy instrument, largely aiming to help farmers to survive under difficult circumstances while in a few cases increasing their production as well. Given the considerable differences between plains and LFAs in terms of the availability of fertile land, irrigation and favorable agro-climatic factors, given further that high income crops cultivated in the LFAs, such as oriental tobacco, are not on the CAP list of crops to be officially promoted, and finally, given the lack of research on alternative crops and enterprises, it is little wonder that the CAP modernization schemes exclude farmers in the LFAs.

Especially since 1981, the Greek extension service has played a largely administrative role with respect to the implementation of the CAP regulations and provision of subsidies. A supplementary role, which often creates tension between extensionists and farmers, is that of controlling the implementation of regulations and verifying the statements farmers make in their bid to receive subsidies and compensation. The lack of a cadastre and/or farmers' register are major institutional deficiencies necessitating such a mode of operation. But extension's role as bureaucrat and controller of the distribution of resources available under the CAP is also supplemented, in terms of working time, by providing technical advice upon request. It must also be noted that the absence of any well-defined perspective on the farmer as entrepreneur allows for personal estimates of farmers' status and needs in the course of scheme implementation. This situation is enhanced by the way the service operates in terms of communication. Local personnel and middle-upper echelons communicate within the typical framework found in any bureaucratic organization (Pettigrew, 1979), which allows for considerable degrees of freedom at the local level (Lipsky, 1980; Wagemans, 1987).

This, of course, does not mean that extensionists accept their burden without reaction. On a number of occasions they have expressed their negative feelings about the kind of work they are required to do, asking for their present role to be abandoned and to be given an active role in local development through involvement in programmes and communication with, and training of, farmers.

Working in offices allows only for a certain degree of interaction with a limited number of clients. According to the normative mode by which the extension service functions, these clients tend to be the so-called 'progressive farmers' (Röling, 1986, 1988), whose profile corresponds to those of the 'dynamic' farmers in the plains areas. Dynamic farmers from the plains areas certainly have more access to the service's resources, not only because the CAP framework enables their participation in the development schemes, but also because they belong to informal social networks shared by service staff, which in turn facilitate their access (Papadopoulos, 1995). This situation can explain the farmers' perceptions of extensionists. Dynamic farmers in the plains, who already have large, efficient enterprises, are mainly in need of information concerning technical developments and modernization schemes. Such information can be relatively easily provided by extensionists, to whom access is easy in social terms. Therefore, dynamic farmers in the plains tend to be satisfied with the extension service.

The situation changes for dynamic farmers located in the LFAs. Such producers try to keep up with modernization in agriculture usually under less favourable circumstances. They do not feel satisfied with extensionists and accuse them of being bureaucrats, of not being easily accessible for advice, and of making no significant contribution to development. Distance from the local offices, lack of appropriate social networks, difficulty in meeting regulatory requirements, the administrative burdens, as well as their preoccupation with dynamic farmers in the plains, can explain the reactions of the dynamic farmers in the LFAs.

The same is true for the 'intermediate' farmers in the plains. These farmers would like to see extensionists becoming more active, both in the office, that is, more helpful with respect to the implementation of regulations in terms of bureaucratic procedures, and in the field in terms of technical advice about recently introduced crops and techniques. As far as the intermediate and less dynamic farmers in the LFAs are concerned, they seem to be satisfied with the mode of operation of the extension service. This can be attributed to their age and state of farming. With no modernization prospects, they are satisfied with the opportunity to gain money through the compensatory allowances without which they would find it difficult to survive, and they are consequently not attracted by prospects of agricultural training.

5.8 References

Archibald, K. (1968). *The Utilisation of Social Research and Policy Analysis*, pp. 164–84. Michigan, USA: Ann Arbor.

Bennet, C. (1990). *Cooperative Extension Roles and Relationships for a New Era*, Extension Service, US Dept. of Agriculture. Washington DC, USA.

Chambers, R. (1987). Sustainable livelihoods, environment and development: putting poor rural people first, IDS Discussion Paper, no. 240. United Kingdom: University of Sussex.

Chambers, R. (1993). Methods for analysis by farmers. *Journal for Farming Systems Research – Extension*, **4**(1), 87–101.

Chambers, R. & Conway, G. (1992). Sustainable rural livelihoods: practical concepts for the 21st century. IDS Discussion Paper, no. 296. United Kingdom: University of Sussex.

Chambers, R. & Jiggins, J. (1986). Agricultural research for resource: poor farmers, IDS Discussion Paper, no. 220. United Kingdom: University of Sussex.

Collinson, M. (1981). Farming systems research: diagnosing the problems. *Agricultural Administration*, **8**, 33–50.

Daoutopoulos, G., Theofilou, M. & Ananikas, L. (1989). Rural Youth, *Unit of Agricultural Extension and Rural Sociology*. Greece: Aristotle University of Thessaloniki.

Etling, A.W. & Smith, R.B. (1994). Participatory needs assessment. *Journal for Farming Systems Research – Extension*, **4**(2), 45–56.

Farrington, J. & Martin, A. (1987). Farmer participatory research: a review of concepts and practices, ODI Discussion Paper, no. 19. London.

Hildebrand, P. E. (1982). Farming systems research: issues in research strategy and technology design. *American Journal of Agricultural Economics*, Dec. 1982, 905–6.

Koutsouris, A. (1994). Crucial factors related to the education/training of new entrants into agriculture in Greece. PhD thesis. Unit of Agricultural Extension, Dept. of Agricultural Economics, Agricultural University of Athens.

Labrousse, C. (1983). *Introduction à l' Econometrie*. Paris: Dunod.

Lagarde, J. (1983). *Initiation à l'Analyse des Donees*. Paris: Dunod.

Lightfoot, C. & Noble, R. (1993). A participatory experiment in sustainable agriculture. *Journal for Farming Systems Research – Extension*, **4**(1), 1–9.

Lipsky, M. (1980). *Street Level Bureaucracy: Dilemmas of the Individuals in the in-public Service*, Russel Foundation, NY, USA.

Long, N. (ed.) (1989). Encounters at the interface. *Wageningse Sociologische Studies, no. 27*, Wageningen: Agricultural University.

Ministry of Agriculture (1989). *Proceedings of the 1st Scientific Meeting of Agricultural Extension*, Athens, June, 21–23, 1988.

Murdoch, J. & Clark, J. (1993). Sustainable knowledge. Paper presented in the XVth European Congress of Rural Sociology. 2–6 August 1993, Wageningen, The Netherlands.

Panagiotou, A. (1974). *Comparative Research on Agricultural Development at the Prefecture of Phthiotis: A Typological Approach*, Agricultural University of Athens, Greece.

Panagiotou, A. (1986). Production structures and agricultural development in Greece: a methodological approach to the development of unities of agricultural development. In *Proceedings of the International Congress on Economy and the Rural Sector*, Agricultural Bank of Greece, November 1984, Greece.

Panagiotou, A. & Louloudis, L. (1984). *Higher Agronomic Studies and the Effectiveness of Extension Officers*, Agricultural University of Athens, Greece.

Panagiotou, A. *et al.* (1990). Rural youth: the results of an empirical research in Central Greece. Paper presented in the 1st Panhellenic Conference of Rural Economy, Agricultural University of Athens, Greece.

Papadopoulos, D. (1995). From negotiations to networks: a study of the responsibilities' orbits at the interface between a rural bureaucracy and dynamic farmers in Greece. MSc thesis, MAKS. Wageningen: Agricultural University.

Pettigrew, A. M. (1979). On studying organizational cultures. *Administrative Science Quarterly*, **2**, 570–81.

Potter, J. & Wetherell, M. (1987). *Discourse and Social Psychology: Beyond Attitudes and Behaviour*, London: Sage.

Röling, N. (1986). Extension and the development of human resources: the other tradition in extension education. In *Investing in Rural Extension: Strategies and Goals*, ed. G. E. Jones, pp. 51–64. Essex: Elsevier Publishers.

Röling, N. (1988). *Extension Science. Information Systems for Agricultural Development*. Cambridge: Cambridge University Press.

Röling, N. (1992). Facilitating sustainable agriculture: turning policy models upside down. Paper in the Workshop Beyond Farmer First: Rural People's

Knowledge, Agricultural Research and Extension Practice, October 27–29, 1992. Sussex: IDS, University of Sussex.

Röling, N. (1994). Communication support for sustainable natural resource management. *IDS Bulletin*, **25**, 125–33.

Röling, N. & Engel, P. (1991). The development of the concept of agricultural knowledge information systems (AKIS): Implications for Extension. In *Agricultural Extension: Worldwide Institutional Evolution and Forces for Change*, ed. W.M. Rivera & D.J. Gustafson. Amsterdam: Elsevier Science Publishers.

Röling, N., Ascroft, J. & Chege, F.W. (1981). The diffusion of innovations and the issue of equity in rural development. In *Extension Education and Rural Development*, ed. B.R. Crough & S. Chamala, pp. 225–36. Chichester: John Wiley.

Schwartz, L.A. (1994). The role of the private sector in agricultural extension, ODI Network Paper 48, London.

Scoones, I. & Thompson, J. (1993). Beyond farmer first – rural people's knowledge, agricultural research and extension practice: towards a theoretical framework. Paper presented in the XVth European Congress of Rural Sociology, 2–6 August 1993, Wageningen, The Netherlands.

Volle, M. (1985). *Analyses des Doneés*. Paris: Economica.

Wagemans, M.C.H. (1987). Voor de verandering. Een op ervaringen gebaseerde studie naar de spanning tussen theorie en de praktijk van het besturen. Dissertation: Wageningen Agricultural University.

6 Integrated arable farming in the Netherlands[1]

WILLEM VAN WEPEREN, JET PROOST AND NIELS G. RÖLING

6.1 Introduction

The Dutch Government has been experimenting since 1979 with integrated arable farming (IAF) on its experimental farms in Nagele, Vredepeel and Borgerswold. This work was initiated in response to pressure to carry out research on more sustainable farming systems. Three systems were compared:

1. the (very) high external input system that has been conventional in the Netherlands since the late 1950s;
2. a biological farming system using no chemical inputs at all; and
3. an IAF system that replaces chemical inputs of pesticides and minerals as much as possible with mechanical and biological products and processes, but does not ban them entirely. In other words, 'integrated' is used in the same sense as in 'integrated pest management' (IPM) (see Röling & van de Fliert, this volume), which integrates mechanical, biological and chemical pest controls. In fact, IPM is a component of IAF. IAF does not seek solely to increase physical output, but rather to reduce production costs and improve the quality of the produce by reducing the use of inputs that endanger the environment and possibly also human health and by increasing the use of knowledge and labour (Vereijken, 1990).

Ideally, IAF requires considerable change in farm practices. In the first place, it relies heavily on farmer observation of what happens on the farm. Such observation requires new methods, instruments and indicators to make things visible. For example, considerable attention must be paid to the analysis of the organic manure used on the farm so as to increase the precision of nutrient application.

In the second place, farmer observation implies interpretation and anticipation. In fact, a farmer must develop considerable conceptual skills to be able to interpret his/her observations and to take forward-looking decisions. An example of such a conceptual change is the adoption of an eco-system perspective, which expresses itself, for example, in the deliberate care of the habitats of the natural enemies of pests. Or, to take another example: instead of only focusing on feeding individual crops, the integrated arable farmer will also think about nutrient

[1] This chapter is based on Van Weperen *et al.* (1995); Röling (1995) and Proost *et al.* (1995).

movements in the soil across the rotation of several crops, and of ways to keep nitrogen available in the upper layers of the soil between crops.

Thirdly, IAF requires the use of more resistant and/or tolerant cultivars, attention to more diverse crop rotations, and use of new tools and equipment for mechanical weeding, for preventing drift when spraying, and for low dosage and spot applications of chemicals. IAF also implies the use of different, more specific and less toxic chemicals. In all, IAF comprises two different sets of change:

1. the adoption of different external technologies, such as new cultivars or machines, and the purchase of services from specialized agencies. This type of change does not really require the farmer him or herself to change much;
2. a transformation in farming practices, i.e. managing the farm as an ecosystem based on observation, interpretation and anticipation. This type of change requires the farmer to go through a great deal of learning of both an experiential and technical kind.

We make this distinction up-front because we shall have cause to come back to it later.

During the period 1986–1990, substantial reduction in chemical use was achieved in the IAF experiments on the three experimental farms. The use of herbicides, fungicides and insecticides was reduced by 50–60%, and nematicides were no longer used at all. What is more, in comparison with conventional and organic arable farming, IAF scored (moderately) better in terms of financial profit. These results were encouraging enough to justify testing under farmers' conditions.

These achievements coincided with the growing public alarm about the effects of conventional intensive agriculture on the environment and with the acceptance of far-reaching environmental laws specifying reduction in the use of pesticides, conservation of bio-diversity, and reduction of manure surpluses. The laws were promulgated with a focus on achievement targets.

One example is the Multi-Year Crop Protection Plan (MYCPP) of 1991, which specifies that pesticide use must have been reduced by 50% over 1984–1988 levels by the year 2000, with intermediate steps which are monitored by Parliament. But the Plan does not specify *how* farmers are to achieve these targets. This is left up to the industry, within the existing market conditions. In other words, no special subsidies or other assistance to farmers is provided, except for funds for experimental research and innovation projects. The government did mount a special extension campaign to support the MYCPP. Later on in the chapter, we summarize the evaluation of this campaign. (The reactions of farmers to the Nature Policy Plan are discussed by Van Woerkum & Aarts and by Wagemans & Boerma, this volume.)

Given the political context, culminating in the various environmental laws, the experimental development of IAF methods was a godsend. Here was a ready-made alternative, albeit insufficiently tested on different soil types, which perhaps could solve the problem raised by the new laws. Adoption of IAF by 100% of the farmers was foreseen by 2000 (Wijnands *et al.*, 1992).

The policy-driven need to introduce viable alternatives to conventional high external input agriculture has led to several initiatives by informal groups of farmers, farmers' organizations (e.g. the Young Farmers' Association), drinking water companies, various local government agencies such as provinces and communities, and by the Ministry of Agriculture, Nature and Fisheries. Somers (this volume) reports on some of these initiatives. The present chapter reports on another: the 'Integrated Arable Farming Innovation Project' of the Ministry which was concluded in mid-1994 (Wijnands *et al.*, 1995). In this project 38 arable farmers from various parts of the Netherlands volunteered to experiment with the incorporation of integrated farming techniques in their current farming practices. These 38 farmers represented *all* those who responded to an open call in the Dutch farming media for volunteers. After the conclusion of the project, 500 other arable farmers became involved in a new project called 'Arable farming 2000'. This project is expected to be concluded by mid-1996. The lessons learned by the 38 innovators are incorporated into this second project.

Meanwhile, the environmental laws mentioned above continue to exert pressure on all arable farmers, so that many of the IAF techniques and methods are being adopted fairly widely, as we shall see later when we discuss the impact on industrial potato production in the northern part of the Netherlands.

The chapter focuses on the transformation that farmers experience when changing to IAF. To this end, three sources of data are examined. The first source is the 38 participants in the 'Innovation Project Integrated Arable Farming'. They were asked to complete a questionnaire about their experiences during the last meeting, in Lunteren in 1994, of all those involved in the Project. The following section reports on these experiences. The second source is a case study which focuses on the changes in arable farming in the industrial potato producing area. We report on this case with a view to examining the impact of the environmental laws on a wider scale. And third, we give an overview of farmers' ideas and attitudes regarding integrated farming based on two national surveys in 1991 and 1994 to evaluate the MYCPP.

In the final section, we formulate some hypotheses on the basis of a comparison of the three studies.

6.2 The experience of the 38 innovators

The 'Integrated Arable Farming Innovation Project' consisted of government funding to support research by various institutes, and intensive extension guidance to the participating farmers. The farmers were divided into five regional (soil type) groups, each about eight farmers. Each group was guided throughout the project by a specialized advisor employed by the privatized Agricultural Extension Service (DLV). The farmers were trained and backstopped by the Experimental Station for Arable Farming and Broad Acre Vegetable Crops (PAGV) and the related Information and Knowledge Centre. The project was managed by a guidance committee consisting of all these and other involved institutions. The project funding comprised a small sum for farmers to defray costs of demonstrations and meetings, and insurance against crop failure as a result of the project. The objective of the project was twofold:

1. to take the IAF approach to regular farmers, and to monitor the results, experiences and feasibility under farmers' conditions; at the same time, to develop region-specific IAF technologies.
2. to introduce IAF in the farming community in order to enhance transfer of knowledge and practical experience, to evaluate farmers' interest and adoption rates, and identify bottle-necks in the technology.

The 38 volunteers signed an agreement declaring their willingness to adhere to IAF practices for the duration of the project and made farm plans in consultation with the specialist advisor. These plans detailed various farming activities, such as the choice of crops and cultivars, crop rotation, fertilizer application, crop protection, weeding and mechanization. The DLV advisor coached the farmers in the utilization of the new practices and visited the participants about twice each month during the growing season. In addition, special study group meetings were organized in the different regions. During the project, the focus shifted from introducing the total package that was developed at the experimental stations, to allowing farmers to adopt those components of the package that best suited them.

6.2.1 Some tangible results of the project

The results of the project were carefully monitored and analysed throughout. Pesticides and minerals used, were measured and recorded for each farm, as were yields and economic results. Wijnands *et al.* (1995) give the following summary in their overview of the technical and economic results of the project:

> The use of pesticides in the last two project years was compared with farm-specific use in 1987–89, the two years prior to the project. The use of herbicides, fungicides, insecticides and growth-regulators was reduced by 60%, 57%, 58% and 66%, respectively, to 1.2 kg, 2.0 kg, 0.1 kg, and 0.0 kg active ingredient per ha. The potato cyst nematode could be increasingly controlled by intensive sampling, testing of cultivars and adapted choice of cultivars. This led to a reduction in the use of nematicides by 75%.
>
> With respect to minerals, integrated farm management led to a dramatic reduction, especially with respect to the surplus of phosphorus. Thus the average surplus of P_2O_5 and K_2O at the farm level decreased by 55 kg and 70 kg, respectively, to 25 kg and 25 kg per ha. The basis for P application was animal manure (an average of 80% of the total P_2O_5 input).
>
> The surplus in nitrogen was decreased by an average of 35 kg to 115 kg per ha. But the percentage of farms on which the average amount of N which remained after harvest in the respective project years was more than 70 kg per ha (at a depth of between 0 and 100 cm), was 77%, 74%, 78% and 18% (wet autumn). Fertilizer management with respect to nitrogen, therefore, did not lead to sufficient

control of potential N losses through leaching. This issue requires more attention, especially with respect to adaptation of N application to crop-specific demand, and with respect to the use of animal manure, especially in the autumn after the harvest.

For 1992–93, the total cost of fertilizers and pesticides ranged from Dfl775 per ha in the north-east to only Dfl330 per ha in the south-east. In the clay areas, Dfl500–700 per ha was spent on pesticides and minerals. This represented a saving of about Dfl250–300 per ha for the clay areas and about Dfl400 per ha in the north-east and south-east. The economic results for the first three years of the project, at farm-level, demonstrate that the 38 innovating farms compare favourably with the farms surrounding them, and that their relative position has not deteriorated.

The norms of the Multi-Year Crop Protection Plan for the year 2000 were achieved for all categories of pesticides. In some of the regions and for some of the categories, the results were far below these norms. A bottleneck was weed control in the north-east, where risks of night frost (in potatoes) and wind erosion impeded a drastic replacement of herbicides by mechanical weed control.

In all, IAF also seems able, in practice, to lead to good results with respect to the input of pesticides and restoring the equilibrium of the mineral balance of P_2O_5. The supplies of N which remained after the harvest were, however, still too high to sufficiently prevent N losses as a result of leaching. This issue asks for further attention.

With respect to economics, IAF had no demonstrable negative effect on farm profitability. Nothing, therefore, needs to impede wide-spread adoption. However, expert guidance and sufficient motivation are indispensable.

The project's results as reported by Wijnands, provide a necessary backdrop for understanding the way in which the 38 volunteers experienced the project. We turn to this subject next.

6.2.2 Farmers' motivation to participate in the project

The 38 participants found the project attractive for various reasons. IAF makes a bigger claim on farmers' professional skills. The project assisted them, as conventional farmers, to incorporate some ecological principles into their farming practice. It gave them the opportunity to experiment with new practices under the guidance of an extension worker.

IAF was felt to be in line with changes the respondents saw already taking place in the arable sector. To some, IAF presented a genuine alternative to conventional arable farming. This was partly based on their rather negative perception of conventional arable farming, and partly on their perception that IAF was a practical and feasible scenario for the future and therefore it was wise to take part in its development. The fact that they would be coached in this innovation process added to the attractiveness of the project.

IAF was seen to challenge the farmer's professional skills. The emphasis is much more on the interpretation of data, on risk assessment and making decisions on the basis of personal observations (instead of on standard recommendations).

The respondents experienced this different type of entrepreneurship as rewarding. It was seen as reducing the more or less obligatory and routine character of, for example, standardized spraying schemes, and to increase work satisfaction. This observation is remarkably similar to the advantages of adopting IPM mentioned by small-scale Indonesian rice farmers (Röling & van de Fliert, this volume).

The opportunity to experiment with new practices under guidance enhanced the attraction of the project. Farmers over the life of the project increasingly wanted to reduce the amount of chemical input into crop production, but they often lacked the knowledge and courage actually to do so. Participation in the project assured the farmers of a guiding hand during the trial of new practices. An essential incentive for the farmers to adopt IAF was the chance to be coached during the innovation process. This is a remarkable result, given that we are dealing with volunteers, all of whom were already motivated to make the change.

In fact, participants distinguished themselves by their motivation and responsiveness to IAF, but risk was said to be an important consideration in whether or not to take part in the project. The offer of professional guidance and insurance against crop failure reduced that risk considerably. Ecological considerations were equally important in deciding whether to participate. The tangible results of the project, especially in terms of reductions achieved in the use of chemical inputs, stimulated participants to proceed on a course that many of them had already embarked upon..

6.2.3 A learning process

The comments of participants strongly suggest that the shift towards integrated arable farming systems is a learning process. Participation made them realize that their former way of farming was not sustainable. Furthermore, integrated farm management demanded several skills which were largely new and had to be acquired. Many confessed that, in the beginning, they felt quite inept when using the new skills, but that their confidence slowly grew.

During the project, attitudes also changed. The shift to IAF led to increased pleasure in farming, to increased self-confidence and to a more critical stance towards information. They wanted to gain knowledge and practical experience that could be applied in their specific situation. The chance to do IAF changed their perception of what it means to be a farmer. Many felt they had acquired more control over their farms and felt surer about their actions.

Farmers became aware that there were feasible alternatives to the conventional methods they had taken for granted. Mechanical weed control actually worked and satisfactory results could be obtained with less spraying. They learned to apply new observational skills, using them to become (re-)acquainted with the various stages in crop growth and to assess crop damage and determine the proper time for spraying. For determining the right moment for nitrogen application, the so-called 'nitrogen window' was used, an instrument which complements the observational capacities of the farmer. Observation also played a role in assessing the consequences of using too much fertilizer. For example, the analysis of samples of water from field drains made visible the extent of nitrogen leaching. In short, the shift towards integrated arable farming stimulated the development of observational, interpretive and predictive skills for decision making.

Not all participants perceived the change in the same way. For some, it was a drastic change, for others IAF fitted closely with their preferred farming style. Individuals developed distinct preferences for some components of IAF and were less happy with others. In general, the criteria used for judging a specific component were its effectiveness, amount of labour needed, cost, stability and economic profitability. The methods used most widely and approved most generally were mechanical weed control, spraying techniques that reduced pesticide use, and new fertilization strategies.

The most important constraint in transferring to IAF was said by some farmers to be that the effort required was not rewarded by a better price for the product. They also felt they had to take much more risk, and were uncertain about the intentions of government, especially with respect to the possibility that IAF would be forced on them in the future. Others often said that the participants were being used to prove that IAF was feasible and hence could be imposed subsequently on all farmers. The lack of more positive feedback from their neighbours also had a constraining effect on the participants.

6.2.4 The role of extension

Being coached while making the shift towards IAF was regarded by all as very positive and inspiring. The initial expectation was that the extensionist would provide information and act as consultant and motivator. In general, counselling visits were paid once or twice a month during the growing season. What appeared important was that the advisor was available at critical moments in the cultivation cycle.

The discussion and learning groups into which the 38 participants were divided had a positive influence on their decisions to carry through innovations in farm management. The group meetings motivated them to try out new things. Another significant function of the study clubs was the exchange of information and knowledge from practical experience. The groups also exerted a mild form of social control, gently pushing group members to stick to their decisions.

6.2.5 The impact of the innovation farms on other farmers

The impact of innovating farms on other farms in the area was regarded as positive though limited. According to the participants, the effect of the publicity and excursions was that their story was better known further afield than it was close to home. Many participants think that the project and its results need more publicity. So far, one cannot say that an innovation wave has been triggered by the project.

The participants were pessimistic about IAF's future potential under existing price and policy arrangements. The main stumbling block was seen to be the unfavourable balance between the labour needed and the prices obtained for produce. There was no agreement as to whether IAF should be promoted as a substitute for conventional farming. Some regarded this as restricting a farmer's freedom to choose his own way.

6.2.6 Implications for agricultural extension

Many farmers have experienced the change to IAF as a series of add-on innovations. Most importantly, it has also been a process of learning a new approach to farming that requires a new set of skills (see also Somers, this volume). This observation has far-reaching implications. We shall return to them in the last section of the chapter. Suffice it here to point out that the role of the extension worker in this type of transition is quite different from promoting the adoption of specific innovations or from the farm advisor who sells his expertise on demand. Indeed, the project experienced difficulties in training extension workers in their new role, especially in giving up the role of expert and embracing that of facilitator of learning.

This is not to say that technical expertise became unimportant during the project. The fact that the advisors were, on the whole, knowledgeable and could fall back directly on the experimental station, did a great deal to ease the farmers' sense of risk in shifting to IAF. The frequency and timing of their visits was apparently crucial. The farmers considered the advisor as a confidant with whom they could share their uncertainties, discuss issues, and come to a decision, but additional skills were required. These skills included the ability to guide farmers in crossing the new boundaries of farming with fewer inputs. That is, he had to be able to understand the learning process the farmer was going through, and stimulate the farmer to continue learning. Similarly, it is one thing to deliver an expert presentation to a group of farmers, quite another to set in motion a process of lively exchange and group formation.

There was another important lesson regarding methodology. The introduction of methods, procedures and equipment to make visible the environmental impacts of farming practices had a powerful effect on farmer learning. An example is the procedure which allows the farmer himself to measure the N content of drain water after using fertilizer. The effects obtained from the reduced use of pesticides was thus very rewarding. The creation of such feedback loops was also found to be very effective in the Dutch dairy sector, among users of the mineral book-keeping system (van Weperen, 1994). The chapters by Hamilton (this volume) and Röling and van de Fliert (this volume) provide further evidence along the same lines. Perhaps this observation warrants the conclusion that extension efforts should start working with practices that have a larger visible impact.

The respondents appreciated the positive atmosphere of the study groups which acted as venues for exchanging experiences and for making (risky) changes acceptable (see also Darré, 1985). Since few sources of farm-specific knowledge were available to the farmer, this proved extremely useful in the development of IAF at the farm level, since IAF relies more on the farmer's adaptation and application of general principles to the farm-specific situation, than on the routine application of ready-made technologies. The small groups provided a good climate for pursuing socially accepted experimentation and generating a sense of solidarity, and (some) competition. This was particularly important in the early stages of the project when some of the participants were ostracized by neighbouring farmers.

6.2.7 Conclusions

Even though the 38 participants were the only arable farmers out of a total of some 40 000 in the Netherlands to volunteer for the project and can be assumed, therefore, to have felt a considerable affinity with IAF, most still appeared to go through a difficult learning process that took a number of years to feel comfortable with. It seems to be a necessary condition for the successful introduction of IAF. The farmers experienced the guidance of highly trained advisors as essential in making the risky and complex transition. Will a widespread transformation from conventional arable farming to IAF require the same fundamental transformation for all who undertake it? If so, we might ask whether such a transformation can be brought about by autonomous diffusion processes, or whether each farmer will require intensive guidance through the learning process. If the latter is the case, perhaps attention should already be given to IAF in primary and secondary agricultural training.

A second conclusion refers to prices and consumers. The respondents felt that the effort they had made to become qualified in IAF was equivalent to acquiring a new diploma. They also felt that the tangible achievements in terms of a better environment and higher quality products were not rewarded by higher prices for their products. Dutch consumers, buyers for supermarkets, and other institutions did not seem prepared to pay for the increased product quality or environmental protection. It must be noted here that the lobbying power of conventional farmers allows them to prevent any public advertising that compares the more ecological IAF products with those of conventional farming. Thus the public does not get the information needed to make a different consumer choice.

6.3 A case study of the industrial potato production area

6.3.1 Industrial potato farming

Every year, groups of foreign students on the International Course on Rural Extension (ICRE) engage in a week-long rapid appraisal of an aspect of Dutch agriculture. Two of the present authors acted as workshop leaders for a group which visited the industrial (starch) potato production area in the north of the

Netherlands in 1995. The case study is based on this experience.[2] We report it here because, as will become clear, the industrial potato producing areas represent a form of arable farming that is under severe pressure to clean up its environmental impact, yet is also experiencing increasing economic hardship. It provides a good case for examining the extent to which IAF is being taken up in the Netherlands.

The terms of reference for the group were to study how pesticide use in potato production is being affected by the MYCPP, which requires pesticide use in industrial potato cultivation to be cut by 60% by the year 2000 compared to the level in the baseline period of 1984–1988. Soil disinfectants and pesticides, applied particularly in intensive potato and sugar beet production, account for about 70% of the pesticides used in arable farming, equivalent, at the inception of the Plan in 1990, to about 19 kg of active ingredients per ha. Before the introduction of the Plan, chemical soil decontamination was compulsory.

Potato production for starch is a relatively insignificant industry in the Netherlands, but much larger in Germany and Eastern Europe. Its viability in the Netherlands seems low compared to other European countries, where farms are not as specialized, usually engage in some off-farm employment, and are not so heavily indebted as Dutch farms (Knickel, 1994).

Potatoes for the starch industry are grown on the sandy soils remaining after digging down the peat cover. The underlying sands are mixed with the top layer of the peat. The boggy moorlands converted in this manner once covered vast expanses of Europe. The exploitation of the peat itself occurred largely in the nineteenth century, but new communities of farmers have settled on the sandy soils made usable by the discovery of chemical fertilizers.

Today, industrial potato growers in the Netherlands are under pressure. Their farms are relatively small (the average farm size being 40 ha) compared to the 70 ha estimated to be needed for a viable farm (about the maximum a farm family without hired labour can manage). To compensate for the small size and the low price of cereals, potatoes and sugar beet are grown relatively frequently (every second and fourth year) in the crop rotation. This necessitates protective measures against eelworms (nematodes). In addition, intensive potato production involves frequent applications of fungicides against phytophthora (there are no resistant varieties as yet), while potatoes and sugar beet both require frequent applications of herbicides before the canopy has formed. The word 'requires' reflects conventional practice: industrial potatoes can be produced biologically, as some farmers in the area demonstrate.

Although the MYCPP raises strong emotions, it is depressed prices rather than the Plan itself that puts pressure on arable farmers (Wossink, 1993). Incomes

[2] The authors would like to thank the International Agricultural Centre (IAC) in Wageningen, and especially its ICRE team, for making their participation possible. We were fortunate to have Mr Teo Dunning of Larenstein Agricultural Collegeas the main workshop leader of the group. He knows the area very well and helped us greatly to understand the issues. We would also like to thank the farmers and the representatives of the major institutions actors involved, such as DLV, for their contributions and explanations to the group, from which this article benefits. The authors remain entirely responsible for any misunderstanding and mis-representation.

for the whole arable sector have been negative for some years. Large-scale, highly educated young farmers still look towards the future with confidence, and have accepted the inevitability of pesticide reduction, but the older, smaller scale and more vulnerable farmers are resistant. One said:

> I grew up on a farm of ten hectares. There were a large number of children and all of us helped on the farm. Now my husband and I farm 65 hectares. We work incredibly hard to make ends meet, but it is difficult. We recently started a broiler production unit to augment our income, but I don't know whether we will make it. How much better it would be if there were four farming families on our land!'

Learning IAF practices takes an extra effort and time that such farmers can scarcely afford.

The further reduction in farm numbers expected, and decrease in government funding for institutions supporting agriculture, place tremendous stress on the various agencies involved in farm support. Their dwindling operational coverage has forced formerly separate farmers' unions (Catholic, Protestant and non-denominational) to amalgamate into one union – the NLTO). Socio-economic advisory services, SEV, likewise are about to amalgamate, with an inevitable shake-out in the numbers of advisors. The future of the regional adaptive research centre and experimental farms is uncertain. Borgerwold biological experimental station (one of the stations where the original work on IAF was done) is certain to close. But the Hildebrands Laboratory, largely funded by the co-operative potato-processing company AVEBE, which buys most of the industrial potatoes in the area, will probably continue to carry out work on resistant varieties. As with so many companies that emerged from Dutch agriculture, AVEBE has long since developed interests outside the Netherlands, which makes it independent of the relatively small production of the farmers in the area.

6.3.2 Extension in the industrial potato growing area

Extension services in the context described are the responsibility of a large number of often competing actors, including farmers, who either consider extension as their main task or as a necessary component for economic or organizational survival. The distinction between extension and technology development has become blurred, as we shall see.

The co-operative union (ACM), which supplies most farmers with pesticides, has joined with NEFYTO (the Netherlands pesticide producers' organization) and the NLTO to negotiate the 'details' of the Crop Protection Plan, including the covenant which stipulates that farmers can revert to pesticide use if matters run out of hand, and to provide specialized extension. Most farmers are members of the ACM, and pesticide suppliers tend to be the advisors most trusted by Dutch arable farmers (Proost, Van Keulen & Schönherr, 1995). The AVEBE co-operative potato processing company provides advice about varied selection and cultivation.

The privatized extension service, DLV, which we shall discuss more fully below, focuses on knowledge-intensive changes in farming practice, including the change to IAF. However, since DLV became privatized, its main interest is in selling its services. Hence they would rather charge to provide farmers with an analysis of the environmental status of the farm, instead of enabling them to carry out such analyses themselves. DLV is contracted by the government to guide the introduction of IAF in the context of the 'Arable farming 2000' project. However, in the course of our rapid appraisal we perceived that most farmers prefer technical fix options, which are provided by co-operatives such as ACM and AVEBE.

The farmers' union (NLTO) employs socio-economic extension agents (SEV), and provides its members with a weekly magazine, *Oogst* (meaning harvest), which provides technical, organizational and policy information.

The 'Terra College', a government-supported centre for primary and secondary agricultural education and apprenticeship, provides (almost wholly self-financing) short courses to farmers, especially during the winter months. DLV advisors often are hired as teachers on these courses and they also write articles for various other agricultural magazines, such as the commercial weekly *De Boerderij* (the farm).

We should also mention farmers' study clubs. The success of these clubs is based on the careful recording by each member of their inputs, costs and results, and an open discussion and comparison of these. The clubs invite speakers, sometimes engage in experiments, and are the prime example in the Netherlands of farmer-to-farmer extension and technology development. They are better developed in intensive glasshouse horticulture than in arable farming, but arable farmers do have their own study clubs and such clubs have been deliberately set up to help implement 'Arable Farming 2000'. DLV is often involved in these study clubs as part of its contract under the AF 2000 programme.

6.3.3 Focus on DLV

We focus here on DLV, specifically the arable farming team in Emmen, which is responsible for the starch potato area. At the time of writing, it had 14 staff and covered about 3000 arable farmers spread over four provinces, or 120 000 ha of arable land.

The original (1990) idea under the government's privatization plans was that DLV would become gradually privatized, with half the cost of the 700 DLV staff employed nationwide by 1999 paid for by government and the remainder paid from a general levy on farm produce (15%) and from DLV profits (35%). In fact, by 1995, the team in Emmen were already paying 45% of the costs from profits, of which 30% was commercial profit and 15% income from provincial and national government contracts. The provincial authorities in the north, which is an area relatively rich in 'nature' and earns much from tourism, are keen to support integrated agriculture and are ready to pay DLV to introduce IAF.

DLV's interests are beginning to diverge from those of the Ministry of Agriculture, in that DLV now concentrates explicitly on benefits to farmers, and must be partly responsible for the advice it gives in terms of results. The Ministry's policies are increasingly focused on nature values and environmental benefits, in addition to agriculture and food production. This has been a source of tension between the Ministry and farmers since the late 1970s, when public extension workers refused even to explain the Ministry's policy to farmers for fear of losing the clients' trust. In the present circumstances, DLV staff sometimes advise farmers against following Ministry policies. DLV is in an equivocal position, given that a large part of its income still derives from direct payments from special contracts with the Ministry, which usually demand policy implementation.

The DLV team sells advice on request, often by telephone. 'We advise on problems and give solutions'. That is its main business, but it also has some 200 clients out of a total of 3000, who receive regular advice throughout the year on the basis of a farm plan, which costs about Dfl2000 ($1400) per annum. DLV's regular clients are the bigger farmers with around 70–75 ha.

DLV has been contracted by the Ministry to guide the 50 farmers in the area who participate in the 'Arable Farming 2000' project. It organizes open days and helps farmers to record pesticide and mineral use and costs, so that results of different practices can be examined and related to farm profitability and other vital statistics. The experience has led to an important opportunity for DVL: it stores information on various practices, use of chemicals, data on yields, etc. in a database, which allows it to identify best farmer practice and so better advise other farmers. As a typical example: in 1994, DLV recommended the use of 1.5 kg of fungicide per ha on wheat costing Dfl200 ($140)/ha/annum. It tried in 1995 to reduce the amount on the 'Arable Farming 2000' farms to one kilogram, split into two applications, before and after the conventional application time. The results allowed DLV in 1996 to recommend two applications of 0.4 kg/ha/annum, resulting in a saving to the farmer of about 50%.

This example, and the opportunity provided by marketing the experience of 'Arable Farming 2000' farms, show that the distinction between extension and technology development is becoming increasingly blurred. The blurring can be expected to increase as government invests less in experimental farms.

6.3.4 The impact of the crop protection plan

So far, many arable farmers in the industrial potato producing area have been able to reduce pesticide use by an estimated 70% on nematicides and 50% on herbicides compared to that in the base period of 1884–88. They have done this by adopting nematode-resistant and tolerant potato varieties, by using nematicides selectively on the basis of soil sample testing carried out by specialised agencies, by using highly specific and potent herbicides and more effective application techniques requiring less active ingredients per application, and by exchanging land with dairy farmers so as to enlarge rotation area. Recently ploughed grass land has had no potato-threatening eelworm populations. Farmers also engage in additional high income activities, such as broiler production, broad-acre vegetables, the production of brewers' barley, or try to enlarge their farms in order to cope with economic and environmental pressures.

Progress towards the norms of the Multi-Year Crop Protection Plan has been achieved largely through add-on adoption of ready-made technologies (new herbicides, resistant and tolerant potato varieties, commercial soil testing), which have been developed and/or commercially offered by agencies. Transformation on the basis of knowledge and labour-intensive adaptations of farming practice (prediction based on systematic observation of crops, use of mechanical weed control, etc.) has been more limited. In fact, mechanical weed control is losing out against the new herbicides. IAF, as we described earlier, has been only partially adopted, notwithstanding attempts by government to introduce it through projects such as 'Arable Farming 2000'.

All the measures farmers have used so far have drawbacks: the new potato varieties are losing their resistance and tolerance, as existing nematodes adapt and new ones spread into the Netherlands. Access to dairy farms for exchange of land is limited to the happy few, and the 'fourth crop' option is under heavy competitive pressure. Buying additional land, or buying out your siblings, is expensive.

6.4 The evaluation of the impact of the multi-year crop protection plan

The following targets were set for the MYCPP:

1. *Reduction of total kilograms of active ingredient pesticides* used in agriculture, 35% by the year 1995 and 50% by the year 2000. Detailed reduction targets were set for each crop sector and pesticide category. In all sectors, soil disinfectants can only be obtained by prescription and the number of applications is limited. Measures taken under the 1986 Soil Protection Act will also limit the use of chemicals. Mechanical weed control is encouraged, as is the use of field monitoring and improved application equipment. A certificate of competence for the application of pesticides is now required.
2. *Decreased dependence on pesticides* (a qualitative goal). Reduction of dependence on chemicals is stimulated by encouraging integrated farming and organic agriculture. In addition, healthy planting material, careful handling of imported planting materials, crop rotation, and the use of pest-resistant crop varieties will help farmers realize reduction targets.
3. *Reduction of emissions (total kilograms) of active ingredients of agricultural pesticides:* by 50% into air, 75% into groundwater and soil, and 90% into surface water by the year 2000. Reduced use of pesticides will result logically in the reduction of emissions into the environment. Closed cultivation systems are being introduced for glasshouses and mushroom growing.

Implementation of the MYCPP makes considerable demands on the agricultural sector. Extra investment is estimated to be 2.3 billion guilders for the years

1990–2000. The majority of Dutch farmers are expected to have sufficient means to make the changes needed, although a certain number will have to stop farming.

The MYCPP implementation process is characterized by consultation and negotiation with all the parties involved: farmers, local and regional authorities, the pesticide industry, input suppliers, environmental organizations, researchers, and extension specialists. Government, the pesticide industry, and farmers' organizations have set their commitment to implement the reduction measures in the form of a contract. The contract is not legally binding but is strong enough to commit all parties to voluntary action and changes in their practices. Negotiations between these stakeholders have resulted in the rejection or postponement of policy instruments such as pesticide taxes and the banning of certain environmentally harmful pesticides. If the official reduction targets are not reached, coercive restrictive regulations can be expected.

Considerable importance is given to extension as a means to reach voluntary changes in farmers' production practices. Special programs for extension and research were financed under the MYCPP. The combination of coercive regulations and the stimulation of creativity within the agricultural sector has proved to be effective in the implementation of environmental measures. With the start of MYCPP implementation in 1991, the Department of Communication and Innovation Studies of the Agricultural University at Wageningen conducted a survey, financed by the Ministry of Agriculture, Natural Resource Management and Fisheries, of about 900 farmers in seven different crop sectors. The aim of the survey was to find out how farmers perceive national crop protection legislation and to look at what actions they are taking. This survey was repeated in 1994 to examine what progress had been made in terms of both attitudes and practices.

6.4.1 Results

Farmers are concerned about water, soil and air pollution through agricultural practices. They recognize the dangers of chemical use for the environment, health, and agriculture in the form of resistance to pesticides.

In 1991, around 15% of respondents showed a negative attitude towards environmental issues. They did not recognize the contribution of agriculture to environmental pollution and showed little readiness to take action. By 1994, this group had reduced to 1%. In both 1991 and 1994, farmers identified the following three major problems for their sector: decrease in prices for their produce; increasing production costs because of environmental measures related to crop protection; and a more restrictive registration policy for new pesticides. Asked about the problems associated with pesticide use, 50% of the growers saw environmental hazards as the most important. They are also concerned about the future viability of their enterprises because of the rising costs of pesticides, the decline in the image of the agricultural sector, and the fear of excessive pesticide residues (in relation to export standards).

Their readiness to cut back on the use of pesticides is considerable. In

1991, farmers expected to reach an average reduction of 28%. By 1995, a 41% reduction had been achieved. Glasshouse growers seem to be more willing to experiment with alternative crop protection methods, especially biological control. In 1994, almost all glasshouse vegetable growers were using biological control.

Extension

When the MYCPP came into force, the extension outlook seemed rather positive. Extension agents deal with target groups that are well aware of what is at stake, possible solutions to technical problems are known, and there is willingness to take action. Growers are well informed about possible solutions via agricultural magazines. Salesmen and advisors working for agricultural input suppliers also play an important role in the dissemination of crop protection information. Farmers find them knowledegable and reliable. DLV advisors are consulted for a second opinion, to check the supplier's advice.

In 1994, farmers appeared eager for new technical information relating to the environment. They preferred 'instant' packages of information ('tell me what to do'). About 50% mentioned their membership of a study club. The Netherlands has a strong study group tradition, especially in the horticultural sector. Participation in these groups is beneficial to farmers because of the discussions they have, the new knowledge that comes up, the trials and research being undertaken, etc. Together, farmers observe, experiment and exchange ideas and experiences. As part of a group they are stimulated to try new techniques. Two-thirds of growers keep pesticide records. Many farmers' groups compare results and this increases a farmer's insight into his own spraying practices and the possibilities for change. Asked about the changes that record keeping brought about, farmers stated that they gained more insight into pest management and were able to reduce pesticide use. Some claim to have changed the pesticides they used. This is a positive stimulus for extension to continue to promote record keeping, because half of the participants still say it hasn't brought any changes in their farming.

One might expect farmers in close and frequent contact with extension agents to reduce pesticide use more than those who seldom or never contact an extension agent. However, it is interesting to note that, in the 1994 survey, no significant correlation was found between the frequency of consulting an agricultural advisor and pesticide reduction.

Special programmes

Farmers expect to increase profitability by reducing pesticide use. Higher prices to consumers are not therefore a prerequisite for improving their environmental performance. Several environmental marketing and labelling programmes have been introduced since 1991 and these seem to be a good starting point for farmers to change their crop protection practices. Although voluntary labelling programmes involve costs to the producers, the majority of growers claim that higher prices for their produce are not necessarily a condition for their participation.

Other special programmes like 'Arable Farming 2000' are discussed in more detail in this chapter. In future, it is very important that the information from these projects is disseminated to all interested parties: farmers, agricultural advisors, researchers, politicians, etc.

Research

Research is no longer seen as a useful source of innovation for decreasing pesticide use. In 1991, farmers had high expectations for more resistant varieties. After three years, however, they feel disappointed. The development and introduction of these varieties will take more time than expected. In general, farmers have become discouraged about results from agricultural research. In 1991, horticulturists, arable farmers, and bulb and mushroom growers in particular had high expectations from research, but, in 1995, many growers complained about useless research recommendations. They do not seem to realize that the adaptations they make in their crop protection practices come mainly from research.

Government interference

Although they recognize pesticide-related problems, farmers feel ambivalent about government interference. They do not evaluate government policy very positively: they find it to be ill-considered and inconsistent. They accept government laws and rules, especially to prevent a neighbour's malpractice, but they also feel that government interference should not diminish their entrepreneurial freedom. Over the years farmers have increasingly preferred being able to take initiatives themselves.

The future

Although farmers are positive about contributing to pesticide reduction, it is not clear that their efforts will achieve the reductions that policy aims for. Many farmers still think in terms of partial solutions: small, simple, and relatively cheap changes in their farming practices. For example, lower dose spraying is a popular idea: these changes will mean a pesticide reduction, but not to the extent foreseen in the MYCPP. Some useful alternatives are not widely implemented, either because they are more information-intensive (for instance, 'guided' pest management), because they involve more risk (mechanical weeding), or because some habits are difficult to change (farm sanitation).

One of the major conclusions from the 1994 survey is that, although farmers think they contribute considerably to meeting reduction targets, their results are, in fact, limited. This may produce a situation in which farmers feel proud of their endeavours but national reduction targets are not achieved. Very few farmers know of the target to reduce dependence on pesticides, even though it is a key issue for the future development of the agricultural sector, and for plant health management in particular. A significant shift towards integrated farming will require more drastic change. IAF initiatives are only small scale so far.

6.5 Conclusions

Listening to the experiences of the 38 participants in the Innovation Project, conducting a case study in one farming area, and reviewing the results of two national surveys of the impact of the Multi-Year Crop Protection Plan, we believe we are in a position to formulate some hypotheses for further research. We have distinguished between two components of IAF:

1. the adoption of technologies such as new cultivars or machines, and the purchase of services from specialized agencies; and
2. a transformation in farming practice, i.e. managing the farm as an eco-system based on observation, interpretation and anticipation.

Progress towards MYCPP norms among starch-potato growers has so far mainly occurred as a result of the first component. A host of organizations (such as the pesticide producers organized in NEFYTO, co-operative input distributors, commercial pesticide salesmen, the potato-processing co-operative, farm equipment producers and sellers, private soil-testing laboratories and the privatized extension service DLV) have vied with each other to provide the products and services that allow farmers to adhere to the norms of the MYCPP through adopting add-on innovations while making minimal changes in their conventional farming practices. Experts agree that further progress cannot be maintained by this strategy alone. A more fundamental transformation strategy is needed if farmers are to make more substantial reductions in input use. At present, it is unclear whether the adoption of add-on innovations to adhere to the environmental policies is a first stage in a more fundamental transformation, or whether it is a dead-end strategy.

Of the 3000 arable farmers in the area, only 50 participants in 'Arable Farming 2000' are receiving the kind of intensive guidance from DLV which the 38 participants in the 'IAF Innovation Project' claimed was essential for making the more fundamental transition to IAF. DLV is deliberately using the experiences of these 50 farmers to develop its own expertise. A possible consequence is that 'Arable Farming 2000' is more likely to lead to greater farmer dependence on add-on innovation and the purchase of specialized services than to increased expertise and self-confidence among the farmers themselves.

Since the early 1960s, the Dutch Government has spent large sums of tax payers' money on research and extension to increase the productivity and competitiveness of Dutch agriculture. Currently, the environmental consequences of this is being addressed largely through regulation. There is no money for the type of mass education campaign which our own study appears to indicate is needed for a transition to IAF. The nature of the changes required for increasing agricultural productivity and competitiveness seems fundamentally different from the changes demanded by a transition to IAF. We can contrast the two types as follows (Cochrane, 1958):

1. Sustained increase in the productivity and competitiveness of Dutch agriculture has been achieved on the basis of technology-propelled

development. Technology-propelled development roughly works as follows. Progressive farmers adopt innovations which allow them to capture pioneer profits. As more farmers adopt the same innovations, the relative advantages of adoption decrease (as the benefits of technological change are passed on to consumers). But as farmers need to adopt to stay in the market place, further change occurs, together with the marginalization of those unable or unwilling to innovate. This process is consistent with the diffusion of innovations studied and described for Dutch agriculture by Van den Ban (1963).

2. The process of achieving sustainability, today an equally important policy goal (Conway, 1994), appears to follow a different type of change process. It is not market driven, in the sense that commercial and competitive advantages are not immediately evident. Market forces do not propel large-scale transition to IAF. The change to IAF is motivated initially from a need to be allowed by government policy and public sentiment to continue to farm; to adhere to the law and, to a lesser extent, out of consideration for environmental values. The change incurs costs which are only to a limited extent repaid. The change replaces external inputs with inputs of labour and the need to acquire new knowledge and farming insights. It requires investment in learning and a great deal of additional effort. Unless the consumer is willing to pay for higher quality and 'greener' production, there seem to be few incentives for such a change to take place on a large scale.

In the current policy climate, there is a tendency to consider only the farmer when discussing the transition to IAF. However, the case study of industrial potato production makes clear that farmers are embedded in networks of actors, including commercial companies, co-operatives, private consultants, banks, and so on. These actor networks (Latour, 1987; Callon & Law, 1989) have emerged in support of the high productivity and competitiveness of Dutch agriculture under conventional farming practices. The immediate interests of these actors lies in maintaining conventional high external input agriculture. As the Dutch arable farming sector shrinks, the actors involved compete intensively among themselves. The collective influence of such actor networks maintains high input farming, resists transition to more ecological agriculture and, at best, promotes adherence to environmental law based on the adoption of add-on external technologies. If this hypothesis is correct, then strategies to promote a transition to IAF will not only require support for the farmer through the learning process but also the restructuring of actor networks that prop up conventional high external input farming.

6.6 References

Callon, M. & Law, J. (1989). On the construction of socio-technical networks: content and context revisited. *Knowledge in Society: Studies in the Sociology of Science Past and Present*, vol. **8**, pp. 57–83. JAI Press.

Cochrane, W.W. (1958). *Farm Prices, Myte and Reality*. University of Minneapolis; Minnesota Press.

Conway, G.R. (1994). Sustainability in agricultural development: trade-offs between productivity, stability and equitability. *Journal for Farming Systems Research-Extension*, **4**(2), 1–14.

Darré, J.P. (1985). *La parole et la technique. L'univers de penseé des éléveurs du ternois*. Paris: Harmattan.

Knickel, K. (1994). A systems approach to better understanding of policy impact: the vulnerability of family farms in Western Europe. *Proceedings of the International Symposium on Systems-Oriented Research in Agriculture and Rural Development*, pp. 966–73. Montpellier: INRA.

Latour, B. (1987). *Science in Action*. Cambridge (Ma): Harvard University Press.

Proost, M.D.C., Van Keulen, H. & Schönherr, I.A. (1995). Gewasbescherming met een toekomst: de visie van agrarische ondernemers. Een doelbereikingsmeting ten behoeve van voorlichting. Department of Communication and Innovation Studies, Wageningen Agricultural University. Research report to the Ministry of Agriculture, Nature and Fisheries in The Hague. 79 pp.

Röling, N. (1995). Who needs extension anyway? Irreverent thoughts on the basis of developments in starch potato production. Thessaloniki (Greece): Paper for 12th European Seminar on Extension Education (ESEE), 28/8–2/9, 1995.

Van den Ban, A.W. (1963). *Boer en Landbouwvoorlichting*. Assen: Van Gorcum.

Van der Ley, H. & Proost, M.D.C. (1992). Gewasbescherming met een toekomst: de visie van agrarische ondernemers. Een doelgroepverkennend onderzoek ten behoeve van voorlichting. Department of Communication and Innovation Studies, Wageningen Agricultural University. Research report to the Ministry of Agriculture, Nature and Fisheries in The Hague. 85 pp.

Van Weperen, W. (1994). Balancing the minerals Moving boundaries. A study on the use of mineralbookkeeping among dairy farmers in the Achterhoek. MSc thesis, Wageningen Agricultural University, Wageningen.

Van Weperen, W., Röling, N., Van Bon, K. & Mur, P. (1995). Het veranderingsproces. Ervaringen van akkerbouwers bij het omschakelen naar een geïntegreerde bedrijfsvoering. Informatie en Kenniscentrum Akker- en Tuinbouw, Afdeling Akkerbouw en Groenteteelt in de Vollegrond, Lelystad.

Vereijken, P. (1990). *Geïntegreerde akkerbouw naar de praktijk. Strategie voor bedrijf en milieu*. Report no. 50. Lelystad: PAGV.

Wijnands, F.G., Janssen, S.R.M., Van Asperen, P. & Van Bon, K.B. (1992). *Innovatiebedrijven geïntegreerde akkerbouw. Opzet en eerste resultaten*. Report no. 144. Lelystad: PAGV.

Wijnands, F.G., Van Asperen, P., Van Dongen, G., Janssens, S., Schröder, J. & Van Bon, K. (1995). *Innovatiebedrijven geïntegreerde akkerbouw: beknopt overzicht technische en economische resultaten*. Report no. 196. Lelystad: PAGV.

Wossink, A. (1993). *Analysis of Future Agricultural Change. A Farm Economics Approach Applied to Dutch Arable Farming*. Wageningen: Agricultural University, published doctoral dissertation.

Part III: Farmer learning, its facilitation and supportive institutions

7 Learning about sustainable agriculture: the case of Dutch arable farmers[1]

NADET SOMERS

7.1 Introduction

Twenty-two years ago, Benvenuti (1961) published his thesis 'Farming in Cultural Change'. Like other rural sociologists of his time, Benvenuti focused on farmers' capacities to adapt from a 'traditional cultural pattern' to modern agriculture. Agricultural incomes lagged behind those of other sectors and thus intensification, rationalization and enlargement of scale in order to increase labour productivity seemed an answer. However, such solutions implied a rapid decline in the number of farmers and far-reaching structural adaptation. In this context, the extension services not only aimed at farm development and management but also at influencing the attitudes and behaviours of the rural population in order to prepare them for rapid structural change.

Rural sociologists in the 1960s in the Netherlands perceived the agricultural transition as a process of acculturation to a totally different lifestyle, to a different style of farm management, and to a new complex of norms and values, attitudes and behaviours. Perhaps never since this period has it been more clearly understood that processes of agricultural adaptation include social, socio-psychological and cultural traits as well as technical and economic aspects. Rural sociologists pronounced a special role for extension in supporting the process of change. Extension had to be aware of and try to influence the deeper cultural layers which, to a large extent, determine actual behaviour.

Today, again, agriculture has to transform itself as a result of the successful transformation of the former period. Public opinion polls show a growing awareness of the severe problems caused by the 'successes' of modern agriculture. The use of large quantities of chemical fertilizers and pesticides threatens the quality of soil and water. Consumers place higher and more diverse demands on the quality of food and the way it is produced. At the same time, farmers are subject to many different policy measures from numerous departments and local authorities. The demands of consumers and environmentalists demand creative solutions. But farmers feel restricted in their options as a consequence of poorly coordinated policies. They doubt that all the norms set by policy can be attained. As in the 1960s, the transformation process in the 1990s generates resentment, resistance, shifting norms and values and changing professional images.

[1] This contribution is based on the conclusions from the study 'Knowledge development for sustainable agriculture: an explanatory study of several experimental projects in arable farming' (Somers & Röling, 1993).

This chapter will limit itself to the following questions: a) how can we characterize the introduction of more sustainable arable farming? b) what does this mean in terms of learning for the arable farmer? and c) what do these aspects mean for extension services that are expected to foster the introduction of a more sustainable agriculture.

Research into these questions is in its early stages and aims more to stimulate discussion than to present any fully rounded explanation. The empirical evidence is based on a preliminary study covering four experimental projects in arable farming in the Netherlands (Somers & Röling, 1993).

7.2 Four experimental projects

The four projects studied differ greatly in their form and particpants (see Table 7.1) and in their goals and implementation (Table 7.2). Although the first three: integrated arable farming, agriculture and environment, and a local farmers' bread project can be considered as projects limited in scope and time, the fourth, biological agriculture, can be seen more broadly as an alternative to 'integrated' and 'established' agriculture.

The projects differ most in their perception of the essence of sustainable agriculture and its practical implications. Participants' perceptions of 'sustainable' agriculture range from 'economically healthy' to conformity with anthroposophic ideals. Yet, despite their different perspectives, participating farmers can be typified as forerunners, who are willing (for different reasons) to learn more about and take risks in the development of more sustainable methods. These forerunners are not necessarily the larger-scale and younger farmers. Although the smallest farms and part-time farmers are missing from the study, the projects encompass a wide variety of farm-sizes and ages of farmers.

We focus on two aspects that are central to the issues of this chapter: aspects of craftsmanship and entrepreneurship and the type of knowledge that has to be generated and transmitted. At first sight, the requirements for a farmer in sustainable agriculture seem not too different from those required of what is perceived as 'the good' farmer (entrepreneur or manager) in established agriculture. Both show a strong motivation to adopt innovations, are open to new information, have technical and book-keeping skills, and are steady and stress-proof. However, a closer look reveals that sustainable agriculture imposes additional demands. Not only does it change the kind of decisions a farmer has to make concerning farm planning and management but it also changes his position *vis-à-vis* external factors. Depending on the stage in the transformation process, a farmer is likely to find himself obliged to develop new knowledge and skills, change his goals and manage risks differently.

To be more explicit, in integrated arable farming, a crucial aspect is determining the critical pest or disease level before spraying; in biological agriculture, the option of spraying is not available. For the farmer, this means, for example, that

- time is spent on field inspection and observation, which has implications for working schedules;

Table 7.1. *Participants and forms of projects*

	Integrated arable farming	Agriculture and environment	Farmers' bread	Biological agriculture
Organizations involved	Ministry of Agriculture, Landbouwschap*, research and extension institutes	Province of Brabant, Landbouwschap*, local farmers' organization, local extension institute	Farmers' initiative in the province of Zeeland, local farmers' organization, environmental and consumer organizations, millers, bakers	440 farmers, organized in an ecological and a biological–dynamic organization (both national); organic research institute and on-farm research; extension team for organic farming
Form of project	38 innovation farms in five regions function as models for study groups. Innovation farms intensively monitored by extension specialists. The study groups are conducted by extension workers of the local extension teams	Two study groups in which 25 and 15 farmers participate. The study groups are conducted by extension workers of the local extension teams	36 farmers, two millers and 75 bakers, supported by one product manager and (informally) one extension worker	

Note: * The Landbouwschap is the apex organization of the three farmers' organizations.

- he must be able to interpret the indicators for a specific situation; the interpretation can differ from one part of the field to another;
- he must be able to relate his observations to indicators;
- he must rely less on external advice and more on knowledge of his own specific farming conditions;
- he must rely less on external farm inputs and more on preventive methods such as shifts in rotation schemes and choice of varieties;
- he must resist the (negative) opinions of his neighbours when his crops develop differently to theirs;

Table 7.2. *Goals of sustainable farming and implications for agricultural practices*

	Integrated arable farming	Agriculture and environment	Farmers' bread	Biological agriculture
Goals of sustainable arable farming	IAF aims at integration of economic and environmental goals: reducing environmental damage while maintaining economic returns	The same goals as integrated arable farming	The same goals as integrated arable farming	Biological agriculture aims at a long-term relationship among human beings, nature and environment by which all three stay healthy
Implications for agricultural practices	In practising IAF the farmer applies a broad crop rotation, resistant varieties, efficient use of fertilizer and strategies for reducing the use of pesticides	The same practices as IAF	For wheat, no chemical fertilizer is allowed. Spraying with a chemical pesticide allowed only once, before the crop emerges	The biological farmer uses no chemical fertilizer and pesticides. He/she tries to stimulate a stable and active soil eco-system. Crop care consists of prevention of diseases

- he must resist the anxieties of neighbours who may be afraid that their crops will be infected by a crop that is not fully 'protected' by chemicals.

In general, ecologically minded farmers need a different type of skill and knowledge, one that is more orientated to their own capacity to make informed judgements. They rely less on external inputs. They are able to live with social disapproval. They see these as a necessary part of advancing broader and riskier transformations. They find sustainable methods difficult and risky. By participating in a project, however, and by experimenting collectively, they gain more confidence.

7.3 Characteristics of innovations and implications for the introduction of sustainable arable farming

Rogers (1983), from a synthesis of a world-wide database on the adoption of innovations, distinguishes five attributes which might describe an innovation and predict its rate of adoption.

1. *Relative advantage:* is it perceived as better than the idea or technology that it supersedes?
2. *Compatibility:* is it perceived as consistent with existing values, past experiences, and the needs of potential adopters?
3. *Complexity:* is it perceived as relatively difficult to understand and to use? The perceived degree of complexity is negatively related to its rate of adoption.
4. *Trialability:* can it be experimented with on a limited basis by the end user?
5. *Observability:* are the results observable. The observability of an innovation is positively related to its rate of adoption?

Let us evaluate the more sustainable methods used against these attributes.

1. Participating farmers derived some financial advantage from adopting more sustainable methods. Arable farmers became aware of possibilities for financial savings, especially by keeping records of chemical inputs and by determining their actual utilization by the crops. Up to a point, they were able to reduce the input of fertilizers and pesticides without jeopardizing the level of production. In reality, they were simply cutting back on unnecessary over-application. In general, reduction in input use is becoming a common strategy among arable farmers for coping with deteriorating financial circumstances. Input-saving is not possible in all farm sectors; many arable farmers will not risk reducing inputs if it raises fears over a loss of harvests. Moreover, many farmers dislike methods which involve more risk and more intensive labour. Field inspection, which takes a prominent place in sustainable methods, requires time. Also weeding is, in several crops, a labour-intensive task. Many of the farmers interviewed feared that the increased labour demand would not be compensated by higher product prices. Extensionists cannot, as long as 'hard' economic figures are lacking, point to positive economic returns, unlike many of the innovations which they promoted in the past.
2. Sustainable agriculture is not wholly consistent with existing norms and values, or the previous experiences and current needs of farmers. The term actually has a negative political connotation among many farmers. Indeed, inconsistencies in policy measures have brought about distrust of policies concerning environmental issues in general. The situation differs among farm sectors, but arable farmers in particular have been hurt by declining prices for subsidized crops, as also for free market crops, which until recently made up for the losses on subsidised crops. In such a situation, arable farmers are willing to do something 'environmental' as long as this goes along with financial advantages (a win–win situation). When, as in the present economic context, stringent policy measures are enacted, the legitimacy and thereby the effectiveness of environmental regulations are reduced. This is especially the case when policy makers and farmers cannot agree on the severity of the environmental problem,

on the role of arable farming in causing the problem, or on the effectiveness of policy measures. The problem of free-riders is a realistic one in such a situation (van der Ley & Proost, 1992; Röling, 1993), but here generating perversely in the sense that some are riding on non-compliance. For younger farmers, who have grown up with pesticides, the new requirements imply a reorientation of their image of the professional farmer. An extension service that is meant to foster the introduction of sustainable practices cannot ignore the accompanying changes in norms and values.

3. Sustainable agriculture is perceived by farmers as relatively difficult to understand and apply. Farmers need more information, for example, about the different seed varieties and the specific characteristics of their fields. They need to know how to determine critical levels of disease and to relate these to former experience and the knowledge of their own farm-specific circumstances. Integrated and biological farming requires an understanding of the management of a complex whole rather than simply knowing how to grow individual crops. Extensionists also need such understanding, but scientific results as well as practical knowledge are thin on the ground. This raises the interesting question of where extensionists are going to get their comparative insights into the practical application of general principles. Farmers and extensionists have to develop many things for themselves. In the projects, farmers were experimenting with methods of reducing chemical inputs, especially in potato production. Extension workers assisted the farmers in designing experiments and interpretating the findings. The lack of scientific knowledge is seen by extensionists as an obstacle to the introduction of sustainable agriculture. The inadequacies of the science base can also be seen as an opportunity to develop a different professional image.
4. A favourable characteristic of sustainable agriculture is the opportunity to learn in a stepwise fashion. In general, interest is growing in new varieties and techniques that help to reduce chemical inputs. The financial savings are an incentive for farmers to experiment with the new technologies and practices. However, the willingness of arable farmers to invest in new equipment and technology is small, due to the unfavourable financial situation.
5. The results of sustainable arable farming were still not visible at the time the study was conducted. Some insight into the possibilities for reducing the input of chemicals were available, but financial results were available for only one year and only for a small group of farmers. Extension workers experience the lack of 'hard' facts as a handicap in their efforts to introduce more sustainable agriculture.

We can conclude, in the light of Roger's criteria, that the adoption of new elements into what remains structurally a conventional farming framework, demands a complex transformation process. The respondents in the study estimate that policy measures must be introduced to accelerate the speed of innovations. It is possible to speak of an interaction whereby the basic principles developed by research institutes are made regionally specific and a shared and gradual learning process occurs among farmers and extensionists. Positive experiences with one aspect of integrated or biological agriculture can lead to a different type of farm management and a growing willingness to try out other aspects. It would be more accurate to speak of a process of gradual learning with shifting goals and perceptions than of a process of adoption of an innovation.

7.4 Characteristics of the learning process

The opportunity to experiment and learn about more sustainable methods can be a point of departure for an extension service that wants to promote the introduction of sustainable agriculture. The projects show that by experimenting, arable farmers improve their knowledge and increase their motivation to engage in sustainable agriculture. The quality of the learning process seems to be an important aspect but we were unable to pursue this in our limited study. We found the following learning processes taking place.

- using indicators to make environmental problems visible;
- using record-keeping to make economic results transparent;
- regular field observation and measurement as the basis for decision making;
- inference from observations based on social 'theoretical' knowledge;
- exchange of information and experience among co-learners;
- use of scientific principles.

The move towards sustainable agriculture has implications for the kind of knowledge needed and the institutional structure within which knowledge is generated. For instance, farmers need more information concerning mechanical weed control, the functioning of minerals in the soil, the field observation and measurement of problems and conditions and the development of resistant varieties. Farmers do not require definitive and ready-to-implement solutions, but principles, ideas and suggestions that can be tested in their own specific situation. The knowledge supporting sustainable agriculture is locally specific knowledge with which farmers can themselves experiment.

This, in turn, implies that sustainable agriculture can best be served by an intensive interaction between scientific knowledge and the knowledge generated by farmers. Our study shows that farmers' perceptions of the risks involved in the use of sustainable methods demands a local try-out of standards that are developed elsewhere. Experts can assist farmers in starting experiments and in analysing the results. In the Netherlands on-farm research under farmer management deserves more attention than it has nowadays.

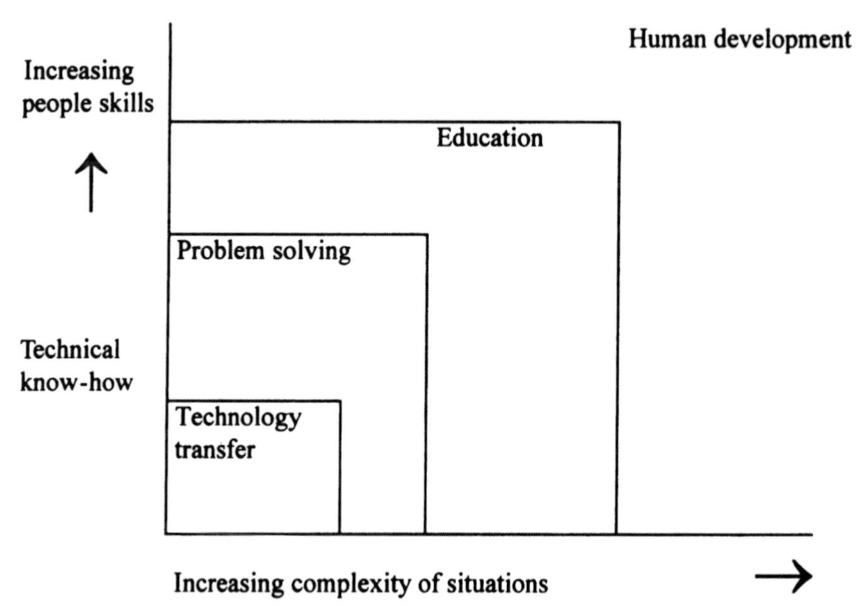

Fig. 7.1. Role of extension in shifting contexts. (*Source:* Coutts, 1994, p. 7.)

7.5 Concluding remarks about the role of extension

We conclude that sustainable agriculture implies more than a shift in farming practices. A change in attitude and knowledge about how to grow a 'good crop' is inherent in the process of adaptation. Without putting too much emphasis on the concept of 'cultural pattern' which prevailed in the 1960s, we want to focus on the process of social–cultural adaptation that farmers are asked to go through today. What a challenge this means for extension!

It seems that the focus must be less on bringing about specific behavioural change and more on raising consciousness. Group work can stimulate individual and collective reflection on measurements, experimentation with new methods, and the acceptance of new norms of appropriate behaviour. New values to support 'good' practice develop in a group. Extension roles shift from advisory activity based on technical expertise, to the role of facilitator of a learning process for both farmer and extensionist. Fig. 7.1 reflects the hypothesis that, when situations become increasingly complex and make higher demands on the knowledge and skills of farmers, the type of extension will shift.

We have learnt from the projects that the introduction of sustainable agriculture is not a matter of simply changing a certain behaviour or adopting a certain technology. The growing confidence of growers to be involved in, and contribute to, the process of developing a more sustainable agriculture seems to be the key. The relationship with the farmer is not so much instrumental or strategic as communicative. This role poses new demands on the skills and attitude of extensionists.

Like farmers, extensionists must also learn to live and work in a context of cultural change.

7.8 References

Benvenuti, B. (1961). *Farming in Cultural Change*. Assen: van Gorcum & Comp. N.V.

Bosker, M.E.C. (1992). De boer als motor, de voorlichter als coördinator. (een case-studie van het project 'Akil Bruchsal'), dissertation, Wageningen: Landbouwuniversiteit, vakgroep Voorlichtingskunde.

Coutts, J. (1994). *Process, Paper Policy and Practice. A Case Study of a Formal Extension Policy in Queensland, Australia, 1987–1994*. Wageningen: Agricultural University, Published doctoral dissertation.

Rogers, E.M. (1983). *Diffusion of Innovations*, 3rd edn. New York: The Free Press.

Röling, N.G. (1993). Agricultural knowledge and environmental regulation in the Netherlands, *Sociologia Ruralis*, **33**, 261–80.

Röling, N.G. (1994). Facilitating sustainable agriculture: turning policy models upside down. In *Beyond Farmer First: Rural People's Knowledge, Agricultural Research and Extension Practice*. London: IT Publications.

Somers, B.M. (1991). Small farmers and agricultural extension. (dissertation) Agricultural University, Wageningen.

Somers, B.M. & Röling, N.G. (1993). *Kennisontwikkeling voor Duurzame Landbouw: een Verkennende Studie aan de Hand van Enkele Experimentele Projecten in de Akkerbouw*. Den Haag: NRLO, Wageningen: Landbouwuniversiteit, vakgroep voorlichtingskunde.

van de Fliert, E. (1993). Integrated pest management: farmer field schools generate sustainable practices. PhD dissertation. Wageningen: Agricultural University.

Van der Ley, H.A. & Proost, M.D.C. (1992). *Gewasbescherming met een Toekomst: de Visie van Agrarische Ondernemers* (Een doelgroepverkennend onderzoek ten behoeve van voorlichting). Wageningen: Landbouwuniversiteit, vakgroep Voorlichtingskunde.

8 The diffusion of eco-farming in Germany

ALEXANDER GERBER AND VOLKER HOFFMANN

8.1 Introduction

Over the last decades, agricultural intensification has greatly contributed to environmental pollution in Germany (e.g. Diercks, 1993; Bach, 1987; Scheller, 1993; Burdick, 1994; Köpcke, 1994). The emergence of ecological farming, or eco-farming as we shall call it here, in 1924, at first in the form of bio-dynamic farming, coincided with the onset of agricultural intensification. Nowadays, eco-farming is considered to be the farming system which best fulfills the requirements of sustainability (Rat der Sachverständigen für Umweltfragen, 1985).

Eco-farming, with the development of its knowledge system and its diffusion, provides an example of the driving and inhibiting forces that govern any sustainable innovation that affects and involves society. The knowledge system supporting eco-farming includes:

- all the know-how and facilities necessary for producing, processing, marketing and consuming products within the eco-farming system;
- the epistemology and the influence of the socio-cultural context; and
- the institutions supporting the promotion of these processes.

This chapter provides an overview of the emergence and diffusion of eco-farming in Germany, from its inception in the anthroposophical movement to its present market-driven development. It describes the epistemology of eco-farming, and the different phases of its diffusion since 1924. It then describes the typical nature of the knowledge system and other institutional arrangements supporting eco-farming. The chapter ends with some statements about the future of eco-farming which are suggested by the historical overview.

8.2 The epistemology of eco-farming

Eco-farming voluntarily restricts itself to the use of certain management options. This is a special feature of its knowledge system and a main reason for its environmental compatibility. These voluntary restrictions owe their inspiration to an expanded understanding of science.

Mainstream natural science, even today, assumes that living nature is only a complicated structure of unliving matter (Mengel, 1991). Implicitly, it is presumed that the total can be described as the sum of its parts and as having mono-causal relationships. Eco-farming principles hold that only the relationships between matter in a biological system can be described by chemical and physical laws, not the organism as it functions as a whole. An organism has a history which influences its behaviour, has activities and variability, and organizes its living processes, all of which indicate a higher order. It was Johann Wolfgang von Goethe, who first tried to describe this scientifically. His 'organic' theory was taken up and developed further by Rudolf Steiner, the founder of bio-dynamic farming, the oldest variant of eco-farming.

Departing from the organic understanding of nature and alluding to the ordering principles of biological systems, Steiner (1924) postulated farming to be the shaping and managing of the farm as an organism. As such, the farm should develop an 'agricultural individuality' in each respective location, with the associated economic and social conditions of that location, and it should be understood as such. This is recognized as a basic principle in all types of eco-farming. 'Eco-farming follows the principle that the farm is a goal-oriented organization of agricultural production which is to a large extent self-sufficient and internally balanced' (Köpcke, 1994).

Conventional modern farming suppresses undesirable elements on the farm through the use of pesticides, and introduces desired ones through the use of chemically synthesized fertilizers. Eco-farming tries, instead, to intervene by stimulating natural processes and consciously makes use of ecological relationships (Schaumann, 1977).

8.3 Phases of the diffusion of eco-farming

The diffusion of eco-farming to date is seen as having three main phases: The first, a generating and consolidating phase, is followed by two phases of expansion.

8.3.1 First phase (1924–1970): the generation of eco-farming through bio-dynamic farming

Some farmers, veterinarians and researchers who were members of the anthroposophic movement were disturbed and unsettled by the early development of modern 'industrialized' agriculture. The findings of Justus von Liebig on the importance of nutrients led to a search for substitutes for natural and organic matter and to the synthetic production of nitrogen. At this early date, the anthroposophists had already observed some problematic developments, for example, decreasing fertility in cattle, increasing soil acidity after fertilization, a decrease in product quality, as well as the decreasing ability to sustain legumes over several years. So they approached Steiner and asked him to advance a viewpoint for the well-being of agriculture derived from his philosophy and research, which had already led to concrete achievements in other fields, such as medicine, pedagogy and the arts. The 'Agricultural Course', held by Steiner at Whitsuntide in 1924, is seen as the start of bio-dynamic farming.

Klett (1994) gives another reason for the early emergence and diffusion of the bio-dynamic system of farming, which he saw, perhaps somewhat romantically, as rooted in the history of human consciousness: 'Until that time, agriculture had been practised more or less intuitively. It was still part of the conserving impetus of tradition, and even though it was fading, the moral imperative of work made itself felt through the inherited force of the community'.

The new trends in agriculture asked for a more rationally based professionalism. At the time, the need to transform what was considered an 'intuition-driven' peasantry, and to take into account the findings of natural science, were generally recognized, especially by the Anthroposophic School founded by Steiner. It was interested in integrating scientific findings into the broader perspective on what it means to be human. This interest, and especially its adoption on large holdings located mainly in middle and eastern Germany, led to an extension of bio-dynamic farming on several thousands of hectares before the Second World War.

Directly after the 'Agricultural Course', interested farmers and researchers grouped together in the *Versuchsring* (Association for Experimentation), to examine and further develop Steiner's guidelines. Many of the principles of bio-dynamic agriculture came from this source and are still valid today (König, 1994). These early investigations were mostly based on empirical observations of single or very few cases, and often could not satisfy scientific or biometrical criteria (Gerber, 1994).

The typical organization, still partially existing today, of the advisory work for bio-dynamic farming developed quite early. Experienced practitioners gave up their own farm work to advise others, or took over advisory tasks after retiring.

In 1941, bio-dynamic farming was prohibited by the Nazi regime, leading to its near complete breakdown. Its restructuring and expansion after the war was much slower than at the time of its creation. In Western Germany, it faced very different and more unfavourable conditions, and in Eastern Germany it was more or less impractical.

The association of people farming bio-dynamically took the name *Forschungsring für biologisch-dynamischen Landbau e. V.* (Research Association for Bio-dynamic Agriculture), and in 1954 the administration of its trade mark was transferred to the *Demeter-Bund e. V.* In 1950, the Institute for Bio-dynamic Research was founded and today remains the only research institution within the framework of the federations for eco-farming in Germany.

From the beginning, bio-dynamic farmers were faced with the problem of operating within their own self-imposed restrictions, while at the same time establishing a position in the market. It was necessary for economic survival to find a distinct market for their products and to ensure that the consumer recognized that this food was produced by special methods. This led to the introduction of the trademark *Demeter* in 1927. Because their special quality was not obvious from the appearance or a comparative analysis of the products, the production method itself became the criterion for determining whether food was produced according to the trade mark regulations (Schaumann, 1995). It became necessary, therefore, to provide guidelines for production as well as controls for compliance with them.

The first guidelines for producing *Demeter*-food were published in 1956.

Since then, similar bio-dynamic farming guidelines have been introduced all over the world. The principle of producing according to guidelines and submitting to controls for compliance with them was later adopted by all other eco-farming groups and today also constitutes the basis for the European Union's regulations on eco-farming, to which we shall return later.

The most striking features differentiating bio-dynamic farming from other eco-farming systems are the application of bio-dynamic preparations and the consideration of cosmic rhythms. The production, application and effects of bio-dynamic preparations have been described and discussed at length (Steiner, 1924; Sattler & von Wistinghausen, 1985; Dewes, 1994; Gerber, 1994).

8.3.2 Second phase (1970–1988): first expansion as a reaction to ecological problems

The slow expansion of bio-dynamic farming after the Second World War and its initial adherence to anthroposophical principles might lead one to assume that only farmers interested in anthroposophy were its adopters. In 1960, about 3200 ha, and in 1970 about 3920 ha, were farmed according to the bio-dynamic guidelines (Brugger, 1990).

From the 1970s onwards, things changed, however. The negative consequences of industrialized farming became increasingly visible, and the emerging 'ecological consciousness' did not leave agriculture itself untouched. Recognizing that modern agriculture was a major source of pollution, some farmers decided that the situation had become critical. The ensuing review of farm practices was, in turn, followed by an increasing number of farm conversions.

At that time, protection of the environment was the most widespread reason given by farmers for the conversion of their farms (Fischer, 1982). In many cases, the decision to convert the farm was reinforced by animal health problems and, consequently, high veterinary costs. Quite often the conversion was preceded by a key event (Grassinger, 1984). Sickness in the family played an outstanding role, but visits to farms on which conversion had been completed are also mentioned as a key event. Other reasons have been the desire to lower costs and to gain autonomy through the reduction of purchased inputs, as well as problems with crop rotation or soil fertility (Brugger, 1990).

For many, conversion was made easier by the fact that some pioneers, mainly inspired by Christian religious motives (Ziechaus-Hartelt, 1991), founded the *Bioland* Federation. Its origin was the concern of a Swiss, Hans Müller (1894–1969), for ensuring the survival and autonomy of small farm holdings. Towards this end, he aimed at economically viable farms that were not dependent on purchased inputs. The crucial issue for him was maintaining soil fertility through careful treatment and intensive use of farmyard manure. From this emerged the bio-organic farming system. With the *Bioland* Federation, an alternative to bio-dynamic farming came into existence, which was less demanding to practice and, in the eyes of many, less ideologically charged. From then onwards, both types of farming, as

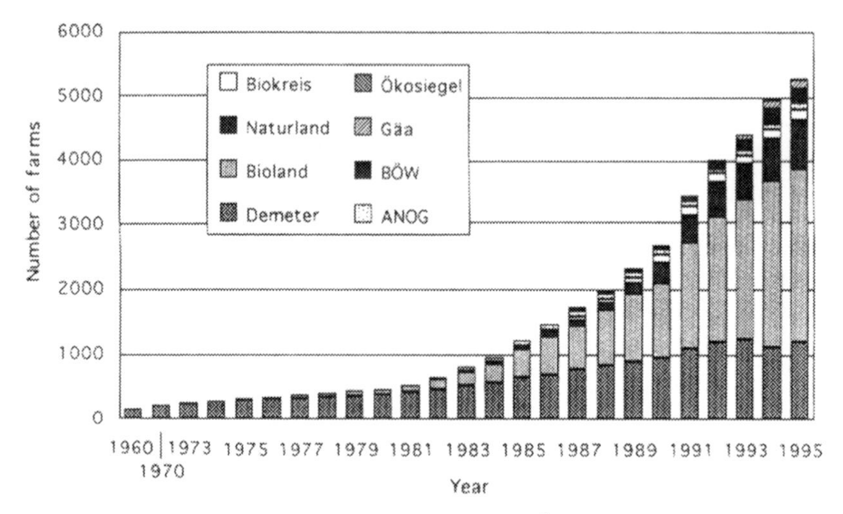

Fig. 8.1. Growth in the number of farms in AGÖL federations. AGÖL=Arbeitsgemeinschaft Ökologischer Landbau, is the central organization of all the German federations of Eco-Farming. *Source:* AGÖL; Hamm, FH Neubrandenburg.

well as the whole eco-farming movement expanded. The number of farms and the area cultivated increased greatly and many additional associations of eco-farming developed (Figs. 8.1 and 8.2).

All federations of eco-farming have compliance with some minimum standards in common, laid down in production guidelines. In this second phase, these are the basic guidelines of the IFOAM (International Federation of Organic Agriculture Movement). Due to the special marketing situation for their products, farmers converting to eco-farming are more or less required to become members of one of the federations of eco-farming.

8.3.3 Third phase (since 1989): second expansion through government promotion programmes

At the onset of this phase in 1989, AGÖL was founded as the central organization for eco-farming in Germany. It coordinates and executes political lobbying for eco-farming, works out general guidelines valid for all member federations, and tries to elaborate common strategies for consolidation and expansion of eco-farming in Germany.

Due to generally declining incomes in agriculture during this phase, eco-farming has gained interest as an economic alternative to conventional farming. Now, economic considerations are often the main reason for converting (Zink, 1986). Reasonable subsidies for eco-farming through the EU programme for agricultural extensification must be seen as a decisive factor in its economic attractive-

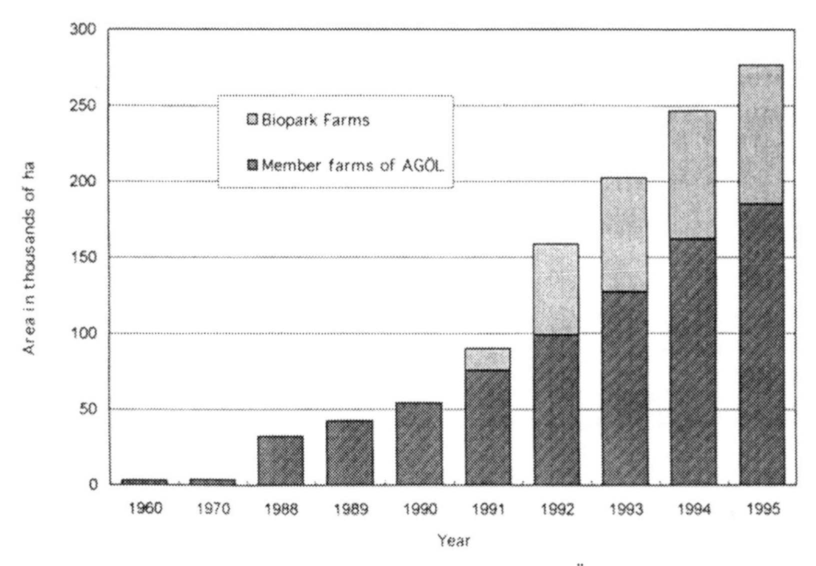

Fig. 8.2. Expansion of the cultivated area under AGÖL guidelines. *Source:* AGÖL; Hamm, FH Neubrandenburg, own investigations. Biopark: eco-farming federation farming according to EU and AGÖL guidelines, being not yet an AGÖL member.

ness. The EU's main motive has been the reduction of agricultural surplus production. One of the variants of the programme accepts the conversion according to guidelines (hereafter referred to as EU regulations) for eco-farming. This promotional programme was especially attractive to farmers in disadvantaged locations, because they already applied extensive farming methods, and thus participation in the programme did not require severe changes from them (Dabbert & Braun, 1993). Especially in the five new German States, with many substandard farm sites, the programme often offered agriculturists a way out of the difficulties created by German unification. Within the framework of the extensification programme, a much larger area has been converted to eco-farming in the East than in the West, though the West has a much larger land surface (Hamm, 1994).

Two problems of the extensification programme merit mention:

- The programme intends to reduce total food production. Therefore it subsidizes the production of ecologically produced food. This has disturbed the market for eco-products by creating an excess supply. Consequently, producer prices have been pushed down;
- The farms which were converted before 1989 have suffered especially from this programme. They could not participate in the programme but encounter subsidized competition in the market.

In the framework of the EU agrarian reform, the extensification programme is being replaced by national programmes promoting all farms practising eco-farming.

Table 8.1. *Number of ecological farms at the date of the 1994 EU control, broken down by different criteria of definition*

Farms controlled by EU regulation	Member farms of AGÖL and Biopark	Farms controlled according to EU regulations, without membership in a federation	Farms participating in the EU extensification programme	Maximum number of extensification programme farms, being controlled according to EU regulation
5866	5669	197	13 200	3563

Note:
Biopark:Federation in Eastern Germany farming according to EU regulations, being not yet an AGÖL member.
Source: Bundesanstalt für Landwirtschaft; AGÖL; Hamm, FH Neubrandenburg, own investigations.

Table 8.1 shows clearly that only a minority of farms participating in the extensification programme chose the eco-option, either by becoming members of a federation or being controlled by EU regulations for eco-farming. The regulation of 1993 clearly defines which farms can be counted as part of the eco-farming system. Table 8.1 is based on this definition. In 1994, only 1.6% of Germany's agricultural land and 1.0% of all German farms fell under the official EU definition of eco-farming.

In addition to eco-farming gaining official recognition in this phase through the EU regulation and extensification programme, advisory work in eco-farming also received increasing governmental support. External pressure on the eco-farming system influences its organization, its self-definition and its knowledge system. This external pressure comes from increasing supply in constrained marketing channels with corresponding price-cuts; increasing quality requirements from processors and consumers; increasing conversions to eco-farming for economic reasons; the growing specialization of eco-farms; and subsidized competition as a result of the EU extensification programme.

There is, in general, a tendency for eco-farming principles to become weaker. Indications of this are found in the discussions within the eco-farming system. The important issues being debated, for example, are concerned with gaining access to conventional food marketing channels and important customers such as canteens. Other discussions concern attaining more income through better prices or subsidies (Hermanstorfer, 1994), and establishing less strict guidelines for processing, although the requirements for '*health food*' (In German: Vollwertkost. Literal translation: full value food) have, so far, not been affected (Koerber *et al.*, 1993).

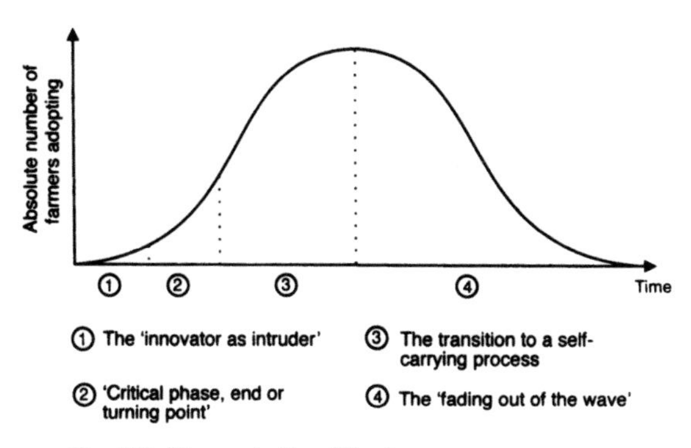

Fig. 8.3. Phases in the diffusion process.

8.4 Assessment of the diffusion of eco-farming based on diffusion theory

According to Albrecht (1994) and others, successful diffusion occurs in waves with successive phases (see Fig. 8.3). The first phase, called 'the innovator as intruder', starts with the innovator perceiving a problem, either through seeing a decline in conditions or through becoming aware of new, superior options. Being the first to adopt the new solution to a problem, he normally runs a risk by not being quite sure whether the innovation will lead to success. As a rule, adopting an innovation leads to a certain isolation from the rest of the community. Innovators are laughed at or even rejected. This can be seen as a psychological defence mechanism which protects people from having to take the new option seriously. Very seldom does this rejection lead to an abortion of the innovation. Generally, the innovator needs to profit from the new solution and would suffer if it failed.

In the second 'critical phase', if it proved successful for the innovator, the innovation is adopted by other members of the community who are in similar circumstances. If this second group is successful, it proves that the innovation will work and is acceptable for all those for whom it is relevant. Further adopters will follow, as the risk of adoption is seen to be lower now. The theory assumes that the process will continue autonomously without external support after a critical number of adopters have successfully adopted. At the village level, this ratio is thought to be about 10–20 % of the potential adopters.

The third phase is now initiated, the 'transition to a self-carrying process', the innovation becomes something of a norm. In the fourth and final phase, the 'fading out of the wave', further adoption slows down.

If we consider the phases of the diffusion of eco-farming described so far, we can draw the following conclusions:

1. The expansion of eco-farming seems to be a typical diffusion process. As eco-farming expanded, it went through the typical first and second phases: some pioneers generated and introduced it as a result of their special

awareness of ecological problems. Conventional farmers and the whole 'agricultural scene' react with strong rejection and discrimination. After it becomes obvious that it works in practice, it is adopted more and more, even by formerly sceptical colleagues.

2. The expansion of eco-farming is at the same time a totally atypical diffusion process. Normally, innovations are introduced as achievements of technical progress and therefore are compatible to general norms of the society. They are developed by science and are introduced into practice by extension work. The main obstacle to adoption is a lack of experience and knowledge, not incompatibility. Eco-farming was not developed by established science and afterwards introduced into practice. On the contrary, it was first developed by practitioners, and only at a rather late stage did science show interest in it. As previously shown, eco-farming is based on an understanding of science which goes beyond the general consensus in the scientific community and society.

These may be some of the reasons why eco-farming, more than 70 years after its beginning and despite its undoubted advantages, has a very modest diffusion rate covering only 1.6% of the total agricultural area. The speed of diffusion is dependent to a great extent on the features of the innovation. The features of eco-farming are rather unfavourable for quick diffusion (see also Somers, this volume).

Compatibility with existing norms (Albrecht, 1994) is a feature of eco-farming of special importance. It is critical for the success of farm conversions that the population accept the system and that consumers are willing to buy eco-products at higher prices. The introduction of the innovation in this case not only consists of a change in total farm organization, but also depends on its acceptance by at least part of society. Eco-farming is not an innovation at farm level only but ultimately an innovation for society as a whole.

Many of the methods of eco-farming, are not directly *visible* – such as the application of bio-dynamic preparations or *measurable* – such as food quality. In general, the workload is higher, the decisions depend on highly complex relations, and the conversion process is accompanied by initial and drastic declines in yields before the desired balance is established. According to AGÖL guidelines, farmers are not permitted to convert in part or on a small scale. These features make adoption of the innovation highly unfavourable. Hence eco-farming is not easily *divisible.*

While the creation of financial incentives by the EU and national governments could be seen as a favourable condition for further diffusion, it has also created the problems mentioned earlier.

The information system has played an important part in the diffusion of eco-farming in phases 2 and 3. In addition to the mass media (television, radio and the press) professional journals, technical newspapers and magazines, as well as personal contacts have played major roles (Albrecht, 1994). Personal contacts have been decisive in the conversion to eco-farming (Grassinger, 1984), while for consumers it has been a mixture of personal and media influence. In general, eco-farming has generated much more public discussion than it merits, given the percentage of farmers actually involved in it.

Summing up so far, we observe that eco-farming is in the 'critical phase' of diffusion and not yet in a self-carrying process. Comparable examples of eco-farming's diffusion processes are not known to the authors. For us, it is quite difficult to see the future role eco-farming will play in agriculture in Germany and Europe. The federations of eco-farming aim to cover 10% of the agricultural area by the year 2000 (Hansen, 1991). Hamm (1994) predicts 6–7% over the same period.

8.5 Examples of knowledge and communication systems supporting eco-farming

The conditions which brought about the development of eco-farming went hand in hand with the creation and adaptation of special knowledge systems and communication strategies among farmers, as well as between farmers and consumers. Some typical examples of this are provided in the following.

8.5.1 Advisory work

Farmers who want to convert to eco-farming need a lot of advice (Benecke *et al.*, 1988; van Weperen *et al.*, this volume). As a consequence of developing outside the normal channels of agricultural policy in phase 1 and 2 of its diffusion, eco-farming did not receive advisory or financial support from government agencies. Hence, farmer-organized extension and advisory work has always been of great importance within the federations. This 'collegial extension' has taken three forms:

- experienced practitioners fulfilled extension tasks;
- personal, informal interaction among farmers;
- regular meetings for exchange of experience among farmers from the same region.

The last two points are still important elements of advisory work in eco-farming. Group extension (exchange of experience) occurs mainly within the bio-dynamic and the Bioland federations, with closed regional groups and regular meetings but each with a different organizational set-up. For Bioland, the regional groups are an important part of their federation structure. Group extension has advantages in making individual knowledge accessible to others and in enhancing co-operation among farmers on all types of tasks and problems. This approach is limited by the group's restricted internal knowledge, unequal composition, group size, a lack of feedback about errors, and by overloaded group leaders (Luley, 1996).

In the third phase of expansion, eco-farming received official recognition as a farming system of merit, along with support for advisory work. The organization

Table 8.2. *Organization of advisory work in German eco-farming*

Organization of advisory work	Federal State	Government share in % of finance	
		Personal cost (DM/year)	Running cost (DM/year)
Extension-circles	Schleswig Holstein	45	0
	Niedersachsen	50	0
	Brandenburg	90	90
	Rheinland Pfalz[1]	50	12 000
	Baden Württemberg[2]	50	100
	Bayern	80	50
Official extension	Rheinland Pfalz[B,D,AAF]	100	100
	Hessen[B,D,BÖW]	100	100
	Thüringen[AAF]	100	100
Federation extension	Mecklenburg-Vorpommem[N]		
	Niedersachsen[N]		
	Nordrhein-Westfalen[A,B,D]	Through project support	
	Sachsen-Anhalt[N]		
	Rheinland-Pfalz[N]		
	Hessen[B]		
	Sachsen[D,G]	Three advisers of Gäa are financed	
	Baden-Württemberg[B,D,N]		
	bundesweit[A]		

Source: Luley (1995) and own investigations. 1=Advisory work in marketing, 2=Extension circles for bio-dynamic farms and circles for eco-farming subjects across all federations; B=*Bioland*; D=*Demeter*, AAF=across all federations; BÖW=Federal ecological viticulture; N=*Naturland*; A=ANOG, G=Gäa.

of extension in eco-farming and the percentage of financial support from government in the different Federal States of Germany are shown in Table 8.2.

Three distinct forms of extension organisation exist in eco-farming:

- *Federation extension*: advisory work through advisers of the federation (partly government subsidized);
- *Extension-circles*: associations of producers which employ advisers to assist its members (usually the government contributes part of the financing);
- *Official extension*: advisers employed by government or Ministry of Agriculture charged with extension tasks in eco-farming.

8.5.2 Information services

Information about eco-farming has, for a long time, only been available to farmers and consumers inside the eco-farming system. Journals, magazines and newsletters of the federations had an important role to play. In phase 1 and 2, one independently edited journal, *Garten organisch, organischer Landbau*, contributed much. In phase 3, however, it changed ownership and changed its profile towards home gardens. Meanwhile, the SÖL (Foundation of Ecology and Farming) was founded and now edits the journal *Ecology and Farming* (*Ökologie und Landbau*) and has two book series about eco-farming. The journal tries to present scientific, practical and political topics of eco-farming, independent of the federations. Recently, basic books have also appeared and are developing into important sources of information for practitioners (Sattler & Wistinghausen, 1985; Neuerburg & Padel, 1992; Siebeneicher, 1993).

In general, the discussion of topics and issues of eco-farming has continually increased in phase 2 and 3, in the professional press as well as in all public media.

8.5.3 Education

The economic success of an ecological farm is, to a high degree, dependent on the qualifications of the farm manager (Schlüter, 1985). Vocational education in agriculture is mainly carried out by government agencies, and since there has so far been no special educational programme for eco-farming, the necessary knowledge and skills are still, for the most part, acquired through self-instruction.

All federations offer one week introductory courses. The only courses of a longer duration are available from the *Forschungsring für biol. dyn. Witschaftsweise*, which offer a 4–week course and a study year of more intensive educational programmes. Inside the bio-dynamic movement specific bio-dynamic apprenticeships are also offered. At some vocational agricultural schools there is an indication that eco-farming is becoming part of the curriculum (Bliefernicht, 1994). In Landshut, Bavaria, there is now a technical high school offering master and technician courses in eco-farming.

Primarily due to the initiative and interest of students, all agricultural faculties at German universities (and university colleges) now have a study programme in eco-farming. The University Kassel/Witzenhausen has been at the forefront, offering eco-farming as a specialization in its postgraduate studies. It has also been trying to use participatory working methods and to make students more active in planning their own study progress. There is also an attempt to take special features of eco-farming into consideration when making didactic arrangements.

8.5.4 Associations between producers and consumers

The necessity of marketing outside conventional channels, has led to strong efforts by producers to 'capture' consumers and to affiliate them with the farms. Equally, consumers have developed an interest in eco-farming for health or environmental reasons. The following modalities for marketing have emerged.

Direct marketing

For many farms, direct marketing is important for earning income. Having consumers come to the farm offers the possibility of informing them on-site about the problems of agriculture. Eco-farming was a leading pioneer in direct marketing, nowadays also increasingly practised by conventional farmers.

Consumer/producer associations

These associations were more important in the beginning when shops and purchase opportunities for eco-food were scarce. In these consumer/producer associations, the consumer agreed to buy certain amounts of products throughout the growing season.

Product networks

These networks aim to achieve a sufficient income for producers based on justice and security. A group of producers, processors, traders, and consumers get together and contract for certain quantities and prices. An example of this is the *Märkische Wirtschaftsverbund* in Brandenburg.

Country communities

This form of union between consumers and producers developed mostly in anthroposophical circles. Departing from the idea that each individual has the right to a piece of land, in these communities the consumers as a group own the farm land. Looking at the available agricultural land in Germany, each citizen would get a quarter of a hectare of land. A 50 ha farm would be owned by an association of about 200 people. They would install a farmer to run the business as a trust for them. The farmer also has the right to propose the farm successor. In this way it is possible, but not obligatory, to hand the farm over to a family member. Because the farm cannot be sold when it belongs to an association, it is withdrawn from land speculation. The classic example of this model is the *Dottenfelder Hof*, described by Bauer (1994). Other similar models are reviewed by Gengenbach and Limbacher (1985).

Aktion Kulturland

This is a foundation based in Hamburg which buys land, mainly in northern Germany, to take it out of speculation. It then gives the land, in trust, to farmers, groups of farmers working together, or to country communities, to be run as explained above.

Financing

The high capital requirements and low profit rates in agriculture often make it difficult to invest. The GLS-Bank, working on anthroposophical principles, offers its customers the possibility of financing eco-farming projects directly, either by direct project financing or through the agricultural fund where interest is paid in kind, in the form of eco-products.

8.5.5 Producer associations

Many producer associations were started during the time when the EU extensification program led to a strong increase in products offered by eco-farming. Producer associations are defined in the *Marktstrukturgesetz,* which is the legal basis for government support of such associations. They are working towards a common market with the advantage of more homogeneous quality in larger quantities. This makes it possible to satisfy the needs of large customers who would then find it easier to introduce new groups of products into the bio-products market. Very often this also leads to closer co-operation with processing firms.

8.5.6 Science

In the field of science, there are two opposite positions towards eco-farming. On the one hand, many investigations point out its superiority with regard to environmental conservation. On the other, there is much scepticism about a system which restricts its range of options and uses technical progress only selectively.

It has mainly been the interest and pressure of students which has led to the establishment of posts for professors or coordinators of eco-farming in the agricultural or horticultural faculties. All of these faculties now have experimental farms, or at least experimental fields, which are ecologically run. However, once eco-farming is integrated into conventional science establishment, it will be affected by the same problems that affect mainstream natural science. Two problem areas are cited:

- eco-farming research produces large quantities of highly specialized and detailed investigations. However, no instruments or procedures have been developed so far to bring them together into a global and holistic view. This phenomenon can be observed in the documentation of the *Third Scientific Congress on Eco-farming*, edited by Dewes and Schmitt (1995).
- The relevance of research in eco-farming must increasingly be assessed with respect to its use in practice. Another question is how to communicate important results to practitioners.

While research in eco-farming formerly worked mainly by comparing systems (ecological versus conventional), recent research is more concerned with optimizing the eco-farming system itself.

Figure 8.4 gives a final overview of the levels and forms of communication inside the different elements of the knowledge system of eco-farming.

8.6 Conclusions

The conclusions are formulated in the form of statements about the future development of eco-farming.

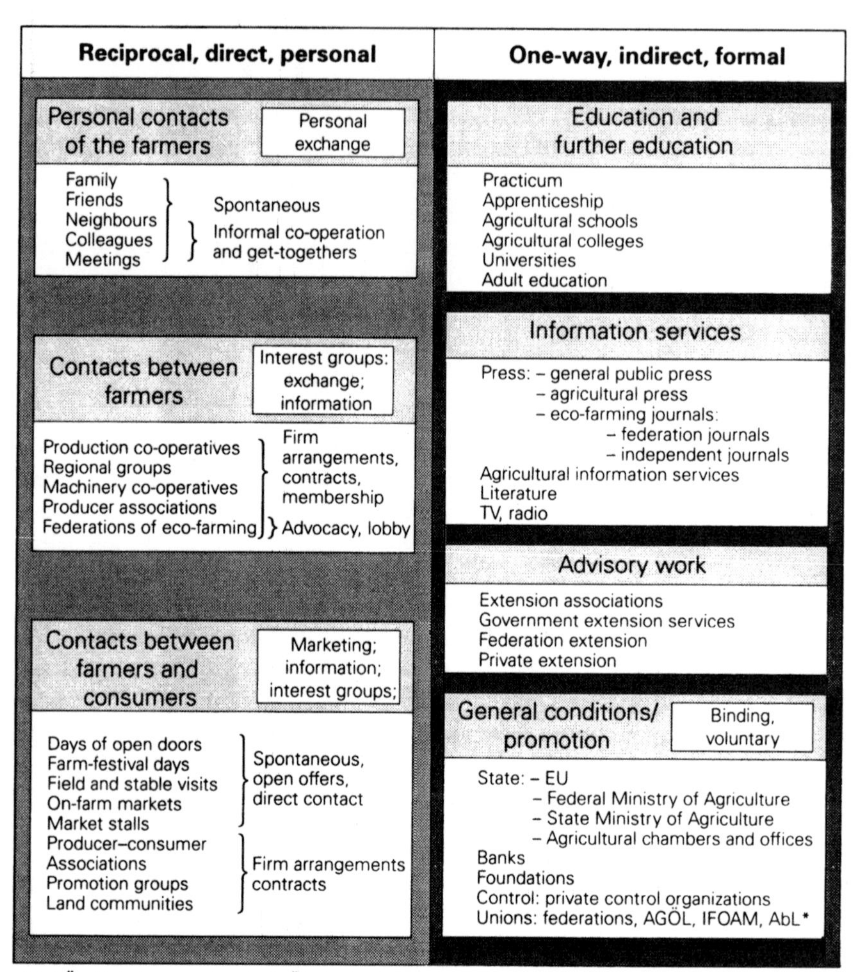

* AGÖL (Arbeitsgemeinschaft Ökologischer Landbau): union of the eco-farming federations; IFOAM: Federation of Organic Agriculture Movements; AbL (Arbeitsgemeinschaft bäuerliche Landwirtschaft): is an alternative farmer's union in opposition to the traditional farmer's union.

Fig. 8.4. Levels and forms of communication in eco-farming.

8.6.1 Eco-farming offers a perspective for a future sustainable agriculture

A further aggravation of problems can be expected in conventional agriculture. It may be through the environmental problems it causes, or in the field of agricultural policy, or through market and structural problems. As solutions are explored, eco-farming, being the most sustainable farming system available, will gain importance in society, science, and policy making. This will create a further expansion of eco-farming, but also endanger the eco-farming movement.

8.6.2 The knowledge system of eco-farming is threatened by increasing external pressures

Until the end of the 1980s, eco-farming existed in a niche governed by its own laws. Nowadays, with its growing importance, eco-farming is increasingly exposed to social aspirations, political interference, scientific research and a more competitive market.

These outside influences bring into question more and more the governing principles of the movement. These principles, which have been successful so far, are also being discussed among the members and federations of eco-farming. Indications of this are:

- conversions to eco-farming are now mainly motivated by economic concerns;
- this leads to a new tendency of pushing the guidelines to their very limits. Eco-farming is only perceived as a system with certain restrictions which must be respected, but which can also be used for maximum economic profit. This attitude contradicts the idea of the farm as an organism and of system integration;
- the enhanced specialization of farms;
- the processing of eco-farming products – the corresponding system to eco-farming in the food sector – shows a weakening of the criteria for 'health food'.

8.6.3 Eco-farming will only remain an alternative for action if it defends its knowledge system

If the development pointed out in statement 2 continues, eco-farming and Integrated Farming will become increasingly similar and finally merge. Integrated farming allows pesticide and mineral fertilizer use, at reduced or minimal levels. By this, eco-farming would lose its attractiveness as an alternative for action and lose its importance as well.

To prevent this happening, an intense discussion should take place inside eco-farming to determine its characteristic objectives and aims. If eco-farming is to remain a perspective for future agriculture, these objectives and aims should then be presented professionally and aggressively to the public in order to counteract the strong external pressures. More work in public relations, marketing and political lobbying is necessary to strengthen the position of eco-farming. The Federations and AGÖL, therefore, should co-operate more closely. Increased efforts should also be made in the field of consumer information.

8.6.4 Eco-farming will only be sustainable if its knowledge system is further developed

Eco-farming has emerged from practice. Its development has created and maintained some dogmatic positions which cannot be supported by either science or practical experience. The rigid application of these practices disregards the necessary adaptation to locations and situations. Two examples are given.

- Tilling and mixing of the topsoil only, is recommended by Bioland, but by not ploughing, soil structure and fertility can sometimes be negatively affected, and, in addition, leads to weed problems.
- The widespread use of composting solid farm manure in bio-dynamic farming leads to a high loss in potassium and organic matter (Mayer 1995).

Eco-farming should allow the freedom to question all established principles, and to have the liberty to adjust them to special conditions and new experiences. Otherwise, the system will some day die from mental sclerosis.

8.6.5 The knowledge system of eco-farming offers an ethical model for action

Eco-farming offers alternatives for action, and at the same time indicates the weaknesses of other knowledge systems. Eco-farming is based on ethical maxims for action from an enlarged epistemology. As a result, the handling of scientific and technological progress by applying self-imposed restrictions can be seen as responsible behaviour, in the sense of Jonas (1979). Consequently, eco-farming offers a model to other disciplines for solving problems in society and science.

8.7 References

Albrecht, H. (1994). Die Verbreitung von Neuerungen – Der Diffusionsprozeß. In *Beratung als Lebenshilfe*, ed. V. Hoffmann. pp. 21–22. Weikersheim: Verlag Josef Margraf.

Bach, M. (1987). Die potentielle Nitratbelastung des Sickerwassers durch die Landbewirtschaftung in der BRD. *Göttinger Bodenkundliche Berichte*, p. 93.

Bauer, D. (1994). Landwirtschaftsgemeinschaft Dottenfelder Hof. Ein Modell für die Zukunft. *Ökologie und Landbau*, **89**, 28–31.

Benecke, J., Kiesewetter, B. & Urbauer, H. (eds.) (1988). *Bauern stellen um.* Alternative Konzepte, p. 62. Karlsruhe: C.F. Müller Verlag.

Bliefernicht, K. (1994). Ökologischer Landbau in der Berufsschule. In *Lebendige Erde*, **3**, 212.

Brugger, (1990). *Landbau – Alternativ und Konventionell.* Bonn: AID-Heft 1070.

Burdick, B. (ed.) (1994). *Klimaänderung und Landbau – Die Agrarwirtschaft als Täter und Opfer.* Alternative Konzepte, p. 85, Karlsruhe: C.F. Müller Verlag.

Dabbert, S. & Braun, J. (1993). Auswirkungen des EG-Extensivierungsprogramms auf die Umstellung auf Ökologischen Landbau in Baden-Württemberg. *Agrarwirtschaft*, **42**(2), 90–9.

Dewes, T. (1994). Die Wirkung der biologisch–dynamischen Präparate. In *Ökologischer Landbau – Perspektive für die Zukunft!*, ed. J. Mayer. Bad Dürkheim: Sonderausgabe Nr. 58 Stiftung Ökologie und Landbau.

Dewes, T. & Schmitt, L. (eds.). (1995) *Beiträge zur 3.* Wissenschaftstagung zum Ökologischen Landbau vom 21. bis 23. Februar 1995 an der Christian-Albrechts-Universität zu Kiel. p. 7. Gießen: Wissenschaftlicher Fachverlag.

Diercks, R. (1993). *Alternativen im Landbau.* Stuttgart: Ulmer Verlag.

Fischer, R. (1982). *Der andere Landbau. Hundert Bio-Bauern und Gärtner berichten über ihre Beweggründe, Arbeitsweisen und Erfahrungen.* Zürich: Verlag Buchhandlung Madlinger-Schwab.

Gengenbach, H. & Limbacher, M. (eds.) (1985). *Kooperation oder Konkurs?* Stuttgart: Verlag Freies Geistesleben.

Gerber, A. (1994). Einfluß einer Flächenspritzung des biologisch-dynamischen Baldrianpräparates (507) auf das Wachstum von Sommerweizen und Winterroggen, unter besonderer Berücksichtigung der P-Versorgung. Thesis, Institute for Plant Nutrition, University of Hohenheim.

Grassinger, P. (1984). Psychologische Aspekte der Betriebsumstellung auf Ökologischen Landbau. Thesis, Institute for Psychology, University of Tübingen.

Hamm, U. (1994). Perspektiven des Ökologischen Landbaus aus marktwirtschaftlicher Sicht. In *Ökologischer Landbau – Perspektive für die Zukunft!*, ed. J. Mayer. Bad Dürkheim: Sonderausgabe Nr.58 Stiftung Ökologie und Landbau.

Hansen, H. (1991). 20 Jahre Bioland – Bioland ins Jahr 2000. *Bioland*, **2**, 3.

Hermanstorfer, U. (1994). Wie weiter mit der Landwirtschaft? In *Rundbrief Dreigliederung des Sozialen Organismus.* Stuttgart: Netzwerk Dreigliederung.

Jonas, H. (1979). *Das Prinzip Verantwortung.* Versuch einer Ethik für die technologische Zivilisation. Frankfurt/Main: Suhrkamp.

Klett, M. (1994). Bewußtseinsgeschichtliche Aspekte zur Entwicklung des biologisch-dynamischen Landbaus im 20. Jahrhundert. *Lebendige Erde*, **5**, 338.

Koerber, K.W., Männle, Th. & Leitzmann, C. (1993). *Vollwert-Ernährung.* Heidelberg: KF Haug Verlag.

König, U.J. (1994). 70 Jahre biologisch–dynamische Forschung. *Lebendige Erde*, **5**, 327.

Köpcke, U. (1994). Nährstoffkreislauf und Nährstoffmanagement unter dem Aspekt des Betriebsorganismus. In *Ökologischer Landbau – Perspektive für die Zukunft!*, ed. J. Mayer *et al.* Bad Dürkheim: Sonderausgabe Nr. 58 Stiftung Ökologie und Landbau.

Luley, H. (1995). Beratungsringe im Ökologischen Landbau – Arbeitsweise und Verbreitung. In *Beiträge zur 3. Wissenschaftstagung zum Ökologischen Landbau vom 21. bis 23. Februar 1995 an der Christian-Albrechts-Universität zu Kiel*, ed. T. Dewes & L. Schmitt. Gießen: Wissenschaftlicher Fachverlag, vol. 7, pp. 29–32.

Luley, H. (1996). *Information, Beratung und fachliche Weiterbildung in Zusammenschlüssen Ökologisch wirtschaftender Erzeuger.* Weikersheim: Margraf Verlag.

Mayer, J. (1995). Erfassung von Bilanzen und Versorgungsgrad für die Nährstoffe Kalium und Phosphor in einem biologisch-dynamisch wirtschaftenden Betrieb. Thesis. Institute for Plant Nutrition, University of Hohenheim.

Mengel, K. (1991). *Ernährung und Stoffwechsel der Pflanze.* p. 10. Jena: Verlag Gustav Fischer.

Neuerburg, W. & Padel, S. (1992). *Organisch-biologischer Landbau in der Praxis.* München: BLV Verlag.

Rat der Sachverständigen für Umweltfragen, (1985). *Umweltprobleme der Landwirtschaft.* p. 337. Stuttgart: Kohlhammer Verlag.

Rogers, E.M. (1983). *Diffusion of Innovations.* New York: Free Press.

Sattler, F. & von Wistinghausen, E. (1985). *Der landwirtschaftliche Betrieb.* Biologisch-Dynamisch. Stuttgart: Ulmer Verlag.

Schaumann, W. (1977). Der Biologisch-Dynamische Landbau. In *Ökologischer Landbau – eine europäische Aufgabe,* ed. Stiftung Ökologie und Landbau. Alternative Konzepte 21. Karlsruhe: Verlag C.F. Müller.

Schaumann, W. (1995). Der wissenschaftliche und praktische Entwicklungsweg des Ökologischen Landbaus und seine Zukunftsaspekte. In *Beiträge zur 3. Wissenschaftstagung zum Ökologischen Landbau vom 21. bis 23. Februar 1995 an der Christian-Albrechts-Universität zu Kiel,* ed. T. Dewes & L. Schmitt. p. 7. Gießen: Wissenschaftlicher Fachverlag.

Scheller, E. (1993). *Die Stickstoff-Versorgung der Pflanzen aus dem Stickstoff-Stoffwechsel des Bodens.* Weikersheim: Verlag Josef Margraf.

Schlüter, C. (1985). Arbeits- und betriebswirtschaftliche Verhältnisse in Betrieben des alternativen Landbaus. *Agrar und Umweltforschung Baden-Württemberg Bd.* 10. Stuttgart: Ulmer Verlag.

Siebeneicher, G.E. (ed.) (1993). *Handbuch für den biologischen Landbau.* Augsburg: Naturbuch Verlag.

Steiner, R. (1924). *Geisteswissenschaftliche Grundlagen zum Gedeihen der Landwirtschaft,* Dornach 1979: R. Steiner Verlag.

Ziechaus-Hartelt, C. (1991). Bioland – ein Verband entwickelt sich. *Bioland,* **2**, 13–14.

Zink, F. (1986). Naturel et Agroalimentaire en Republique Federale d'Allemagne. PhD, Paris-Grignon: Institut National Agronomique.

9 Introducing integrated pest management in rice in Indonesia: a pioneering attempt to facilitate large-scale change

NIELS G. RÖLING AND ELSKE VAN DE FLIERT[1]

9.1 Introduction

Integrated pest management (IPM) is an important component in sustainable agriculture. The paper describes the National IPM Programme which the Indonesian Government has been implementing since May 1989. It is the first large-scale attempt to systematically introduce more sustainable agricultural practices as a national, public sector effort. As of October 1995, the Programme had trained an estimated 229 000 farmers in season-long farmer field schools. In doing so, the Programme had learned immensely important lessons for all of us who are interested in what it takes, in practice, to foster sustainable agriculture. This chapter is an attempt to capture some of these lessons, although the authors fully realize that they can only describe some of the highlights.

The Programme was managed from the start from a professional but an informal educationist perspective and not from a technical perspective. This is in no way meant to sneer at technical scientists, such as entomologists. In fact, our experience is that many of them, such as irrigation specialists, are particularly sensitive to the human dimension of agricultural change. It was the entomologists in charge of organizing IPM in Indonesia who made a deliberate choice to involve lay

[1] Part of this chapter appeared in *Agriculture and Human Values* Vol 2 (2+3), Spring and Summer, 1994: 96–108 as 'Transforming extension for sustainable agriculture: the case of Integrated Pest Management in rice in Indonesia'. That article benefited from assistance from Mrs Jennifer Dunn (who edited a 50 page 1991 mission report into a first draft), from Dr Patricia Matteson who made helpful comments on the third draft, and from Dr Russ Dilts, the team leader of the FAO TA team of Indonesia's IPM Programme, who made equally helpful comments on the fourth draft. Dr Haynes and Dr Lori-Ann Thrupp, editor and guest editor of *Agriculture and Human Values*, also made important contributions. The article was originally based on a report written by Niels Röling during a field trip to the project in 1991. At the time, the second author was carrying out an evaluation of IPM Farmer Field Schools in a district in Central Java for the FAO. This evaluation became the basis of her doctoral dissertation (van de Fliert, 1993). The article was later updated and changed on the basis of her work, and again in 1996 on the basis of the first author's participation in the World Bank's Mid-Term Review of the Programme in September 1995. The Mid-Term Review was particularly useful in learning more about the efforts to scale up the impact of the programme, especially through the use of farmer-trainers.

The authors wish to express their gratitude for having been exposed to this exciting, innovative and pioneering programme of the Indonesian Government, FAO and the World Bank. They wish to thank them for the lessons learned 'on behalf of all of us'. They want to especially thank FAO's TA team in Jakarta, under the inspiring leadership of Dr Russ Dilts, and FAO's Regional team in Manilla, under the equally inspiring leadership of Dr Peter Kenmore.

educationists in its management. And that makes the Programme especially interesting for this book: facilitation strategies were deliberately designed, and ways to overcome the hurdles of scaling up the successful pilot attempt were experimentally established.

The Programme is a temporary structure that will be continued for a limited number of years. Its first pilot phase (1989–92) was financed by donations from USAID to BAPPENAS, the national planning agency, that were originally meant for pesticide subsidies. The second phase (1993–98), carried out by the Ministry of Agriculture, is supported by a World Bank loan. Both phases were implemented with assistance from a FAO technical team that consists of 'technical' and educational scientists.

The Project was preceded by a Presidential Decree, in 1986, which banned 57 pesticide brands for rice cultivation, and declared IPM the national pest control policy. A second policy measure gradually reduced the subsidy on pesticides, previously 85%, to zero in January 1990. These regulatory measures created a favourable climate for the implementation of the National IPM Programme.

Farmer training was considered necessary, in addition to the regulatory measures. As will become clear, these measures were, in themselves, sufficient to reduce the use of pesticides and reduce the threat to food security. The IPM Project gives the following grounds for elaborate additional farmer training:

- The conventional, high external-input approach to agriculture is heavily ingrained in the national system and can be expected to have considerable momentum. The pesticide industry and the input distribution apparatus, including some extension workers who support their income through pesticide sales, can be expected to exert continuous pressure on farmers to use pesticides, replacing banned substances with permitted ones. Evidence to this effect is the promotion of pesticides with relatively expensive carbofuran granules permitted by the Presidential Decree. Only a critically aware farming populace can provide the necessary counter-pressure;
- Local pest outbreaks can easily be used to scare farmers and local officials into massive use of pesticides and to undermine IPM. The IPM Programme has experience with organizing collective activities to prevent unnecessary pesticide use, even in outbreak situations;
- An impact study conducted for the Programme in 1991 among over 2 000 IPM Field School graduates in five provinces showed that trained farmers used 50% less pesticides than untrained ones, especially with respect to the banned substances (Pincus, 1991). This was later confirmed by other studies (e.g. van de Fliert, 1993).

However, the main reason is that IPM, with its reliance on knowledge-intensive local agro-ecosystem management, requires that farmers are 'experts' in their own fields, capable of observation, experimentation, anticipation, joint deliberation and considered decision making.

The National IPM Programme provides an ideal case to contrast the facilitation of sustainable agriculture with that of promoting high external-input agriculture. IPM is being introduced into a farming system, irrigated rice, in which the Green Revolution has been deeply embedded for the past 20 years.

9.2 The Green Revolution in Indonesia

The present generation of Indonesian rice farmers has grown up with the Green Revolution. From 1968, when famine threatened the Indonesian people, high-yielding varieties (HYV) of rice and agro-chemicals were introduced, often by force. Usually, village officials exerted pressure in various ways to promote Green Revolution technology. In some areas, crops of farmers not growing the new HYVs were cut down by village officials, or planting of HYVs and use of fertilizers were enforced by the army. Inputs were distributed through village administration, which allowed easy control. Moral pressure to co-operate in intensification programmes was high. When farmers purchased input packages on credit through the Village Unit Co-operative (KUD), they had to take the entire package (seeds, fertilisers and pesticides prescribed) as part of the blanket recommendation covering the entire rice-producing area.

Decisions to apply pesticides were often made by officials, and entire areas would be sprayed by plane. Farmers were not allowed to take paddy-land out of production if they wanted to grow more profitable crops. In fact, the policy focused on production in order to obtain self-sufficiency, and farmers who might have been more interested in profit found themselves at odds with the goals of the Ministry of Agriculture.

The coercive nature of the introduction of the new technologies should, however, not be overstressed: farmers soon discovered the benefits and readily adopted the package, at least partly. The present situation is more relaxed. Many erstwhile compulsory measures are soon carried out voluntarily. Although farmers feel that rice production with the use of the new technologies is riskier and more of a hassle, they say they are much better off.

However, rice farming is still considered official business. Farmers are treated as if they are the lowest level of civil servants and considered passive acceptors of official wisdom. Extension's task is to tell them what to do. The agricultural extension system, based on the training-and-visit (T&V) model, organized farmers into official farmer groups. For extension convenience, grouping is based on adjoining rice areas and consequently puts farmers from different neighbourhoods and sometimes villages together. Logically, many of these artificial groups do not function in practice. A carefully calibrated scale, based on ten criteria, is used to 'grade' farmers and allow them to advance, civil service style, to 'progressive farmer'. Few, so far, have attained this lofty position.

In all, the Green Revolution seems to have been effective. Indonesia attained self-sufficiency in rice in 1983, after having been the world's largest importer for many years. Price relationships are carefully managed so that most farmers continue to make a minimal living, while rice remains cheap and allows low urban wages. Especially from a national point of view, the approach can be considered a success. The political turmoil which coincided with famine in the 1960s has ensured that food security remains a political priority. This was reinforced by a shortfall in rice production of 2 million tonnes in 1994 as a result of drought. By then, the effectiveness of IPM in reducing 'outbreaks' had paradoxically reduced the priority of IPM, so that the shortfall was able to provide renewed impetus to large-scale promotional campaigns of high-input technologies. However, we are running ahead of an orderly story.

9.3 Beyond the Green Revolution

The Green Revolution seems to have run its course in Indonesia and it is time for the next 'wave' of innovation that will allow farmers to increase efficiency, while maintaining and improving what has been achieved. The productivity of (irrigated) rice is plateauing at about five tonnes per hectare. In addition, the present high external-input farming has a number of problems:

- Serious environmental and human health effects. Such effects include loss of food sources such as fish, frogs and ducks, and poisoning of drinking water supplies. Health effects from high exposure to biocides is observed by farmers who usually do not wear any protective clothing.
- Threats to food security through vast yield losses as a result of mass resurgence of such pests as brown planthoppers, stemborers, and rice leaffolders. These outbreaks are the invariable result of indiscriminate pesticide use (van den Bosch, 1980; Kenmore, 1980; Gallagher, 1988). The broad-spectrum pesticides commonly used by farmers kill both pests and natural enemies. This results in massive pest outbreaks, since pest populations build up faster than natural enemies. A second problem caused by the indiscriminate use of pesticides is that farmers tend to make concentrations too high but cannot afford to spray the required volume per hectare. Spraying is often very uneven, and therefore ineffective. What is worse is that this creates ideal circumstances for the development of pesticide resistance.
- Many traditional rice varieties appear to have been lost, and with them a store of genetic diversity which took literally thousands of years to develop. Large areas covered by crops of the same genetic make-up create conditions for pests and diseases to spread rapidly.
- Continuous rice cropping in irrigated areas leads to a situation in which pests always find sufficient food.
- Indigenous knowledge (e.g. Warren, Slikkeveer & Brokensha, 1991) about some of the components of rice farming seems strangely lacking. Indigenous knowledge of names and life cycles of many pest insects and their natural enemies is virtually absent, partly because many current major pests were not important previously, but also because of the input-orien-

tated technology advocated by extension. Most current farmers have grown up during the Green Revolution or have forgotten indigenous practice.

9.4 The introduction of IPM in Indonesia

Efforts to introduce IPM started as early as 1979, after Indonesia experienced its first nation-wide brown planthopper outbreak in 1975–77. Attempts followed the transfer of technology approach which had been so successful in the Green Revolution. Technical assistance was provided by FAO's Inter-country IPM Programme.[2] IPM training activities focused on packages and prescriptions, and were incorporated in routine extension meetings. No clear impact of these activities has ever been reported.

The problems associated with pesticide use culminated in a major threat to food security in 1985–86. During two seasons, an estimated 275 000 hectares of rice were destroyed by the brown planthopper. There are two contrasting stories about the way the problem was appreciated at national level (Box 9.1).

Box 9.1. Two versions of the management information that led to IPM

First version. Brown planthopper damage was not apparent at first in the national pest infestation records as a result of a principle called 'Asal Bapak Senang'. This means something like: as long as one's superior feels good about it. The story is as follows. The infestation records are based on sampling reports of special field officers (pest observers who are under the Directorate of Crop Protection). These reports are amalgated stepwise as they move up through the administrative hierarchy, from subdistrict level via district and province to the national level. Since records of severe pest outbreaks occurring in one's jurisdiction were considered a potential embarrassment to one's superior, brown planthopper damage during the crisis years became progressively smaller at every administrative level. It was only when the home villagers of the President came to him for help, and pressure from different quarters had led to an independent survey by BAPPENAS, the national planning agency, that the extent of actual damage became apparent. A politically dangerous situation had been created. It was this crisis which led to a Presidential Decree in 1986, declaring IPM the national pest control strategy.

Second version. The yield loss due to brown planthopper damage was systematically over-reported. Fields with some affected areas could be counted as affected fields, and areas in which some fields were affected could be counted in terms of the total acreage of the area. The reason for this systematic over-reporting was to obtain higher pesticide allocations.

[2] The FAO 'Inter-country Programme for the Development and Application of Integrated Pest Control in Rice in South and South East Asia' started with a pilot IPM (training) programme in 1978–80 in the Philippines. For almost the next 20 years, funded by the governments of the Netherlands, Australia and the Arabic Gulf Fund, the programme expanded to 15 participating countries in South and South-East Asia, where it initiated IPM training and research activities, and supported national programmes.

In looking at IPM activities in Indonesia, we must, as we said above, make a crucial distinction between *farmer training* efforts, on which most of this chapter focuses, and its *regulatory* and *fiscal* context. The Presidential Decree, INPRES 3/86, introduced an impressive array of policy measures which provided important conditions for the training effort, including:

- IPM declared the national pest control strategy
- prohibition of 57 broad-spectrum insecticides for rice, leaving ten brands of narrow spectrum insecticides, most of them considered especially effective against brown planthoppers;
- creation of 1 500 new pest observer positions within the Directorate of Crop Protection, bringing the total up to 2 900;
- enforced use of resistant rice varieties;
- in several irrigated areas, enforced introduction of one (dry) secondary food crop after two irrigated rice crops, prohibiting continuous wet rice farming;
- crash action through so-called 'POSKOs', or commando posts, involving specially trained farmers to give mass applications of narrow-spectrum insecticides, if necessary;
- the removal of the 85% subsidy on the price of pesticides (a separate measure, not part of INPRES 3/86).

As a reaction, IPM training was intensified through the T&V extension system (Matteson, Gallagher & Kenmore, 1993). The Government requested the World Bank to use the US$4.19 million remaining for the second phase of the National Agricultural Extension Project (NAEP II) to be used for IPM training. Senior pest observers were trained as 'IPM master trainers', and the new pest observer recruits and selected village extension workers were given a six-day crash training programme. The trained pest observers and extension workers, in turn, had to train farmers. FAO's Inter-country IPM Programme provided technical assistance.

In this crash programme, a tremendous effort was made to develop trainers' guides, flip charts, slide–audio modules, leaflets and pamphlets of which 150 000 copies were distributed by NAEP II. Travel money, honoraria, vehicles, subsistence and pocket money for farmers and other monies were paid. The entire budget, which would have totalled US$7 million if calculated on an annual basis, was spent in 7 months. Although the activities had Presidential priority and were facilitated by the Ministries of Finance & Planning and Economic Affairs, only 8.5% of the allocated resources were delivered to the field to train less than 10% of the farmers targeted (10 300 persons). Where farmers were reached, trainers used top-down approaches and did not use the field or farmers' own experiences. Only 25% of the training groups actually entered a rice field. Farmers trained reported not to have learned many new things, and their decision making remained dependent upon the officials.

The 1990 Review Mission summed up the experience:

> A rigid system equipped to move simplistic messages to a large number of passive farmers could not absorb the energy of IPM's field skills training. A transformation from within was needed to meet the new challenges from outside.

In other words, a T&V-type 'transfer of technology' approach to IPM farmer training did not work. The same conclusion has been drawn from similar attempts in other countries (Agudelo & Kaimowitz 1989; Matteson *et al.*, 1993).

Despite the meagre result from this crash IPM training Programme, the *policy measures* resulting from INPRES 3/86 were enough to:

- end the threat to food security from massive brown planthopper resurgence, induced by the destruction of biological control agents;
- save an annual outlay for the insecticide subsidy of between US$ 110 and US$120 million a year;
- vastly reduce pesticide imports; and
- make farming more cost effective. Contrary to popular belief fanned by the pesticide industry, careful experimentation has shown yields to be unaffected by reductions in pesticide use. Environmental and health effects at farm and macro-level are less easily measurable, but assumed to be substantial.

In 1989, the time was ripe for the approval of an IPM project to start the large-scale implementation of a revised IPM *farmer training* approach in major irrigated rice growing areas, based on principles of informal or non-formal education. Having learned from 10 years of experience in IPM training and implementation in various Asian countries, the Indonesian model embarked on a new course, with respect to both technology and training. From mechanical instructions for field sampling and spraying based on centrally determined economic threshold levels, IPM shifted to more ecological principles. These different principles required a different approach to farmer training, as will be shown below, but we must first describe the institutional actors in the National IPM Programme.

9.5 Introducing some of the actors involved

The pilot phase prior to 1992 was run by both expatriate and local experts and located within BAPPENAS, the planning agency, and not the Ministry of Agriculture. This gave the Project much more flexibility, especially since it was strongly supported by the then Director General of BAPPENAS. There were no senior counterparts in the traditional sense of the word. A Steering Committee, an Advisory Board, and a Working Group with members from various government institutions and universities were called into being to assist project management. For management and curricula development purposes, special secretariats were established in Jakarta and Yogyakarta.

The Project worked intensively within the country's existing framework, putting strong emphasis on creating linkages and contracting for specific jobs, such as curriculum development and training. In addition to training regular extension staff and farmers, the Project supported research activities such as a field laboratory in West Java focusing on white stemborer problems, a health impact study in a pesticide-intensive area in Central Java, insect habitat studies (generally establishing the incredible robustness of the rice eco-system), studies on IPM in secondary food crops, and several training evaluation and impact studies. An express training and action programme was organized in West Java to control a severe white stemborer outbreak. Airplane spraying ordered by some senior officials could be prevented only at the last moment through effective mass action by school children and others to collect stemborer egg masses. The mass action was followed by a marked reduction in pest pressure, which boosted the national standing of IPM considerably.

The Project continued to actively learn from its experience. A good example is the 'threshold' concept for chemical control decision making (Box 9.2).

Box 9.2 The evolution of the 'threshold' concept

It started as a *technical* damage count, considering only the number of pests per rice hill or square meter. The threshold then moved to the more sophisticated concept of economic threshold level (ETL) that weighed expected yield loss (in terms of damage (kg/ha) multiplied by price per kg) against estimated pest control cost. Applying this concept under farmer conditions, however, appeared complex and confusing, and therefore not workable. As a reaction to this, the '*experience* threshold' was started, which develops as farmers learn and experience, and focuses on the procedure of decision making. The former entomologist of the Programme, Dr Kevin Gallagher (1990), came to the conclusion that what mattered was that farmers made a sound decision, based on information collected from the field, whatever the actual decision taken. Thus the Project came full circle: from prescribing a concrete concept, the only thing that mattered in the end was the process. Meanwhile, however, it takes time for such learning experiences to penetrate the curricula, while staff who have been trained in previous concepts continue to propagate them.

At the regional level, the pilot phase operated from 12 field training facilities (FTF). Existing in-service agricultural training centres in the main rice growing provinces were partly transformed into IPM FTFs. Primary trainers at the FTFs were the field leaders I (21 in total), assisted by field leaders II (219 in total). They work in the Directorate for Plant Protection, but were loaned to the project. Most of the field leaders belong to the group of pest observers who were upgraded in the crash IPM training in 1986 to become IPM master trainers. This cadre of highly qualified trainers is a vital pivot in developing the multiplier for the training of the 10 million farm families in the major 'rice bowls' in Indonesia. Field leaders assisted in designing the final farmer training curriculum and field guides.

Pest observers are field level staff, also employed by the Directorate of Crop Protection, whose numbers were doubled by INPRES 3/86. As indicated by their name, the pest observers' task is to monitor and report pest outbreaks observed on fixed sample fields. For this purpose a network of Pest Control Laboratories has been established throughout the country with Japanese assistance (Japan is a major exporter of pesticides). The pest observers were the farmer trainers used by the Programme.

The Programme has trained 2112 pest observers out of a total of 2900 in the 12 Programme provinces. After secondary agricultural school, all pest observers were initially trained in the crash IPM Programme in 1986–87. Since then, they have received 15 months' special intensive training through the Programme. This training is a second vital pivot in the multiplier mentioned earlier. It will be described in greater detail later (Section 9.4). During the pilot phase of the Programme, the trained pest observers functioned as IPM trainers at the Regional Extension Centres (REC), to train both farmers and extension staff, with each REC having one or two pest observers. All staff involved in IPM training are temporarily assigned to the IPM Programme and receive some topping up of their salaries.

Given the experiences in the earlier crash training programme, it seemed obvious from the start that the field level extension workers (FLEWs) would not be suitable candidates. Numbers were doubled by INPRES 3/86. As indicated by their name, the pest observers' task is to monitor and report pest outbreaks observed on fixed input distribution activities, which conflicts with the nature of IPM. Yet, the Programme faces a dilemma. It is impossible for some 2 100 pest observers to train 10 million rice farmers effectively. There are 18 000 FLEWs. Hence the project has systematically involved FLEWs from the start, by making them responsible for farmer selection, by giving them short training courses and by involving them in actual farmer training (see below) and now (1995) by giving them the regular 15 months IPM training through the extension training academies.

Rice farmers are obviously the intended beneficiaries of IPM training activities. At first sight, they seem similar in their cultivation practices. However, farmers differ a great deal in terms of their use of inputs, farm size, tenure status, the type of off-farm jobs they engage in, and their activity in farmer groups and so on (van de Fliert, 1993). The villages in which farmers live also show great diversity due to geographical and infrastructural isolation, leadership, history, and other factors. As described above, farmers are formally organized in farmer groups, which are officially used as a unit for selection of IPM training participants, but these groups differ a great deal in the extent to which they are 'active'.

Vested interests in pesticide use are not immediately apparent beyond the agro-chemical companies. But inputs are an enormous industry with a turn-over valued at some US$1.5 billion per year. Involvement in this industry can be found in various sectors and levels, including salesmen, organizations such as government village co-operatives, village officials and extension workers. The influence of these interests on the effects of IPM training and implementation have been substantial.

Research institutes and universities have only been marginally involved so far. The universities have trained the pest observers for a few months to provide them with a diploma which allows them to advance in salary scale. Greater involvement of research institutes and universities is planned for the future.

9.6 Farmer training

The principles developed for rice IPM in Indonesia are:

1. grow a healthy crop;
2. observe the field weekly;
3. conserve natural enemies; and
4. farmers are IPM experts.

Based on these principles, the project emphasizes the following:

- focus on a *healthy crop*, tolerant to local pests and diseases, and able to compensate for pest damage. The focus on healthy crops opens the possibility to give considerable attention to soil fertility management, a crucial avenue for expanding the project from its narrow focus on 'bugs', to a broader focus on sustainable practices;
- a *good knowledge of pests and their natural enemies*, not in terms of their (Latin) names, but in terms of their function in the rice ecosystem, what they do to plants and to each other at what stage of the crop. Such knowledge includes the life cycles of pests and natural enemies and the recognition of their different stages. This knowledge is expected to be updated and improved by farmers' own observation and experimentation, and by farmer-to-farmer exchange of experience;
- *regular and systematic observation* of the field, using systematic procedures (random selection of sample rice hills) to assess the occurrence of pests and natural enemies in relation to the crop's development stage;
- *sound decision making* (whatever the decision) and discussion with other farmers about such decisions;
- *experimentation* with planting times, varieties, soil cultivation practices, fertilization, rotations and biological controls for their effect on pest populations;
- *use of relevant, science-based knowledge*, such as the work of the International Rice Research Institute, IRRI, on the regenerative capacities of rice varieties after pest damage, or the work of the Programme's own experiments, e.g. parasitism of white stemborer egg masses.

It is clear that with such priorities, the farmers' own expertise and mastery is fostered rather than only their adoption of external information. It is remarkable, in this respect, that when asked about advantages of IPM, farmers tend to mention not only the reduction in production costs, but also the confidence in their new-found expertise (van de Fliert, 1993).

This raises an important issue. One can regard IPM training as simply a way to reduce pesticide use. In the present World Bank-financed National IPM Programme, the positive internal rates of return for the project were based on the expected reduction in pesticide use. This focuses the Programme on targets which are expressed in term of the numbers of farmers trained and makes it logical to give the responsibility for IPM to the Directorate of Crop Protection after the termination of the Programme. However, one can also see farmer IPM training as an entry point for moving the entire government approach to agricultural innovation. Away from its present narrow focus on heavy-handed technology transfer, which views the farmer as a passive adopter of uniform imposed technology packages, to a broad focus on enlisting the farmer as an intelligent and expert partner in the challenge to make optimal use of Indonesia's diverse natural resources (see also Castillo, this volume). One could even claim that this is the natural post-Green Revolution approach to capturing additional productivity gains. Such a focus would locate post-Programme responsibility for IPM at a much higher bureaucratic level, say the Directorate General for Crop Production, rather than box it in at the level of the Directorate for Plant Protection. At the time of writing, the choice has been made to box it in.

The basis for the training approach developed in the Indonesian IPM Programme is non-formal education (NFE) (Dilts, 1983, 1990), itself a 'learner-centred' discovery process. It seeks to empower people to solve 'living' problems actively by fostering participation, self-confidence, dialogue, joint decision making and self-determination. Group dynamic exercises are an important part of this approach. IPM training is organized in so-called farmer field schools, described in Box 9.3.

Box 9.3 Key ingredients of the IPM farmer field school

- An IPM farmer field school (FFS) consists of a group of 25 farmers, selected either from one farmer group, or across such groups within one village;
- During the training, farmers work in small sub-groups of five, the optimal size according to NFE experience world-wide;
- Training starts with a 'ballot box' pre-test of knowledge and ends with a post-test. The tests, which have a multiple-choice character, are done in the field and about field problems. The scores of the tests, which are fairly meaningless in themselves, are a motivational device for the participants, and give an important diagnosis on trainees' relative ability to the trainers;
- The farmer field school lasts for the main part of an entire rice growing season so as to follow all stages of crop development. The school meets once a week for 10–12 weeks;
- Each farmer field school has one training field, divided into two plots: one IPM-managed field, and one field with the package recommended by the Agricultural Service, including one preventive granular pesticide application;
- During the training, lecturing is hardly used. The trainers do not allow themselves to be forced into the role of expert. They do not answer questions directly, but try to make farmers think for themselves. 'What did you find?' 'What did it do?' 'What do you think?'. This is called the 'Apa ini?' principle, meaning literally 'What is this?' Answering a question directly is considered a lost opportunity for learning;

- The field school meets somewhere in or close to the field under a tree or in a small shack which provides some shade;
- The main activity, the first in the morning, is to go into the demonstration fields in groups of five and observe sample rice hills, usually chosen randomly along a diagonal across the field. Notes are made of insects, spiders, damage symptoms, weeds and diseases, observed on each hill. The stage of the plant is carefully observed, as is the weather condition. Interesting insects and other creatures are caught and placed in small plastic bags;
- Drawings of what was observed are made in the sub-groups, during the 'agro-ecosystem analysis'. On large sheets of cheap newsprint fixed to a sheet of plywood, using different coloured crayons, farmers draw the rice plant at its present stage of growth, together with pests and natural enemies occurring on it. A conclusion about the status of the crop and possible control measures is drawn by the five members together and written down on the paper;
- The sub-groups' agro-ecosystem analyses are presented to the whole field school group. The conclusions drawn from the field observation with respect to pest control are discussed in the entire group. The field has become the main training material and farmers' own observations the source of knowledge for the group;
- During each session, special subjects are introduced. Their training provided the pest observers with a substantial repertoire of modules carefully developed to avoid lecturing. Special topics relate to occurring field problems, such as rat population dynamics, effects of pesticides on natural enemies, and life cycles of rice field inhabitants;
- Group-dynamic exercises enliven the field school and create a strong sense of belonging to the school;
- Farmers often keep an 'insect zoo', plastic netting around four bamboo poles set around a rice plant in a container. Inside, various pests and predators are introduced, and watched by farmers. Through their own experiments and observations, farmers gain ecological knowledge;
- Active members of groups are encouraged to train other groups. This farmer-to-farmer dissemination is an important strategy for mass replication. We will come back to this later;
- Once during a field school, a 'field day' is organized for local officials, other members of the farmer group, and so on. This has an important diffusion effect. Resources are available to entertain the guests with snacks;
- Farmers participating in the IPM field school receive a compensation of Rp1000 (approximately US$0.5) per day from the Programme to remunerate possible loss of income while spending time in training. Many groups use these monies for buying caps and T-shirts, decorated with the emblem of the Programme and their farmer group name, visibly increasing group spirits. Some groups also use (a part of) the compensation to go on excursions, for instance to experimental stations or training centres;
- The 'unit costs' of farmer field schools, that is the costs for snacks, for renting an experimental field and its treatment, for paper, plastic bags, crayons, etc. is borne by the Programme.

Senior visitors to the field schools marvel at what is happening. Here are farmers and some village officials, the lowest ranked people on the bureaucratic hierarchy, discussing their problems actively and intelligently, drawing often very accomplished and accurate pictures of various insects, speaking in front of others (including such visitors as the Minister of Agriculture, and making considered decisions about pest control. It is small wonder that the Indonesian Farmer Field School model has now been adopted in at least ten other Asian countries, in both government and non-government IPM programmes.

At this juncture it is perhaps opportune to point out that the 'discovery learning' by farmers on the basis of 'agro-ecosystem analysis' which uses their own field observation, is science informed. The agro-ecosystem analysis methodology was developed carefully on the basis of the latest entomological knowledge. Hence this participatory approach does not represent a violation of the 'integrity of science', but rather a new interactive way of deploying science (see also Hamilton, this volume).

9.7 The training of trainers

The type of farmer training we have discussed requires different staff training. After the fiasco with the crash IPM training, T&V style, immediately following the 1986 Presidential Decree, the National IPM Programme opted for a much more fundamental and penetrating approach consistent with the needs of IPM. The training of trainers, as well as extension worker training, is more or less the same as farmer training. The same basic elements, such as field observation for agro-ecosystem analysis, recur at all three levels.

The NFE approach to staff training began in earnest in July 1989 with the training of trainers. Twenty-one field leaders I (formerly the IPM master trainers), 15 pest observer coordinators, five heads of pest control laboratories, and ten trainers from agricultural in-service training centres were trained thoroughly to form the basic IPM trainer cadre.

An important key to the success of the IPM Programme obviously lies in pest observer training. It basically uses the same principles as farmer training, as described below. The training takes 15 months. Its components are described in Box 9.4.

The goal of training the pest observers is to make them confident IPM experts, instill an attitude of self-learning through experimentation, and develop a cadre of effective trainers of farmers and extension workers. Since 'the methods we learn from are the methods we fall back on when we teach others' (Pontius, 1990), the methods used during pest observer training are those they are expected to use with farmers. During their training, pest observers work in their fields every morning, a rare event for civil servants. Special topics are developed and presented into a set of modules which pest observers feel confident to handle with farmers or extension workers. The training is supported by elaborate field manuals.

Box 9.4 Components of pest observer training

- Rice IPM induction training (3.5 months);
- Practical training in IPM farmer field schools (3.5 months);
- Dry season secondary food crops IPM training (3.5 months);
- a diploma course at the university (4.5 months), which allows them to be promoted in the formal system;
- Each training group consists of 50 people, divided into subgroups of five each;
- Rice and secondary food crop IPM training takes place at the field training facilities (FTFs) where the pest observers grow their own crops. They have to become farmers first before they can face farmers in a position as trainers;
- The training curriculum is completely field-orientated. The 'Apa ini?' principle is the basis for learning. Field problems discovered in the practice fields become topics for discussion. Carefully designed field experiments, such as systematically comparing varieties, fertilizer treatments, variations in pesticide treatments, and a range of special topics (modules to be used in farmer training) are introduced and discussed;
- Practical training takes place in the work areas of the pest observers, where they each conduct four IPM farmer field schools during one season. In this training, two field level extension workers (FLEWs) for each pest observer are trained in IPM on-the-job;
- Field leaders facilitate the FTF training, and supervise pest observer training in the field.

During their training, one pest observer has to choose two field-level extension workers to form a team for farmer training. These extension workers are given a 1-week introductory training at the FTF in which they are acquainted with the principles of IPM and with the farmer field school approach. The teams formulate work plans for farmer training. One team conducts two farmer field schools, which implies four field schools for each pest observer. During the implementation of the farmer field school, the pest observer is the main facilitator, whereas the field-level extension worker assists where necessary and, at the same time learns, on-the-job, to become IPM facilitator.

With this set-up (of one training cycle of about a year's duration) one farmer training facility can deliver 50 trained pest observers, 100 field-level extension workers, and 5000 farmer field school alumni. During the pilot phase, such training took place for over 2 years at ten FTFs, and at two additional FTFs that were opened in 1992. In all, by May 1992, i.e. at the end of the so-called 'pilot phase', about 1120 pest observers and 4700 field-level extension workers had been trained, following the informal education approach, resulting in over 200 000 trained farmers.

After completing their training, the pest observers pick up their regular work as pest occurrence monitors, but with the additional responsibility to organise at least two farmer field schools a year. This arrangement sets the scene for one of the typical battles which must be fought to scale up the success of a pilot phase. Given that, with IPM, the chance of pest outbreaks has been greatly reduced, the Programme wants to increase the proportion of the pest observer's time that can be devoted to farmer training. The Directorate of Crop Protection still has not lived down the need for early warning systems for 'outbreaks' which must be fought with all possible means. Hence it is loath to release the pest observers.

9.8 Using farmer trainers: the vital multiplier

From the start of the Programme, the big question was how the training approach could be scaled up to reach the 10 million small rice farmers in Indonesia. Various models were considered, one of which was, as we have seen, to use the 18 000 field-level extension workers of the agricultural extension service. Various permutations of pest observer/ extension worker combinations have been considered and tested by the Programme. And this avenue has by no means been closed off. Field-level extension worker induction training at the academies now includes a fully fledged IPM training of 15 months (see Box 9.4), and the new phase of the National Agricultural Extension Project might well adopt farmer field schools as its basic approach.

Meanwhile, spontaneous enthusiasm of many farmer field school alumni to carry out training for other farmers in their farmer group and beyond, led the Programme to embrace farmer-to-farmer training as the most promising multiplier of Programme success. By 1995, as many farmers had been trained by other farmers as by pest observers. Farmer trainers had become the major vehicle for scaling up. In estimates of what can be achieved with farmer trainers, the (conservative) assumption was made that a farmer trainer will at most train one field school and that farmers trained by farmers will not become farmer trainers themselves.

The mechanism for effectively using farmer trainers was set up by the Programme. Field Schools run by farmers are funded in the same way as those run by pest observers, although at first, there was considerable reluctance on the part of the World Bank to fund this. For each field school which a pest observer facilitates, he/she selects two farmers as likely trainers for other field schools. Experience has shown that it is wise to select one 'organizer' and one 'technical farmer' as an effective training team. The selected pairs of candidate farmer trainers are given a 1-week training-of-trainers workshop. They prepare their field school with the pest observer, and plan it carefully so as to draw on funds for supplies, snacks, their own honorarium, etc. As a result of this careful preparation, the quality of farmer field schools run by farmers is not observably lower than that of schools run by pest observers (internal survey by the TA team).

In the 4 years that this system has been in operation, at the District level, a core of active IPM farmers has emerged which was selected twice: once for the farmer field school and once as farmer trainer. They are highly motivated and enthusiastic about IPM, they are the founding members of local farmers' associations, and they are often able to motivate local government at village, sub-district and district levels to finance farmer field schools. As we shall see, such local financing might be the mainstay for IPM sustainability after the termination of the Programme.

The Programme has recognized the potential of these networks of farmer trainers and is deliberately building what it calls a horizontal 'Management Information System' comprising regular meetings for farmers at the sub-district and district levels, and technical workshops where farmers can exchange experiences. The Programme pays the small sums for travel, snacks, meeting rooms and so on, which are involved in these meetings.

Multiplying Programme impact through farmer trainers might not be enough. Farmer trainers are especially active in their own farmer groups, that is, after a field school facilitated by a pest observer has been held there. Presently, new farmer groups are being identified by the field-level extension workers who (are supposed to) work with the farmer group. During the first years of IPM, the 'easiest', most active farmer groups were chosen. Now this has been 'done', either through the Programme or through District funds. For example, in East Java, an IPM field school has been held in about 5000 of the 39 000 farmer groups (13%). According to field leaders working in that area, there are no longer any active farmer groups. This means a discontinuity because an active and interested farmer group is a necessary condition for mounting effective farmer field schools. A non-active farmer group implies that the relationship between local farmers and the field-level extension worker is absent or disturbed. In those conditions, more time and effort will have to be invested in holding preparation meetings, and village heads will need to be involved in mobilizing field schools. This will raise unit costs of farmer field schools.

The present strategy for scaling up farmer training involves the following components:

- hold at least one field school in every village through direct involvement of a trained facilitator (pest observer or farmer trainer). There are several farmer groups in one village.
- train farmer trainers and support field schools run by them;
- build active local networks among alumni of field schools and especially among farmer trainers as a mechanism of autonomous change.

9.9 The institutional support framework

Nearly all observers agree that Indonesia's National IPM Programme has evolved a new and highly successful approach to farmer practice, learning and facilitation (the first three dimensions of Box 1.3 of Chapter 1 of this volume). But it is one thing to develop such an approach to facilitation, quite another to develop and install an institutional framework for supporting it, and especially to manage the necessary change of the existing apparatus.

At the simplest level, IPM seems to require decentralized teams of trained staff, capable of autonomous, locality-specific decision making, based on local monitoring and experimentation, and rooted in active farmer participation and control. In fact, farmer organization was seen, from the beginning, as a necessary condition for IPM sustainability (R. Dilts, pers. comm.). At a more complex level, the approach requires new flows of financing, and, especially if IPM is to become sustainably ingrained into the Indonesian fabric, a bureaucratic support structure. Can a decentralized system as described, function in such a hierarchical system as Indonesia's public sector?

Below, we shall briefly describe some of the experiences which accompanied the change-over of the Programme from its so-called pilot phase to its present status as a World Bank Programme of the Indonesian Government. One thing that happened was the unexpectedly quick change of affiliation from BAPPENAS, the Planning Agency, to the Ministry of Finance, Directorate General of Crop Production. Without going into detail, the following aspects are worth mentioning (Box 9.5).

Box 9.5 Some institutional implications of becoming part of the Government structure

- At the central level, a special Bureau coordinates the Programme and is in charge of financial flows, coordination with other agencies, annual reporting, etc. This Bureau was created rather quickly and, initially, it found it hard to submit the required estimates in time for the annual Treasury procedures. Given that these procedures, even if implemented according to the book, take at least two months from the end of the financial year (March 31) to get new money into the field, a whole year was more or less lost during which hardly any activities could be mounted.
- An estimated 40% of total Programme funds are required to fund 'costs' of various sorts at the national level.
- In the field, the cadre of highly trained field leaders was installed at sub-province and district levels in all 12 provinces of the Programme. There is a pest observer in virtually every sub-district. Monthly management meetings and technical updating sessions for these staff are held, especially at the district level, though their financing is not secure.
- A special structure has been installed which allows funds to flow directly to the Agricultural Service offices at the sub-district and district levels. This is an achievement. One of the problems is that, in some areas, serious competency battles are going on between the Agricultural Service (in charge of agricultural development in a province) and the Directorate of Plant Protection (of the National Ministry of Agriculture) to which the field leaders and pest observers officially 'belong'. As a result, some field leaders are no longer involved in decision making about IPM and pest observers are curtailed in the time they are given for farmer training.
- The unit cost per farmer field school was cut, and snacks, farmer remuneration, supplies, experimental fields (a key to a successful field school) and special activities, such as field days, networking, training of trainers, and technical meetings were no longer financed. In other cases, the funds arrived late, so that

the field school could not complete the growing cycle. This meant a sharp reduction in the quality of the field schools, with many farmers dropping out after a number of sessions. The focus was on targets, that is, on the number of farmers trained. Officially, the target of 400 000 was achieved by 1995. But this figure was arrived at by multiplying the number of field schools held by 25, the official number of participants. In actual fact, the number of farmers properly trained has probably not increased that much over the number at the end of the pilot phase.

In all, the change in the status of the official World Bank Programme has had very serious implications for the quality and enthusiasm of the staff. One can doubt the capacity of the largely self-referential bureaucracy in Jakarta to understand what it takes for the country to benefit fully from this innovative Programme, that was spawned under the Government's own auspices. Although the Bank's Mid-Term Review has raised some of these issues, and although the Jakarta officials agreed to go along with some of the suggestions, one cannot expect farmer meetings in sub-districts, farmer trainers, insect zoos, snacks for farmer field schools, etc. to continue to be financed from the annual Government allocation to the Directorate of Crop Protection once the World Bank funding is terminated. The best bet for sustaining the IPM 'movement' seems to be to create the conditions at district, sub-district and village levels, through NGO and Local Government funding, for the development of local activist networks of progressive and influential farmers who want to move beyond the Green Revolution and have seen the value of IPM-type training for achieving that.

9.10 References

Agudelo, L.A. & Kaimowitz, D. (1989). Institutional linkages for different types of agricultural technologies: rice in the Eastern plains of Colombia. The Hague: ISNAR/RTTL, linkages discussion paper 1.

Dilts, R. (1983). Critical Theory: A theoretical foundation for non-formal education and action research. PhD dissertation. Amherst, Mass.: University of Massachusetts, Centre for Informal Education.

Dilts, R. (1990). *Foundations of Action Research.* Solo (Indonesia): UNS-IDRC-LPTP Action Research Training Document. Amherst Mass.: University of Massachusetts, Centre for Informal Education.

Gallagher, K. (1988). Effects of host resistance on the micro-evolution of the rice brown planthopper, *Nilaparvata lugens* (Stal). PhD Dissertation. Berkeley: University of California, Graduate Division.

Gallagher, K. (1990). The 'MODEL'. Yogyakarta: PHT Sekretariat.

Kenmore, P.E. (1980). Ecology and outbreaks of a tropical insect pest in the green revolution: the rice brown planthopper, *Nilaparvata lugens* (Stal). PhD Dissertation. Berkeley: University of California, Graduate Division.

Matteson, P.C., Gallagher, K.D. & Kenmore, P.E. (1993). Extension of integrated pest management for planthoppers in Asian irrigated rice. *Ecology and Management of Planthoppers*,ed. R.F. Denno & T. J. Perfect. London: Chapman & Hall.

Pincus, J. (1991). *Farmer Field School Survey: Impact of IPM Training on Farmers' Pest Control Behavior.* Jakarta: IPM National Programme.

Pontius, J. (1990). *Consultancy Report to Training and Development of IPM in Rice-based Cropping Systems.* Jakarta: National IPM Program.

van de Fliert, E. (1993). Integrated Pest Management: Farmer Field Schools Generate Sustainable Practices. A Case Study in Central Java Evaluating IPM Training. Doctoral dissertation. Wageningen: Agricultural University, WAU Papers, 93–3.

Van den Bosch, R. (1980). *The Pesticide Conspiracy.* Garden City, NY: Anchor Books/ Doubleday.

Warren, D.M., Slikkeveer, L.J. & Brokensha, D. (eds.) (1991). *Indigenous Knowledge Systems: the Cultural Dimension of Development.* London: Kegan Paul International.

10 Co-learning tools: powerful instruments of change in Southern Queensland, Australia[1]

GUS HAMILTON

10.1 Introduction

Between 1991 and 1995, the Queensland Department of Primary Industries (QDPI) implemented a major extension activity in southern Queensland which focused on fallow management of farming systems and aimed to improve their sustainability. The project was carried out by a multi-disciplinary project team known as the Viable Farming Systems Group (VFSG). For the duration of the project, the region experienced the most severe drought on record, presenting difficult circumstances to engender change. Despite these conditions, the change in the fallow management of farming systems in southern Queensland was dramatic (see Fig. 10.1). While there were many other groups and individuals pursuing similar objectives to the VFSG during this period (and their contribution to this change is acknowledged), evidence indicates that the activities of the VFSG have been a major force in this change process (Hamilton, 1995).

The extension project was accompanied by extension research. Both the project and its complementary research were emergent in design, a mix of pre-planned, opportunistic and serendipitous activities. The research was mainly qualitative in nature due to the complex nature of the social phenomena being inquired into. The extension research developed our understanding of why the changes that occurred eventuated, what the relevant processes were that contributed to this change and how these processes may be reproduced in the future.

Key factors in achieving the change that occurred were:

- the multi-disciplinary nature of the project team;
- the initial two-phase market research exercise that was to greatly influence the operation and direction of subsequent activities and their resultant impact on change;
- the utilization of a constructivist approach – participatory learning and action research (PLAR) – rather than a positivist approach. Our application of PLAR was grounded in the theoretical perspectives of soft systems thinking, agricultural knowledge and information systems and an actor-orientated perspective;
- pursuit of change as an emergent property rather than a predetermined end point;

[1] This chapter derives from Hamilton, N.A. (1995).

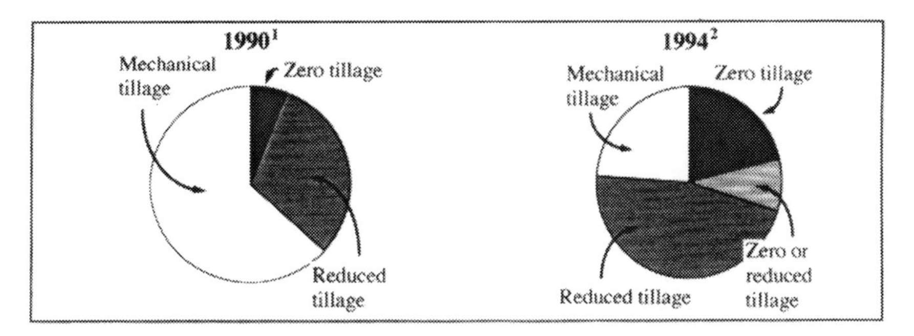

Fig. 10.1. Relative numbers of landholders using various fallow management practices, 1990 vs their stated intention of using various fallow management practices, 1994. (*Sources:* [1]Anon, 1991; [2]Hamilton, 1995.)

- awareness that we were dealing with a 'coupled' system of an extended knowledge system and a hard ecological system;
- the awareness of team members that they had to change their status and role from being technical experts to becoming facilitators of, and equal participants in, joint learning activities, and
- the pursuit of multiple outcomes complementing the multiple perspectives of the individual actors involved; and
- the use of learning tools to develop understanding of the coupled system from the many individual perspectives.

10.2 The location of the project and its agro-ecology

The southern Queensland cropping region of Australia (see Fig. 10.2) has some of the most fertile and productive soils of Australian agriculture. Agriculture in this region is relatively recent (generally less than 100 years), but a combination of growing introduced crop species and the high levels of production and productivity has resulted in an alarming decline in its sustainability.

The region is approximately 400 kilometres from north to south and 450 kilometres from east to west (see Fig. 10.2). There are approximately 6000 landholders and their families (Australian Bureau of Statistics, 1990) engaged in agricultural production, 50% of whom grow crops as part of their farming system.

The region is bisected into two agro-ecological zones (Gentilli, 1972): to the west, a semi-arid zone, commonly referred to as the Western Downs and Maranoa, and to the east, a sub-tropical zone, commonly referred to as the Darling Downs. These agro-ecological zones reflect differing climatic patterns (rainfall and temperature) and differing soils (their water-holding capacity and consequent crop growth).

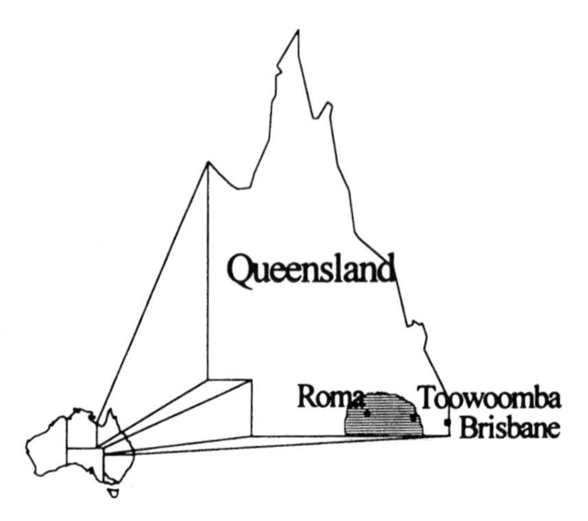

Fig. 10.2. The southern Queensland cropping area.

10.3 Land degradation – a crisis occurring

Figures from the Australian Bureau of Statistics (1990) indicate that there are two million hectares of land cultivated in southern Queensland, supporting approximately 3000 farmers and their families. This region's agriculture is based mainly on continuous cultivation and cereal cropping. Over time, this has led to declining fertility (Dalal *et al.*, 1991), deteriorating soil structure (So, Cook & Dalal, 1988), lower beneficial biological activity (Haas, Evans & Miles, 1957; Dalal & Mayer, 1986), increased incidence of disease and weeds, and increased soil erosion (Lal, 1989).

10.3.1 Soil erosion/storing runoff water in the soil – the major concern

Of the cultivated area, 1.8 million hectares (90%) require protection from soil erosion by earthworks and agronomic practices (Chamala, Coughenour & Keith, 1983). Soil erosion is a result of rainfall running off the soil surface, taking topsoil with it. Soil loss from soil erosion on cultivated lands has been estimated to be minimal (less than 5 t/ha/yr) on the landscapes of low slope and/or low rainfall but rising to 20–50 t/ha/yr on much of the sloping cultivated land of southern inland Queensland. Of the designated 'needs protection' area, only 0.5 million hectares were protected by soil conservation earthworks in 1990 and less than 40% was adequately protected by agronomic measures such as maintaining stubble to combat soil erosion. The other impact of rainfall runoff is the failure to maximize the stored water in the soil for use by subsequent crops.

10.3.2 Soil fertility is the emerging concern

Soil fertility decline is also regarded as a growing problem facing agriculture in southern Queensland. Dalal *et al.* (1991) estimate that in this region, 1.2

million hectares (60%) are affected by soil fertility decline, with a subsequent reduction in crop yield and grain quality valued at an output loss of $AUS[2] 324 M^2/year.

10.3.3 Socio-economic degradation – the hidden concern

The socio-economic degradation of this region reflects the socio-economic degradation of Australian agriculture generally (K. Bowman, pers comm., 1995). This degradation is characterized by declining profitability, a declining contribution to the Australian economy (and consequently, declining political lobbying capacity), and a declining numbers of farmers, growing progressively older.

K. Bowman (pers. comm., 1995) reports that the economic viability of farming in the region has declined dramatically over the past 20 years, likewise reflecting a national trend. He reports that the contribution of agriculture to productivity, employment and export earnings of the region also reflects the rapid decline of agriculture's contribution to national productivity, employment and export earnings (see also Roberts, 1995).

The number of farms and farmers has declined by an estimated 25% over the past decade (Cribb, 1994). Cribb suggests that: '*National environment decline in rural areas* [is a result of] *depopulation, absentee management and the pressure to extract too much from the land* [which] *are the mainspring of land and water degradation*.' The farming population has also been ageing, both in the south Queensland region (K. Bowman, pers. comm.) and nationally (Australian Bureau of Statistics, 1994), indicating a failure of younger farmers to become involved in agriculture.

10.4 Efforts to address the problem

Research into improved fallow management techniques has been conducted extensively over the past three decades. These previous efforts to address problems have been grounded firmly in a transfer of technology paradigm, the dominant approach used in the QDPI at that time. This is a positivist paradigm which focused exclusively on the hard (sub)system of fallow management. Research into the hard system aimed to determine the cause and effect relationships between rainfall, infiltration, stored soil moisture, crop growth, runoff and soil loss. These relationships can be represented adequately by simple cause and effect relationships (see Fig. 10.3).

Water is commonly the most limiting factor in dryland agriculture in Australia. Rainfall is unreliable and this, coupled with high evaporative demand, results in moisture stress during crop growth with subsequent reductions in yield. Few profitable crops result from growing on the basis of in-crop rainfall alone. In order to improve the water supply for crop growth, fallowing during the preceding season(s) is a tactic widely used. The combination of moisture stored prior to crop growth (fallowing) and in-crop rainfall improves crop yields and profits.

[2] $AUS 1 is approximately equivalent to $US 0.75 is approximately equal to Dfl 1.20 (1995).

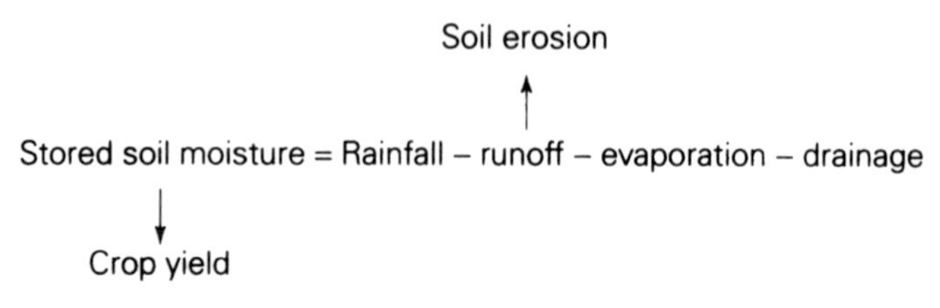

Fig. 10.3. Cause and effect relationships of fallow management variables.

Fallowing is not a particularly efficient process. Research studies in southern Queensland (Littler & Marley, 1978; Freebairn & Wockner, 1986; Radford *et al.*, 1991; Thompson & Freebairn, 1991) indicate that only 20–25% of incident rainfall occurring during a fallow is stored in the soil. The remainder is lost via evaporation (65–70%) and runoff (8–12%). Runoff results in soil erosion. Little can be done to reduce the amount of moisture lost via evaporation. Thus, the focus for improving stored soil moisture is to reduce runoff. Increasing stored soil moisture increases crop yield. Decreasing runoff causes decreased soil erosion. But the studies show that retaining stubble, so that more than 30% of the soil surface is covered over the fallow period, increases infiltration and reduces runoff. Implementation of a system that increases soil moisture storage and decreases runoff improves sustainability. Zero tillage is best able to maintain stubble levels above 30% and, compared to other fallow management options, results in least runoff, least soil loss and most soil moisture storage. The recommendations emanating from this research can be seen to be optimizing this component of the system from a research scientist's perspective by maximizing soil moisture storage and minimizing runoff and soil loss.

It is now an extension problem. The problem and its solution seemed deceptively simple. The research had developed insights into the problems associated with the hard system and determined recommendations to alleviate them. Yet, by 1990, less than 30% of farmers were implementing the findings, and the fallow management practices in use left the soil unprotected from intense rainfall. Consequently, much of the rainfall ran off the cultivations. In the process, soil erosion was worsened and soil moisture storage, so necessary for growing profitable crops, was not optimized.

The problem at this stage was now seen to be an extension problem. In response, the National Landcare Program funded a major project to improve the sustainability of farming systems in southern Queensland by accelerating the rate of adoption of improved fallow management practices. The project initially aimed to use a transfer of technology approach. It was envisaged that, through the use of 'innovative' farmers (those who had already adopted zero tillage) with peer groups of farmers, the farmers who had not yet adopted would become enlightened and change.

Debate, within the project team, the extension environment of the QDPI and within Australia generally, was challenging this approach, and suggesting that participatory action learning was more appropriate. This was confirmed during the market research activities that formed the initial component of the extension project.

10.5 Our first learning experiences – the market research phases

The project commenced with intensive market research activity. This was a consultative process conducted over two phases. The first used rapid rural appraisal (RRA) to produce a wide overview of the problem situation. The second phase used a focus group analysis to validate, and probe in depth, priority issues emerging from the RRA, and to verify or legitimatize project assumptions. The RRA (Hamilton, 1991) revealed:

- important actors (and hence linkages) of the knowledge system that were not encompassed by the original proposal;
- areas of knowledge deficit that required further investigation during the subsequent phases of the market research in order to allow action to be planned and implemented; and,
- areas of knowledge deficit that required action (to be planned and implemented in conjunction with clients) to develop or capture the required knowledge.

The focus group analysis (Blacket & Hamilton, 1992) revealed that:

- the situation in which we were intervening was complex, with an associated extended knowledge system;
- actors had preferred communication channels;
- decision makers required information supporting strategic rather than operational decision making;
- decision makers required an understanding of the fundamental causes of land degradation rather than prescriptions for solutions; and,
- the information required extended beyond technical information to economic and social information and the interactions between these.

The market research phases uncovered some important lessons and changed the project direction. Lessons included:

- The balancing of status between departmental staff and interviewee allowed better information to emerge;
- The use of third-party interviewers demonstrated the value of alternative world views;
- The team members of the VFSG developed enduring social research skills;
- The power of qualitative research processes was demonstrated in the range of emergent issues, breadth of data, contextual details and insights;
- The deemed importance of various stakeholders was altered;
- Project management objectives were changed;

- Long-standing assumptions were challenged;
- Information for strategic decisions was discovered to be a major issue;
- 'Extractive' social research procedures creates a demand amongst participants for follow-up activities.

The market research also introduced us to our first experience of paradigm clash. The outcomes of the different phases of market research produced different reaction amongst the scientific community, the farming community and the project team. The scientists' response was dichotomous: some were supportive and accepted the results. Others were highly (and often aggressively) critical and tended to dismiss the findings as unrigorous because they were not statistically based. On the other hand, those working in the farming community tended to discuss the findings as already being common knowledge.

The criticism at first threatened the cohesion of the team. But after intensive discussion, the team realized that the criticism was basically unfounded, and appreciated that the VFSG was involved in discovering or creating a complementary paradigm. The team came to believe they were on an exciting and important learning path. The market research phases thus led to team cohesion, and an empathy with the range of people involved directly or indirectly in farming. In this respect, the biggest impact was undoubtedly brought about by the greater understanding of the multiple 'world views' concerning viable farming systems through communicating and interacting with respondents.

As a result of the market research, the original project proposal, its assumptions and the transfer of technology approach were discarded, and participatory approaches were taken up as promising mechanisms for initiating and facilitating change. The market research also illuminated areas where VFSG activities could be developed and implemented. These activities focused on developing an understanding of the problem situation, its causes and potential improvements through participatory learning activities.

10.6 Developing tools to facilitate co-learning

The project now aimed to interact, and participate with, clients in groups, jointly discovering what was occurring in their farming system and discussing the information as it emerged and developed. To assist in the discovery process, a range of tools and complementary participative processes were developed. These tools focussed collectively on the same major variables as the market research, although individually they were only able to focus on a subset of these variables. Many of the tools were already in existence as research tools or demonstration tools. The developed process gave control of the tools to the farmers, allowing them to determine treatments and variables they would investigate, encourage them to implement their chosen treatments, make their own observations and discovering for themselves what they considered important. The process facilitated discussion amongst farmers in groups, utilizing knowledge and information contributed by the farmers themselves. This process empowered the participants to understand and learn about the problem situation and its potential solutions.

Over time, the tools developed a status of their own. Farmers became enthusiastic about being involved in the learning activities, and actively sought to have the tools used on their own farms or in their districts. Many farmers also visited activities outside their district when these tools were being used there. The tools developed included the rainfall simulator, the soil corer, *How Wet*, the *Fallow Management Game* and *With and Without*.

10.6.1 The rainfall simulator

The rainfall simulator was already in use and had been used as a demonstration tool during the year prior to the start of the VFSG project. The VFSG developed a suitable process to make it a farmer-controlled PLAR tool. As such, it allowed clients to develop better understanding of the relationship between fallow management practices, rainfall, soil moisture storage and runoff, and to experiment with optimizing soil moisture storage. It only allowed a 'snapshot' in time to be investigated. The process was highly reproducible and with suitable training, could be utilized by a range of facilitators in a range of situations. Being reproducible enabled high penetration of the target audience, with over 1500 farmers attending one or more rainfall simulator learning activities over the 5 years of the project.

The rainfall simulator is a transportable machine that produces 'rainfall' with drop size and energy similar to natural rain. 'Rainfall' is applied to two adjacent plots allowing two treatments to be applied and compared. The plots are separated by barriers to maintain independence between the treatments. Rainfall is usually applied at a constant known rate but is also measured for the duration of the activity. Runoff collected at the bottom of each plot is vacuumed into measuring tanks where it is easily seen and measured.

The rainfall simulator activities, in the context of the VFSG, were intended to be action learning processes used with small groups of up to 20 landholders. This process allows different participants to arrive at different decisions from the same activity (see Box 10.1).

The rainfall simulator activities were underpinned by several key components:

- multi-channel publicity for awareness prior to the event;
- farmers nominating treatments and putting the treatments in place;
- farmers monitoring treatment effects;
- farmers participating in small group discussions;
- scientists facilitating the process of self-learning among participants; and
- scientists adopting the role of expert and contributing knowledge *only* when clients are obviously drawing incorrect inferences from observations, or where the input of scientific knowledge would energize the clients to pursue further discussion and knowledge sharing.

Box 10.1 An influential farmer's involvement in a Rainfall Simulator PLAR activity.

The rainfall simulator allows for multiple outcomes. An influential farmer, who was a member of a major funding body, had expressed doubts about the effectiveness of activities like the rainfall simulator to assist farmers make better decisions and change their farming practice. This made him a prime target to host a rainfall simulator activity. This farmer, with a small group of his peers, tested a range of options with the rainfall simulator. Their fallows generally had fine tilth with no stubble cover due to the drought conditions preventing them from growing stubble, and excessive cultivation. This group tested a range of options including stubble cover (provided by hay), rough tillage, and fine tillage.

At the end of the simulator demonstration, one farmer went home, got his machinery out and roughened the surface up, a result he interpreted from the action learning exercise.

Our targeted influential farmer said, 'Well, you've convinced me'. He went out that afternoon and bought himself a zero till planter, a result he interpreted from the action learning exercise. This farmer became a strong advocate of these types of PLAR tools.

This example demonstrates that these tools allow participants to reach very different outcomes from the same event.

During these activities, paradigm clash again emerged as an issue, but with a difference. In this case, the paradigm clash was an internal struggle for the officers to move from their traditional role of technical expert to the new role of facilitator. When the rainfall simulator was first used by the team members, their natural inclination was to revert to the default extension process, that is, to use it as a demonstration tool to teach farmers improved fallow management techniques. We instituted specific training to assist team members to learn to use the rainfall simulator as an action learning tool with farmer participation. Through this training, team members gained the confidence to utilize the process to allow understanding to emerge amongst farmers, and to allow multiple outcomes to emerge, rather than trying to control the situation and drive it towards the extentionists' predetermined outcomes. Participatory learning and action research became the preferred approach and activities were performed better because of it.

10.6.2 The soil corer

The soil corer is a thin tube, usually 2–2.5 cm in diameter, that is driven into the soil to remove a complete core of soil up to 2 metres. The soil corer existed as a prototype, developed many years previously and used sporadically in the community and by researchers as a demonstration tool on an *ad hoc* basis. The VFSG developed a suitable process to make it a PLAR tool. Mike Foale, senior research officer from APSRU[3] elaborated its design, and arranged for its manufacture. He and the VFSG promoted its dissemination as an extension and farmer-controlled tool. The soil corer allows clients to develop a better understanding of their soil profile and to monitor soil moisture storage over time.

Most farmers have little knowledge of what exists in their paddocks below about 15 cm, the normal depth of planting. Yet the full profile that crop roots can access has huge implications on yield and quality, and the amount of water and nutrients in the total profile is of paramount importance. Most farmers estimate their soil moisture capacity by utilizing a soil probe: a thin metal rod that will penetrate through moist soil, but not dry. While this gives a good indication of depth of wet soil, it gives little indication as to the degree of wetness of the soil. Use of the soil corer allows farmers to see and feel exactly how wet the soil is. As clay-based soils become wet to field capacity, their texture becomes like plasticine. Conversely, when they are close to wilting point, they become hard and crumbly. By removing a complete core, the farmers can see and feel the moisture.

The simplicity of the soil corer and its relatively cheap cost provided farmers with a tool with which to explore their own soil profiles. It gave farmers the capacity to make measurements in the terms used in a range of scientific information and this made that scientific information more meaningful (see Box 10.2). This range of information included the output from crop model.

Box 10.2 A grower's experience using the soil corer

A grower bought himself a soil corer after being introduced to it during a rainfall simulator activity. He reported (unsolicited) that the soil corer had allowed him to test his soil profiles extensively in paddocks across his farm. The moisture profiles of the various paddocks were not what he expected from the varying lengths of fallow each had had. A paddock with a short fallow that he thought would have less moisture stored was found to actually have more. Conversely, a paddock with a long fallow that he had expected to have more moisture, was demonstrated to actually have less than the short fallow paddock. Consequently, he changed his decision to plant the long fallow paddock and planted the short fallow paddock instead.

Crop model simulations basically transform stored soil moisture into a range of yield probabilities. With a measurement of stored soil moisture, farmers could now utilize this information in their decision making. For example, based on the probability of success as determined from crop models, choices could be made between (a) planting immediately or (b) continuing the fallowing process and planting at a later date. The latter choice would aim to accumulate more soil moisture and enhance the probability of growing a viable crop.

[3] APSRU – the Agricultural Production Systems Research Unit, a joint Department of Primary Industries and Commonwealth Scientific and Industrial Research Organization team which aims to develop computer models of the major agricultural crops and farming systems in Queensland.

The alternative was to present the information in a way that would relate to depth of wet soil, which was commonly a known feature by farmers from their work with the soil probe. However, because soil depths vary considerably across the region, and the effective water holding capacity of soils varies considerably, the depth of wet soil (determined by a soil probe) is not a useful variable to relate to information being transmitted by science.

10.6.3 *How Wet*

How Wet was initially developed as a pen and paper exercise to calculate stored moisture in the soil using farmer's rainfall records. Rainfall records are recorded fairly meticulously by most farmers, but are seldom used in decision making or in reflecting upon production outcomes. In an attempt to make this information more useful for farmers in their decision making, and to couple it with information generated by the rainfall simulator and the soil corer, a research scientist (David Freebairn, Principal Research Officer, QDPI, Toowoomba; Freebairn & Wockner, 1986) developed a process which took daily rainfall figures and converted them to stored soil moisture, evaporative loss, and runoff and soil loss.

The initial calculations were done on pen and paper, and converted subsequently to form a computer program. However, when we started to use the computer software with farmers, the 'black box' nature of how the rainfall records were converted to soil water storage, runoff and evaporation made them distrust it. We altered the process so that farmers went through the pen and paper process and understood what happened to individual rainfall events, how much was stored in the surface and lost to evaporation, and how much ran off. We then moved on to entering this same information into the computer program. By this process we opened up the black box and allowed farmers to understand how the final figures were arrived at. When presented in this way, farmers were willing to trust and use the computer program.

The pen and paper process also allowed farmers without computers to participate. Such farmers are still in the majority, so access to the information by simple means is important to extension impact, although it is more time consuming to generate insights in this way. The insights created allowed farmers to access a range of scientific information, presented in terms that farmers could not easily generate on their own farms. *How Wet* has the same capacity to assist farmers in decision making as information derived from the soil corer, but adds an in-time understanding of sequence. Whereas the soil corer gives a picture at a specific point in time, *How Wet* gives a picture across time.

10.6.4 *The Fallow Management Game*

The *Fallow Management Game* was developed in reaction to, and as a by-product of, ongoing development of crop process simulation models. It aimed sequentially to build on and expand the learning activities associated with the rainfall simulator, soil corer and *How Wet*. The intention was to add to the understanding gained with one tool by incorporating the next tool into the learning process.

The *Fallow Management Game* incorporated the software from *How Wet* and turned it into a computer-based game that provides information to users as a normal decision support program. Unlike most computer decision support software, the *Fallow Management Game* is designed to be 'played' within a group with the interaction between players being as important as the interaction between players and the game. When it is used by groups, it also facilitates discussion, information exchange and learning between producers within groups. It further enhances understanding of the farming system through demonstrating the relationships between fallow management and: (a) stubble and stubble cover; (b) infiltration, runoff, evaporation and soil loss using actual rainfall records; (c) nitrogen and moisture accumulation over the fallow period; (d) the costs of accumulating this moisture; (e) the crop growth as a result of moisture accumulation and subsequent rainfall; and, (f) the returns from crop yields.

The *Fallow Management Game* also facilitated the involvement of all members of a normal decision making unit in the learning process without upsetting the often subtle relationships between the individuals in the decision making unit. Farmers also enjoyed playing the game and expressed that 'farming in a fun sense' had an impact on how they responded in practice. Clients play a role in developing the game. By playing the game, farmers identify which variables would be useful to their decision making and these variables are incorporated in updated versions.

10.6.5 *With and Without*

With and Without is a user-friendly comparative economic analysis tool, designed to analyse and test options on a whole farm basis. Standard economic tools did not appear to offer the methods to fully address this situation. In an effort to create applied information from existing knowledge (amongst farmers and extension officers), we turned to a research economist from the QDPI (J. Gaffney, Principal Research Economist). He had designed the *With and Without* tool (Gaffney 1992) to allow an analytical comparison of the total costs and benefits associated with an option with the total costs and benefits without the option (usually the existing situation). Until this time, its main use had been within the QDPI to improve extension officer's understanding, with only the *results* being transferred to farmer, in other words, as a tool for the classic TOT approach. However, the use of the tool in the absence of farmers could not provide the detailed knowledge of actual farming operations, performance levels and likely market prospects relating to the issues the group (on behalf of farmers in the region) wanted analysed. Although much of this knowledge possibly existed in diverse QDPI circles, it was only the farmers in the region who fully understood the complexity of the issues.

We refined the tool and used it in a co-learning situation within a PLAR framework. These co-learning situations were conducted as workshops, where groups of VFSG team members and farmers used *With and Without* to test the implications of major changes in strategy on a whole farm basis. In this situation, the farmers involved learnt the procedure associated with the tool and how to apply it, and were then able to investigate options that interested them. We learnt how farmers thought about farm economics and the various strategies they investigated themselves. These were, of necessity, labour-intensive exercises.

The farmers believed that the information and knowledge generated by the approach led to better informed decisions because it expanded their knowledge base. They considered it an improvement on relying purely on gut feeling and gross margins. Team members developed greater insights into the intricacies of farmer decision making, which assisted them in being more critical of the information they were producing and was believed to add to its effectiveness. For the workshop participants, the information produced was found to be extremely useful. The format allowed them to clarify variables and to put values upon the variables that were meaningful to them. In effect, it allowed them to construct meaningful information for their own use.

The information derived from these workshops was disseminated to the wider farming community in the form of case study reports. Farmers who were not involved in the workshops, but who had read the reports, said in interviews that they had found the information useful (Harris & Gaffney, 1993). However, they did not regard the information as complete and did not agree with all the costs used nor all the benefits gained. But they saw it for what it was, that is, a construction from one viewpoint, highly applicable to the individual case, and this has implications and application for other users as it provides a different viewpoint on a common problem theme. The scientific staff viewed it completely differently. The questions they raised caused a great deal of tension and created wide debate. They questioned the expertise of the people involved, and, consequently the results.

How expert were the farmers? The farmers who participated in the workshops were not what would be commonly termed innovative farmers according to the TOT paradigm. From the constructivist perspective, however, their real-life experience gave them expertise: they had lived the problems. But they were not considered innovative farmers by scientists, and some of the expertise they applied was not regarded as 'expert', the farmers' experiences were played down by scientific staff.

How expert were the officers involved in the process? In crops such as cotton, none of the officers was regarded within the QDPI as a technical expert, but this was not significant as their technical expertise was not what was needed in the workshop. They played a facilitating role appropriate to a PLAR approach, rather than that of technical expert. This distinction was lost on many of the scientific staff. Consequently there was much criticism as to the accuracy and the usefulness of the case study reports. The scientific staff failed to understand that these reports were not an end in themselves but a means to an end. They are constructions put together by individuals for their individual situation, in a format that allows comparison amongst alternatives by others as well as by themselves. They are a starting point to stimulate communication, to assist farmers to seek information to enhance their decision making. For some users, they provide a model of a relevant situation alternative to the scientific model. By enlarging the perspectives brought to bear on a problem situation, the understanding of the problem situation is improved. The scientific staff failed to grasp this concept, and the impact of *With and Without* was diminished because of it.

10.7 Why do these tools work?

These tools were found to be powerful initiators of change. Clients and staff were enthusiastic about being involved in activities where the tools were used. The reasons for their impact are many.

- *The tools were used in an andragogical approach.* That is, participants were regarded as adult learners, with the teacher and learner having equal status. It gave control to the 'learners'. It utilized their highly valued indigenous knowledge, which was given the same status as scientific knowledge. It allowed them the freedom to make their own decisions from an understanding of their own individual situations.
- *The role of the VFSG team changed from that of technical expert to one of facilitation.* The scientist's technical knowledge was not excluded from the process but only introduced to assist farmers to discover further information or to present an alternative perspective on the situation.
- *The process assisted farmers to reflect upon their learning.* This was accomplished by providing a feedback loop, with observations and measurements and discussions of emergent meanings.
- *The activities matched the preferred learning style and perceptual modalities of the participants with the learning activity.* For example, most farmers prefer to learn by participating in a 'hands on' experience – a haptic modality. The most successful activities utilized this 'hands-on' approach.
- The other great advantage of these learning tools is that they allow a *rapid and extensive penetration of the target audience.* Contrasting the various activities carried out demonstrated that once participation was not part of the process, the reliability of the method failed.
- These tools enhance participatory learning by engaging in a form of *'risk-free' action research based on formal representations of real farm situations.* They allow clients to play with their environment, to test various options, to gather information, to enhance our understanding and the client's understanding of the problem situation, and to gain knowledge without any direct risk.
- The *use of the tools in group situations* is also important for their success. They allow participants to understand and explore 'reality' from differing perspectives. Through dialectic discourse and joint observation, alternative perspectives emerge, challenging or complementing existing perspectives

and enhancing understanding. Consequently, more purposeful decision making is made by the clients.

- The tools allowed the *manipulation of impersonal, but contextualized 'reality'*, which could be interrogated at will. In the process, deeper insights into the nature of technology as a social construct was gained.
- They were *highly reproducible processes* for a range of facilitators, client groups, and situations in time and space. Their impact was enhanced by science facilitating the process and the client becoming the researcher. This change in roles enhanced the clients' self-perception of their own status as being knowledgeable, their perception of their ability to generate knowledge and the value of that knowledge. In turn, this enhanced their trust in the research and its results and their understanding of the situation. Such enhanced perception is the driving force that makes these tools and their associated processes powerful agents in creating change. Conversely, fear of a perceived loss of status associated with being a facilitator was a barrier for some team members to adopting such a role.

The approach used contrasts with the transfer of technology approach. The tools provided a forum where the experience and knowledge of scientists, extensionists, and farmers were placed on a more equal footing than is the case in the TOT approach. Information and data were shared, not transferred, and as a consequence the understanding of both farmers and agricultural professionals understanding was enriched by the exchange and opportunity for dialogue. Under a TOT approach lies the belief that an innovation discovered by research will be implemented by the client in the expectation of reproducing the same outcome. Under the PLAR approach lies the expectation that, by understanding socio-economic and technical interactions, farmers will be better able to make informed choices, and innovation will be an emergent result rather than a predetermined end point. For example, we found farmers who dropped zero tillage. They had adopted it as the result of a TOT approach and had implemented it in the belief that there was no flexibility in its application, i.e. fallow management was determined at the commencement of the fallow. After being involved in a PLAR exercise, they understood the processes of water accumulation and loss and recognized fallow as a dynamic environment to manipulate, requiring a series of tactical decisions based on the conditions occurring at the time and the expectation of conditions that may follow. Thus, under PLAR, zero tillage is an emergent end point, not a predetermined end point.

These PLAR tools and their attendant processes reflect the work of Argyris and Schön (1974) who emphasize the importance of double loop, as opposed to single loop learning. PLAR pursues double loop learning by combining action research and participatory learning. Under double loop learning, participants reflect on, and amend, not only their action strategies but also the governing variables behind those strategies. Under single loop learning, the results learnt from implementing an action are used either to avoid reproducing undesirable results or to reproduce desirable ones. The governing variables behind the action strategies are not reflected upon nor amended.

The governing variables of single loop learning are

- to achieve the purpose as the actor defines it;
- to win not lose;
- to suppress negative feelings; and,
- to emphasize rationality.

In double loop learning, the governing variables include

- valid information;
- free and informed choice; and,
- internal commitment.

These very different governing principles lead to strategies under double loop learning that actively seek information and the increased participation of others.

10.8 Paradigm clash

We also discovered, that however powerful the learning process was for the team, it aroused considerable hostility among other members of our intellectual community. Specifically, it challenged the positivist realist framework of R&D and technology development, and the TOT model of technology adoption at farm level. These paradigm clashes arose from the emerging understanding that while positivist science has a valid perspective on the problem situation and its potential solutions, other stakeholders had equally valid, alternative perspectives. The clash highlighted the epistemologial differences between the paradigms and questioned the positivist belief that the investigator and the investigated 'reality' can be independent entities, and that the investigator is capable of investigating the 'reality' without influencing it or being influenced by it. Further, the study questioned the definition of the problem and the solutions proposed by positivist science as free of values and biases. It showed that, if these values and biases were accounted for, the outcomes would be very different.

The conclusion is reached that, for very simple biological systems, positivist reductionist approaches are adequate for explaining simple cause and effect relationships. However, for handling complex situations such as sustainability, soft systems approaches and adult learning approaches are more effective and efficient. With these approaches, individual stakeholders are encouraged to build their own constructions. Meanwhile, their values are accommodated in the process but not explicitly accounted for, and science has a different role in facilitating learning rather than teaching the scientific construct.

These conclusions suggest that, for science to be effective when confronted with a problem situation, it must carefully explore the situation to decide on the appropriate paradigm and approach. Decisions are enhanced if research and extension practitioners develop a fuller understanding of the options available and their underlying assumptions. This study demonstrates that it is pointless to use a positivist approach such as TOT to assist in this decision-making process. Rather, more informed choices are likely to be made and new roles for science to emerge through a participatory exploration of the paradigms in use and their impact on methods and areas of interest.

10.9 An emergent perspective

PLAR processes are powerful initiators of change. We demonstrated that by employing such a PLAR process. More farmers tried one or more options; farmers in a greater range of target categories tried one or more options; and farmers became more purposeful in their decision making. End results were not predetermined by the process. Through improved understanding of the situation they were dealing with, farmers became more purposeful in their decision making, became better decision makers, and arrived at end results that they had determined for themselves.

We came to see the system we were dealing with as a 'coupled system' (Röling, 1994) with the implications: (i) that performance would be diverse rather than standardized; and (ii) that sustainability is an emergent property, not a single objective function or end state. We also came to realize that interactions between decision makers and their farming system were both flexible and locked into their (natural and socio-economic) history. We showed that the way to purposefully move forward is in terms of a co-learning process, by improving insight into systemic relationships and by improving monitoring and visibility of those variables thought to reflect changes in state. We also came to realize that this could only be achieved through the application of constructivist approaches rather than more traditional positivist approaches. We also came to realize, however, that researchers and change agents would not readily adopt such a co-learning and constructivist approach unless and until they themselves began to operate in, and explore, the PLAR mode.

10.10 References

Anon. (1990). Application for Project Funding for the project 'Co-ordinating Conservation Farming Technology with South Queensland Farmer Groups', submitted to the National Soil Conservation program (NSCP), by the Agriculture Branch, Queensland Department of Primary Industries, April, 1990.

Argyris, C. & Schön, D. (1974). *Theory in Practice: Increasing Professional Effectiveness.* San Francisco: Josey-Bass.

Australian Bureau of Statistics (1990). *Agricultural Land Use and Selected Inputs, Queensland, 1988–89.* Brisbane: Australian Bureau of Statistics.

Australian Bureau of Statistics (1994). *Agricultural Land Use and Selected Inputs, Queensland, 1992–93.* Brisbane: Australian Bureau of Statistics.

Blacket, D., Hamilton, N.A. (eds.) (1992). *Understanding Farmer Decision Making on Landuse. Research using Focus Groups in Southern Queensland.* Brisbane: Govt. Printing Office.

Chamala, S., Coughenour, C.M. & Keith, K.J. (1983). *Study of Conservation Cropping on the Darling Downs – A Basis for Extension Programming.* Published report. p. 1. Brisbane: Dept. of Agriculture, University of Queensland.

Cribb, J. (1994). The unsettling of Australia – disturbing trends on the way to 2001'. *The Australian Magazine*, February 12–13, 9–16.

Dalal, R.C. & Mayer, R.J. (1986). Long-term trends in fertility of soils under continuous cultivation and cereal cropping in southern Queensland – I-Overall changes in soil properties and trends in winter cereal yields. *Australian Journal of Soil Research*, **24**, 265–79.

Dalal, R.C., Strong, W.M., Weston, E.J. & Gaffney, J. (1991). Sustaining multiple production systems 2. Soil fertility decline and restoration of cropping lands in sub-tropical Queensland. *Tropical Grasslands* **25**, 173–80.

Freebairn, D.M. & Wockner, G.H. (1986). A Study of soil erosion on vertisols of the eastern Darling Downs, Queensland. I. Effect of surface conditions on soil movement within contour bays. *Australian Journal of Soil Research*, **24**, 135–58.

Gaffney, J. (1992). Overcoming the disillusioned learner problem in farm economics training. In *Future Directions for Agricultural Education. Extension and Training. Workshop* sponsored by NSW Agriculture and Faculty of Agriculture and Rural Development, University of Western Sydney, Hawkesbury. Sept. 30/Oct 1, p. 2.

Gentilli, J. (1972). *Australian Climate Patterns.* p. 33. Melbourne: Thomas Nelson.

Haas, H.J., Evans, C.E. & Miles, E.E. (1957). Nitrogen and carbon changes in Great Plains soils as influenced by cropping and soil treatments. *Technical Bulletin,* No. 1164, Washington: United States Department of Agriculture.

Hamilton, N.A. (ed.) (1991). *Southern Queensland Farming Systems: A Survey of Farmers' Attitudes and Behaviours.* Toowoomba: Queensland Department of Primary Industries Information Series Q191027.

Hamilton, N.A. (1995). *Learning to Learn with Farmers: A Case Study of an Adult Education Extension Project Conducted in Queensland, Australia, 1990–1995,* published doctoral dissertation, Wageningen: Wageningen Agricultural University.

Harris, P. & Gaffney, J. (1993). Partnerships allow effective co-learning in farm economics. In Gearing up for the Future. ed. R. Fell. Presented papers at QDPI Extension Conference 2–4 February, 1993, held at Rockhampton. Brisbane: QDPI.

Harris, P. Hamilton, N.A., Kelly, A. & Field, L. (1995). Evaluation of change in farming practice in southern Queensland and the influence of the Viable Farming Systems Group project (1991–1995) on this change. Final Project Report to the National Landcare Program.

Lal, R. (1989). Conservation tillage for sustainable agriculture: tropics versus temperate environments. *Advances in Agronomy*, **42**, 85–197.

Littler, J.W. & Marley, J.M.T. (1978). Fallowing and winter cereal production on the Darling Downs. Wrk-C108–AB Progress Report, Agriculture Branch. Brisbane: Department of Primary Industries.

Radford, B.J., Gibson, G., Nielsen, R., Butler, D.G., Smith, G.D. & Orange, D.N. (1992). Fallowing practices, soil water storage, plant-available nitrogen, and wheat performance in South West Queensland. *Soil and Tillage Research*, **22**, 73–93.

Roberts, B. (1995). *The Quest for Sustainable Agriculture and Land Use*, p. 49. Sydney: University of New South Wales Press Ltd.

Röling, N. (1994). Creating human platforms to manage natural resources: first results of a research program. In *Proceedings of the International Symposium on Systems-oriented Research in Agriculture and Rural Development*, held at Montpellier, France, 21–25 November.

So, H.B., G.D. Cook & Dalal, R.C. (1988). Structural degradation of vertisols associated with continuous cultivation. In *Proceedings of the International Soil and tillage Research Organisation*, XI International Conference, Edinburgh, pp. 123–8.

Thompson, J.P. & Freebairn, D.M. (1991). Review of fallow management trials – Hermitage Research Station. Internal review to the Viable Farming Systems Group, July.

11 A social harvest reaped from a promise of springtime: user-responsive, participatory agricultural research in Asia[1]

GELIA T. CASTILLO

11.1 Introduction

South, south east and east Asia are home to half the world's people. The region's high productivity and sustainability, over thousands of years, of farming systems based largely on irrigated rice, has supported such numbers. In the past 30 years, the productivity of irrigated rice systems has been boosted by the Green Revolution, based on wide-scale adoption of uniform science-based packages of high-yielding varieties and accompanying inputs. Such packages focus on component technologies while neglecting resource and whole farm development (Dwarakinath, 1994).

The green revolution approach is running out of steam. It largely neglected the non-irrigated areas, the fragile lands. Agriculture in these areas is marked by high diversity, variability, complexity and risk, and major new and appropriate technologies have not been developed. Instead, these fragile lands are threatened by degradation through deforestation, soil erosion, loss of soil fertility, and drought as a result of reduced water retention and climate change. Poverty in its crudest form (lack of access to biomass) is deepened by these trends. Thus the issue of sustainability in these areas cannot be considered only from an ecological point of view. Sustainability is inherently a question of livelihoods.

The local livelihood systems are usually very complex. They build on diversity, spread risk as widely as possible and rely on low external input technologies. People subsist on a great variety of crops in addition to rice, especially sweet potato, cassava, maize and a large number of other crops, including fruits and species of animals. The complexity of the livelihood systems are embedded in social organization in which women play key roles and in complex networks of mutual dependency.

The capacity of science to support these complex livelihood systems and a totally different approach to mobilizing is needed. This chapter explores the type of agricultural research which is emerging in response to the challenge. It is user-responsive, participatory and able to take into account local diversity. It seems to hold promise for supporting sustainable livelihoods in complex and fragile environments under high population pressure.

[1] This chapter is based on a paper originally prepared for a workshop on 'Sustainable Agriculture: Implications for Extension Practice', sponsored by the Department of Communication and Innovation Studies, Wageningen Agricultural University, June 7, 1995.

The chapter reports on the social 'harvest' from this approach, the 'Springtime' of the new approach, an appropriate metaphor for the sense of excitement, energy and hope that has been created. The experiences reported here also bring the realization that the new approach is complex, difficult and demanding. The lessons reported here have been derived from the Asian activities of the UPWARD[2] network, from the experiences of SAPPRAD,[3] and, of course, from relevant experience in rice. Their empirically generated results enrich our knowledge and add to the credibility of the new research paradigm regardless of the labels under which it is pursued. The labels include: farmer-back-to-farmer; farming systems research (FSR); agricultural (soft) systems methodologies (SSM); research with a user's perspective; knowledge systems thinking; and participatory technology development (PTD). As will become clear, the new paradigm does not allow for a clear distinction between the research and the 'transfer' of its results. The research itself is tied up in interventions which involve social networks, new ways of organizing labour, local food security, eco-system management, policy, seed systems and technology development.

The chapter first reminds us of the main attributes of the approach which supported the Green Revolution. It then identifies the main aspects of sustainable livelihoods which have emerged in the new approach. Finally, it describes the research approach itself, providing concrete case studies to illustrate its methods and impact.

11.2 The contours of the conventional research paradigm

A perusal of socio-economic research in agriculture during the Green Revolution era shows practically no diagnostic studies. In the case of rice, for example, adoption of new varieties and other accompanying technologies took place so rapidly that social scientists focused their research attention on adoption behaviour and consequences of adoption. Hence their major role was to monitor and assess impact. After all, given a target of irrigated, favourable areas comparable to research station conditions, scientists seemed to know the architecture, nature, and habit of the rice plant they needed to develop, and went ahead and produced it. Eventually, scientists succeeded in releasing good eating quality rice which was also highly productive. An 'inertia of euphoria' went on for about 25 years and the

[2] User's Perspective With Agricultural Research and Development is a network of Asian agricultural researchers and development specialists committed to the participation of the ultimate users of agricultural technology in the development of production and post-production innovations in root-crop agriculture. UPWARD recognizes that agricultural R&D programmes often fail because low-income users cannot apply results and recommendations generated in experiment stations and untypical farming areas. By building a partnership between the local expertise of users and the 'global' expertise of R&D specialists, UPWARD believes that socially and economically significant innovations can be developed.

[3] SAPPRAD (Southeast Asian Program for Potato Research and Development) also includes sweet potato in its R&D portfolio. The SAPPRAD countries are Thailand, Malaysia, Indonesia, Sri Lanka, Papua New Guinea, and Philippines.

contribution of science and technology to the welfare of rice-eating and rice-growing communities was almost taken for granted. The central focus of social science research was farmer response to the new technologies and the extent to which the technologies had been accepted and adopted.

When the Green Revolution was at its peak, criticisms levelled against it were also at their peak. There was vigorous advocacy for those who have been by-passed by the Green Revolution. Mostly, farmers were not consulted about farming practices and varietal preferences, a criticism applied especially to less favoured crops, less favourable eco-systems, and less favoured farmers. In the meantime, there were repeated tales of new crop varieties and other technologies which were not adopted by farmers. In the end, the Green Revolution turned out to have limits. In 1994 Cassman and Pingali documented the declining factor productivity of irrigated intensive rice production. Rice yields, even in favourable environments, seemed to have 'hit the ceiling' – a phenomenon labelled 'post-Green Revolution blues'.

All of these factors (and many more) brought forth a shift in research paradigm among those who were dissatisfied with the commodity-based, linear approach to technology development, which starts from the research station and then relies on delivery by extension to the farmer. The past two decades have witnessed the flourishing of so-called non-conventional user-sensitive, diagnostic, participatory, location or eco-system-specific technology design and development which almost always begins with local knowledge. In these new approaches there was also place for issues of sustainability, equity, gender-sensitivity and environment friendliness. All these aspects, together with the time-honoured objective of agricultural research to increase yields, has made the research task complex and difficult. Moreover, perhaps because of its newness, substantive research results and real-life technologies and benefits derived from the application of the approach are not yet as robust as the arguments for its application.

The chapter first describes some of the insights which emerge as we learn to look at the complex web of factors in which livelihood sustainability is embedded. The research is based on studies, trip reports, field experiences, and dialogues with users, and on participation in site visits, workshops, and training programmes, with particular respect to sweet potato and potato production systems in less favourable eco-systems. A few cases of rice-based systems have been included to illustrate the application of the research paradigm.

The insights are described under the following headings:

- social networks, community actions and trade relations;
- people–labour dilemmas;
- farmers in multiple agro-ecological niches;
- food availability and the livelihood calendar;
- micro-impacts of macro-policy; and
- seed systems and on-farm genetic diversity.

In the last section, the new technology development processes and practices will be described.

11.3 Social networks, community actions and trade relations

Food security among subsistence farm households is not just a matter of food production but also of investing in, and maintaining, social relations. Keasberry and Rimmelzwaan (1993), in their study of a Philippine Upland Village in transition from subsistence farming to cash cropping, show the role of social relations in food security as experienced by different households. Social relations with relatives, friends and neighbours are involved in the exchange of food, money, labour, land, and livestock. Different types of households can be distinguished:

- those who regard these relations as a short-term solution to temporary food shortage;
- those who mainly receive more food and food-related items than they give and hence are dependent, or those who have others being dependent on them;
- those who manage to maintain equal interdependent food relations or reciprocal dependency;
- rich, cash-crop dependent farmers who consider their socio-economic relations less important for their household food security because they can depend on the sale of crops and other sources of more or less stable income;
- the poorest subsistence households who are the most disadvantaged in all exchange relations because they are not in a position to give and hence are less free to ask.

None of the households is food secure through exclusively cultivating subsistence or cash crops. Most food-secure are the really self-sufficient farmers who have the opportunity to cultivate rice.

The exchange of food is different for rice than for vegetables and other kitchen products. Rice is distinguished as a food product of a higher status. The practices for borrowing rice are similar to small money loans. Rice has to be returned in the same quantity as soon as possible and usually within two weeks. In the borrowing process, the important thing is that the borrower has to feel free to ask, as in the case of relatives, friends and neighbours with whom there has been frequent contact. Cash crops are never loaned in this way, unless farmers have a surplus after sale.

Another study shows the importance of trade relations which are often informal and socially based despite commercialization. Garzon and Arocena-Francisco (1992) traced the relationships among different actors in the sweet-potato marketing chain in the Philippines and identified farmer–trader, trader–trader and farmer–banker/financial institution relations. We describe the farmer–trader relationship under the 'suki' system.

It is a system of regular trading relations between a buyer and a seller based on affinity or consanguinity and/or goodwill. The farmer can deliver to the trader or processor who comprise his 'suki' even without an order and is paid on delivery or before the farmer goes home that same day. This arrangement frees the farmer from worries about selling his product while the trader or processor is saved the trouble of looking for stock in the market. But delivery can also be on assignment with payment some days later, which the farmer might prefer so as to 'save'. Barter exchange is still practised under this system. The trader pays in part for the sweet potato with dried fish, salt, legumes, etc., which he also has in store, and deducts the cost from the value of the sweet potato delivered. Another arrangement is that the trader accepts the produce which a farmer has not been able to sell during the day at a reduced price. The farmer might prefer this to taking home the produce. Depending on the strength of the suki relationship, farmers can usually request cash advances from their suki to meet non-farm expenses for medicines, school fees, etc.

Traders at the local or district level function as a significant source of market information for farmers, particularly with respect to new/preferred varieties of sweet potato. They determine in many ways what varieties farmers should plant in response to market demand. They also know where planting materials can be obtained and often 'broker' links between farmers in this regard. Traders thus unwittingly contribute to the decline in varietal diversity at the farm or community level. It is not unusual for farmers to plant only one or two varieties which are defined as marketable

What the embeddednes of the sustainability of livelihoods in social relations means for the dynamics and heuristics of the new agricultural research paradigm is illustrated by an attempt to control bacterial wilt (BW) at the community level in the Western hills of Nepal (Pradhanang *et al.*, 1994).

For a BW control programme to succeed, the minimum unit of operation is one complete village. Everybody in the village must participate and must accept a moratorium on potato growing in infected lands for 3 years. At the same time, there is a need to continue producing potatoes because of their importance as a food source. The programme included a three-year crop rotation, rogueing of volunteer plants, clean seed multiplication, and education of farmers on disease transmission and crop-hygiene practices. Local farmers' committees were formed to implement the programme. Although the final results are not yet known, some observations are worth noting.

In one village where the programme has been successful, farmers had good reasons for co-operation:

- storage losses in excess of 50% on average;
- prospects of acquiring clean seeds from BW-free areas with subsidized transport;
- availability of BW-free sites for planting potatoes, not unusual in the high hills of Nepal because of abandoned land as a result of labour shortage;
- equity among participants by distributing BW-free land for temporary use on the basis of a lottery by a farmers' 'cropping system improvement committee'. This committee was composed of influential villagers;
- use of Khet (wet-rice) land in valley bottoms which usually remain fallow during the winter after the rice harvest;

- peer pressure: 98% of the village agreed to participate, therefore the rest had to become involved;
- regular visits by the team of BW control scientists.

The approach was not so successful in some other villages. In one, this could be ascribed to lack of farmer co-operation, the large size of the village (500 households) and the scattered potato growing areas, which made monitoring difficult. The village is also a major mountain trekking centre for tourists so that farmers have non-farm livelihood interests which detracted from co-operation with the programme.

In another village, the major difficulty in implementing the programme was the inaccessibility of the seed potato multiplication site in terms of physical distance. This problem could not be overcome even by construction of a suspension bridge across a small river between the village and the site through community participation with support from the Ministry of Local Government.

The education of farmers about the symptoms of BW in the field and stores, and about the means by which BW spreads, is very important. Potatoes taken as gifts at marriage ceremonies were confirmed to be a major cause of disease spread among villages at high elevations (Pradhanang *et al.*, 1994).

11.4 The people–labour inconsistency

From the review of farm and village-level diagnostic studies in unfavourable eco-systems in Laos, Cambodia, Nepal, the Philippines, Thailand, Vietnam, Madagascar, and India, it is clear that labour shortage is a constraint in farm activities and a deterrent to the adoption of such sustainable crop management practices as green manuring, observing economic threshold levels in integrated pest management (IPM), weeding, or nutrient management. Because of this labour constraint, the diagnostic studies suggest that labour requirements, often ignored in the 'romanticism' of low external input agriculture, should be given very careful attention in the design and development of more productive but sustainable technologies.

The real puzzle however, lies, in the phenomenon of 'people plenty – labour short' situations. At the macro-level, most of these countries are characterized by relatively high rates of population growth or at least by large populations. Where are all these people whom the demographers worry about? There are several likely explanations for this puzzling phenomenon:

- complex and labour-demanding but low-yielding production systems;
- competition from off-farm jobs and non-farm livelihood opportunities;
- low energy levels of farming households;
- limited availability of non-human power sources;
- unattractiveness of farming particularly to the young;
- rural–urban migration;
- lack of labour-saving but knowledge-intensive practices;

- a high proportion of the young, school-age population which is deliberately being educated out of farming;
- shifting labour arrangements from co-operative/exchange toward hired labour paid in cash;
- inflexibilities in the gender division of labour.

Broadly speaking, family, exchange, and co-operation are the major sources of labour in unfavourable areas, particularly in countries where de-collectivization has taken place and where non-agricultural employment is limited. However, labour hired for cash or kind has already taken over some specific tasks. Most farm households practise two or three different labour arrangements for different farm operations, different crops and different eco-systems. In industrializing economies, hired labour has become the standard practice, along with mechanization, and leasing in or leasing out of land, as ways of adjusting to labour shortage (Castillo, 1995).

11.5 Farm households in multiple agro-ecological niches

Farm households are often categorized according to commodity or to agro-ecosystem. We have rice, corn, wheat, livestock farmers or we have irrigated, rainfed, flood-prone or upland farmers. Each category of farmer is treated as a particular commodity farmer or a specific ecosystem farmer.

But in real life, it is not unusual for farm households to manage more than one ecosystem and more than one commodity in different parcels of land not always contiguously located. In other words, these are farm households who are engaged in cross-ecosystem, diversified farming. In addition, depending upon their skills and access to non-farm opportunities near urbanizing centres, they can also be engaged in different kinds of non-farm activities and/or receive remittances from household members who have migrated to urban areas. Hence it is more accurate to describe their food security and cash income-generating strategies as 'livelihood systems' rather than cropping systems or even farming systems. Even in typical rice-dependent villages, income from rice is often less than 50% of the total income.

One virtue in using the farm household, rather than the commodity or even the ecosystem perspective is that we begin to see the dynamics of diversified household strategies. We are not dealing each in isolation with rice, fruit or forest trees, livestock, fish, non-food crops, off-farm wage labor, non-farm work, or remittances, but with combinations of the above, sometimes in minuscule amounts but in unimaginable permutations. It is not just adult men, but equally women and children, who participate in this complex production and decision making process. It is not only field agriculture which is important, but also household gardens which usually contribute very essential, but often officially ignored, components of the livelihood system.

An illustration of a very complex household organization in relation to agricultural systems under different environmental conditions is provided by

Bhuktan (1990) in his intensive study of 18 Nepalese rural communities from the central region.

- Although all 204 households studied were landowners, half the land they cultivated belonged to someone else. Usually a household engaged in multiple tenurial arrangements as a risk-adjustment strategy. Land fragmentation was such that a hectare was made up of, on average, five parcels with an average distance between them of 9.86 km.
- On less than 1 ha of this fragmented farm land, the average household grew 35 food and 31 non-food crops belonging to 13 crop types organized temporally and spatially into seven distinct cropping patterns. The 204 households studied together used 56 rice varieties in 65 varietal combinations.
- One of the important criteria that farmers used in selecting a crop was its capacity to accommodate other crops.
- Eleven types of animals were kept, organized into 30 different livestock patterns. Animals were reared for food, fertilizer, power, fuel, medium of exchange, means of repaying debt and collateral for credit. Since farmyard manure is a highly valued source of crop nutrients, households had to organize livestock farming in relation to crop farming.

11.6 Food availability and the livelihood calendar

Any diagnostic analysis of fragile lands, whether for plant and soil science, socio-economics, technology development or extension, or for all of the above, wisely starts with the cropping calendar which details when different farm operations are carried out. There is a time dimension attached to such concepts as dry and wet season, early or late maturing varieties, staggered harvesting, post-rice sweet potato, intercrop, relay, crop rotation, and seasonal disease or pest problems.

The calendar becomes more complicated when it expands into a farming system, or more accurately, a livelihood system calendar. These seasonal calendars assume social significance because all of the activities are managed by the household. The content and sequence of events in the calendar determine the rhythms of life in a community and reflect opportunistic, creative, experimental, and sometimes desperate attempts to 'match' or adjust to rainfall, labor availability, the nature of crop technologies, subsistence needs, market demands, the suitability to growing conditions, and interactions with other components of farming systems (such as livestock, trees, fish, tobacco, sugarcane, etc.) and non-farm livelihood systems.

To understand the sustainability of household livelihoods in different fragile ecosystems, a cropping calendar is, therefore, not enough. We need to put together the following information: labour allocation, food availability and consumption, illness, income, expenditure, gender roles, etc. When we superimpose such a calendar on the basic agriculture calendar, we discover periods of plenty and scarcity, conflicting demands for labour and land use, time for earning and time for borrowing, the seasonality of malnutrition, time for selling and time for spending, and time for socially accepted leisure.

We might also discover that the period of food scarcity coincides with peak labour demand and frequent illness, like respiratory diseases in the cold season, sore eyes in the hot season and diarrhoea during the onset of the rainy season; that gender roles differ at different periods in the calendar such as when adult males engage in non-farm wage labour elsewhere; and that some households sell rice soon after harvest when the price is low and purchase rice before the next harvest when the price is high.

Verdonk and Vrieswijk (1992) illustrate the connections between food availability and the livelihood calendar in Hungduan and Barlig, two municipalities of the Philippine Cordillera, focusing on the role of sweet potato. Almost all Cordillera households are engaged in agriculture where farming is rice based, but as the rice harvest is not enough for the village to last the whole year, other sources of food or income are needed. Swidden cultivation is practised in steeper non-irrigated areas and almost all households have livestock, mainly pigs and chickens. In some places a shift is taking place from rice cultivation to the production of cash crops, mainly vegetables. Although these crops provide higher incomes than rice, due to marketing problems and strong traditions most farmers still plant rice. Even people producing at a subsistence level will sell a small part of their produce to purchase needed items.

Because of the increased need for cash income, people look for non-farm sources such as local paid labour and small enterprises. Insufficient local employment opportunities result in men looking for work elsewhere in the mines and cities. As a consequence, there is an agricultural labour shortage. Farmers employed far from the village lend out their rice fields, practise sharecropping or sell their fields. Labour groups are paid in cash when owners of rice fields are employed outside the village and cannot reciprocate in labour exchange.

Activities in the swiddens take place in-between labour peaks in the rice fields. The swidden farms are cleared immediately after transplanting rice, and planting takes place before rice fields are weeded and cleared. In the swidden farms, sweet potatoes are mono-cropped or intercropped with beans and sometimes maize. The amount of sweet potato grown by a family depends mainly on how poor a family is. More sweet potato is planted by families who own many pigs which have to be fed. For poor families, the sweet potatoes from the swiddens serve as safeguard against rice deficit periods. In those times, the swidden fields are visited regularly and sweet potatoes are harvested for human and animal consumption as needed. The consumption of sweet potato decreases when the mother has a full-time job and is no longer able to look after a swidden field.

11.7 Micro-impacts of macro-policy

Although macro-policies are always blamed or credited for either negative or positive consequences, we do not always understand the dynamics of how macro-policies translate into impact at the micro-level, particularly as far as the small farmers are concerned. Rentian, Xianjie & Jianhua's (1994) analysis of events in one of the five poorest counties in Zhejiang Province, China, tells us one such story. Jingning SHC minority autonomous county is located in a mountain environment where farmers' average income is only a quarter of that for the whole province. China's previous policy insisted that farmers plant a certain acreage of wheat, whether or not they were willing. This county is not suitable for wheat, but farmers had to plant it anyway even if yields were less than half those obtained for the province.

In 1989, some enterprising farmers in the county found the market price of potato highest in April, a time when they would be able to supply potato if they could replace wheat in their cropping system. They therefore started potato farming and selling. During the first year, potato gave them ten times more income than wheat. Farm manure and seed needed for potato cultivation were farm generated and, therefore, did not require such cash inputs as were required for wheat. With the cash income from potatoes, farmers could purchase wheat and other foods, which helped relieve shortages in this poor mountain area. It is little wonder that farmers shifted from wheat to potato in their winter crop land allocation. In addition, they planted potato on marginal lands, roadsides and other areas which officially were not 'arable land' and therefore were not required to be planted with wheat. Gradually, potato moved into the irrigated fields to replace wheat. As the authors put it: 'Wheat "belongs" to Government while potato belongs to farmers'. It represents farmers' own choice. From 1993 onwards, farmers could plant their land with any crop, as they are now allowed to make decision themselves on their production and marketing.

Given the promise of the potato, farmers seek ways of increasing production and of harvesting earlier (the earlier the better). As a result, farmers are keen on good varieties, improved seed quality and seed potato marketing. Farmers learned that seeds from high mountain areas are better than locally stored ones, even of the same variety. The possibility of getting into the seed market offers better opportunities for farmers to earn cash. In order to enhance the seed supply system, new varieties must be introduced into the high mountains, but for the system to be viable, new uses for potato must also be developed beyond the fresh market.

11.8 Seed systems and on-farm management of genetic diversity

One characteristic feature of farming in developing countries is the fact that not many farmers can afford to purchase seeds or planting materials, especially where the formal sectors are concerned. They usually plant seeds from their own previous harvest and/or exchange with neighbours, relatives and friends on a shared and reciprocal basis.

Potatoes provide a typical example. Studies conducted by the International Potato Center in Kenya, Ecuador, the Philippines, Tunisia, and Peru show that informal farmer-based seed systems are most important. Even in a newly industrializing country like South Korea, only 15% of seed potatoes that farmers use are certified; the rest comes from the informal seed system (Crissman, 1989*a*, *b*; Crissman & Uquillas, 1989). Because of this, any diagnostic study of agricultural systems in these environments would be woefully incomplete without a characterization and analysis of seed systems including available varieties, sources, manner of exchange and/or distribution, varietal preferences, seed management, varietal diversity, seed health, etc. Although it has been said that seed quality makes a difference in yield, seed health has not received as much research attention as other aspects of production.

From several studies of sweet potato production systems in the Philippines, the following observations seem relevant to seed systems and on-farm management of varietal diversity (Amihan-Vega, 1994; Mula, 1992; Mula, Fang-Asan & Bennette, 1994; Sandoval, 1994):

- The availability of planting materials is a growing problem which results in the decrease of the area planted. Shortage of cuttings is acute during the long dry spell. In some instances, the reason for planting a particular variety is its availability, not necessarily that it is preferred.
- Pig-rearing increases the local demand for sweet potato roots and vines to the point where the latter are now beginning to be sold for feed or planting materials.
- In Philippine areas, where sweet potato is grown commercially after rice, cuttings have to be obtained from another province and only one or two varieties are planted, usually in response to the local trader's suggestion.
- Where sweet potato is grown in different niches, different varieties with distinct characteristics are maintained and so are the planting materials which are usually available the whole year round. In the case of the Mountain Province, the production systems are: swidden, permanent swidden, paddy field, household garden and stone walls. The household garden used to function as a seedbank where the different varieties of sweet potato were preserved, but this is being crowded out by the need for residential space, hence the swidden is a better niche for seed and variety maintenance.
- Varieties are introduced by villagers who go to other places. Cuttings are usually shared with other households or are tried out first by the carrier and then shared with others.
- In commercializing villages there is less crop diversity and fewer sweet potato varieties cultivated in response to market demand. However, a few rows of preferred traditional varieties are sometimes maintained for home consumption.

The above findings suggest that there is an existing informal system for maintaining varietal diversity. Sandoval (1994) points out that these gene-banks are complemented by 'memory banks' which contextualize the genetic material and provide the human reasons that shaped its selection. 'Genetic information and cultural information are co-evolving systems and one major force in the decline of biotic diversity is human intervention.'

Prain and Piniero (1994) are actively pursuing alternative community gene-banking strategies among families in two villages in Mindanao, the Philippines. They are trying out different 'curator groups' such as individual households, tribal groups, family groups, women's rural improvement clubs, and women's informal groups. In identifying curator groups, two factors were taken into account because of their possible importance:

- gender, because of the frequent association of women with seed and variety management, and because of the degree of cohesion and mutual support found in many women's groups;
- the degree of informality/formality of the groups, because the novelty of 'self-conscious' gene-banking might, on the one hand, benefit from informal flexibility but, on the other, needs some degree of local legitimation.

Two major issues faced by these pioneering efforts are incentives and sustainability. What will sustain the deliberate on-farm conservation strategies when short-term livelihood benefits are lacking? Who will conserve for the purpose of conservation when current use-values of particular genetic materials are not so evident?

With respect to the project on community curatorship of genetic resources (Prain & Triangles, 1992), the involvement of indigenous knowledge occurs via the participation of users in three principal ways: as germplasm consultants who contribute their insights to scientists involved in germplasm collection and classification; as evaluators who receive new genetic material, evaluate it and provide information on it to researchers and disseminate the good material among the local population; and as research curators of collections *in situ*, who not only evaluate technologies coming from global R&D sources but also receive, conserve and evaluate materials collected from small-scale peripheral farming systems.

11.9 Technology development processes and products

In arguing for the new research paradigm, the assumption is that client-orientated, participatory and systems-focused approaches lead to user-responsive and locally adaptable technologies, which would support the sustainability of livelihood systems of less favoured farm households in less favourable ecosystems. As mentioned earlier, illustrations of studies which have actually had the desired impact are not plentiful, but there are some very interesting cases, even if their impacts are not always known at present.

11.9.1 High science-low tech: the case of IPM

In the 1960s and 1970s diffusion/adoption studies were the standard fare of researchers in rural sociology and communication. Training and Visit (T&V) was almost the universal approach to technology transfer. Things were much simpler then. Technologies were discrete components: simple, easily identifiable and meant for widespread adoption. Examples are crop varieties, fertilizers, pesticides, herbicides, seedbed preparation, straight-row planting, etc. The extension worker and the mass media were the 'delivery mechanisms' of the technology.

But new concepts emerged, like sustainability, biodiversity, ecology, integrated nutrient management (instead of just fertilizer application), integrated pest management (instead of simply calendar spraying), and agro-ecological diversity and location-specificity. They challenged the technology transfer model and diffusion-adoption studies went out of style.

Integrated pest management in rice, which was based on research at the International Rice Research Institute (IRRI) (e.g. Kenmore, 1980), was a major breakthrough in terms of technology development and communication. Previous technologies were prescriptive recipes. The adoption of IPM requires decision-making at different stages in the cultivation process. At first, this posed a problem as Goodell, Litsinger & Kenmore (1980) observed: 'Filipino farmers found IPM too intellectually complex as it involved pest identification, varietal identification, threshold determination, dosage and volume calculation and choice of chemicals.'

Because of its complexity and the slowness in adoption of IPM, more research was done on human and social constraints to the implementation of IPM programmes (Escalada & Heong, 1992; van de Fliert, 1993), the relationship among pesticides, rice productivity and farmer's health (Rola & Pingali, 1993), and so on. At the same time, crop protection problems were addressed by programme implementors and policy makers. This led to the evolution of major national programmes for IPM. As Kenmore (1991) put it, 'IPM policy was built on ecological science underlying agronomic principles and fulfilled through an empowering participatory approach to motivating people for action.' Röling and van de Fliert (this volume) provide more detailed information on the new style approach pioneered by IPM.

What is exciting about IPM is that it uses 'high science', for example, in the curricula for farmer training based on sophisticated entomological understanding of population dynamics of rice pests and their predators, while the basic technologies are simple ('low tech') and rely on observation and (collective) inference and anticipation by farmers. One participant in an Indonesian IPM 'farmer field school' said, 'Before, when we followed an agricultural programme, we just did what we were told. Now we really understand what we are doing' (Indonesian IPM Programme, 1993; Kenmore, Gallagher & Ooi, 1995).

11.9.2 User-responsive locally adaptable plant breeding

In Kerala and other parts of the West Coast of India, where non-lodging semi-dwarf varieties (which maximize only seed production) were introduced in the mid-1960s, rice production showed a declining trend in both areas by the early 1980s. Rosamma (1995) reports efforts to evolve varieties which would suit the extremes of the diverse situations under which rice is cultivated in Kerala and which would meet dietary demands as well as livestock production requirements. The chosen strategy involved a shift from the dwarf plant-type towards the semi-tall one so as to increase total productivity, i.e. both grain and straw yield, plus adaptability to specific seasons, locations and dietary preferences. Land races available in the region with specific desirable attributes, improved local strains and high yielding varieties of indigenous and exotic materials were utilized in the breeding programme.

More than 50 000 hectares in Kerala are now planted with these new varieties. They take advantage of the traditional, the improved and the exotic to produce locally responsive varieties to increase total economic yield, not just grain yield. This breeding programme was undertaken at the regional research station in Kerala, not at the central, national experimental station. Incidentally, the International Rice Research Institute recognized the innovativeness and value of this work when C. Rosamma was chosen to be a recipient of the 1995 'Outstanding Young Women in Rice Science Award'.

11.9.3 Learning from farmers

In this era of sustainability concerns and increased value being placed on indigenous knowledge and farmer practices, SAPPRAD has taken seriously the advice 'learn from farmers' (SAPPRAD Newsletter, 1993*a*). As a consequence, it has been sensitive to possible lessons and relevant technologies developed by farmers. Some examples are the following:

- Thai seed cutting technology, which reduces seed tuber requirements (Thongjiem, Kewkaew & Rasco, 1992);
- In-ground potato storage by Filipino farmers which allows tubers to remain undisturbed in the ground for periods up to 6 months (Aromin & Rasco, 1992);
- Use of 'suckers' or excess stems from a tuber-planted potato crop as planting material by a farmer in Sri Lanka. In 1994, four neighbouring farmers adopted the technique. With proper management, the use of suckers can result in 100% establishment and the yield is at least 50% of the yield obtained from tuber seeds (SAPPRAD Newsletter, 1993*b*);
- Comps mounding of sweet potatoes in Papua New Guinea Highlands. Under conditions of high rainfall and porous soils where potassium and nitrogen leaching occurs and where there is very high phosphor fixation, farmers have been able to use sweet potato as the dominant crop in the same land with no fallow for as long as 50 years (Redfern, Daink & Woruba, 1991);
- Application of borax on potatoes by a farmer in Benguet, Philippines. He believes that borax can control scab and late blight of potatoes (Aromin, 1994). A study conducted in farmers' fields showed yield increases in

Granola when boron is applied. This increase might be due to the fact that soils in Benguet are potentially boron deficient (Abosetugn, 1993);

- Ridger-harvesters-in-one were developed based on farmers' advice. When the two implements were introduced separately to sweet potato growers in Zambales, Philippines, it took the farmers less than 30 minutes to conclude that the ridger does a better job of harvesting than the harvester itself.

SAPPRAD analyses, evaluates, experiments, and tries to find the underlying science behind farmers' practice. In this way, the programme is able to share the technology and can explain why it works. Farmer practices then become an integral part of, and not an antithesis to, science. In this respect, SAPPRAD has really taken farmer practice as the farmer's contribution to the enrichment of science.

11.9.4 Sweet potato on a 'new wave' of R&D: a farmer participatory and location specific scheme

In the conventional vertical top-down model of agricultural R&D, the uniform crop varieties are developed at the central experimental station, they undergo multi-locational testing for wide adaptability and then are passed on to the agricultural extension system for introduction to farmers. The latter may accept, reject or adapt the variety. Success is measured by the hectarage devoted to the variety and the production that results from it. While this system seems to have worked well with high input irrigated areas, performance in more diverse agro-ecological settings and cropping systems and more resource-poor farm households is spotty at best. Adoption of experiment station-generated varieties has been poor.

SAPPRAD has tried to be more responsive to diversity. Rasco (1992) suggests that there are narrow limits within which individual varieties are suitable and therefore the concept of 'national' varieties deserves re-examination. In Indonesia, trials in ten environments using 12 promising clones show that none of the clones excelled in more than two environments. In the Philippines, a set of 35 indigenous and introduced cultivars were tried in six provinces. Only three cultivars were in the top five in more than one location.

From these observations and results of trials, Rasco concludes that "there is a need for location-specific evaluation as a basis for location-specific recommendation. A variety selected for high average performance in multilocation trials is likely to be inferior in any specific environment." Furthermore, varieties also have to respond to consumers and users with varied preferences and requirements (Rasco, 1992). Given these data, generated over 2 years, SAPPRAD came up with a new sweet potato breeding scheme which regards farmers as partners. It was also expected to be a mechanism by which improved varieties could be introduced quickly into farmers' fields.

Rasco (1993) compares the new scheme with (a) the typical national variety scheme and (b) the normal way by which varieties are developed in farmers' fields (Box 11.1).

Box 11.1 The new potato breeding scheme compared

Criteria	New scheme approach	Conventional farmers' approach	Normal approach
Broad germ-plasm base	Yes	Yes	No
Location-specific selection	Yes	Yes	Yes
Use-specific selection	Yes	Usually no	Usually no
Farmer participation	All stages	Late stages	All stages
Rate of farmer adoption	High	Usually low	High

Source: Rasco (1993).

Reactions of farmers in Thailand, Indonesia, Sri Lanka, and the Philippines are very exciting. They are presented with a wide range of choices from improved, imported, introduced and local varieties, and actively participate in their evaluation. This may be one way of increasing varietal diversity at the local level where only one or two varieties are often planted at present, because nobody has introduced other possibilities to them.

A significant by-product of this scheme is that farmers themselves are stimulated to conduct their own variety trials. As Rasco (1994*a*, *b*) puts it:

> For so long, the farmer has been viewed by some plant breeders as just a recipient of new varieties. Guided by this thinking, variety development schemes put the farmer at the end of the variety development pipeline, like a customer waiting for a product to be released to the market by a factory. This is probably the reason why plant breeding vocabulary uses the term 'release' which literally means to let go. The term describes the situation wherein plant breeding institutions keep their varieties in the station until they are completely satisfied that they are good enough to be given to farmers.
>
> While the intention is to protect the farmer from indiscriminate introduction of new varieties, they run counter to a more fundamental tenet of plant breeding that variety evaluation should be done using user's criteria and under conditions that are representative of the farmer's fields. Indeed what is a better way of following this tenet than to allow him to do the evaluation on his own farm?

11.9.5 From diagnosis to action research and household gardening development

Mula and Gayao's diagnostic studies (1991) show that sweet potato is an important component of the house-garden system in the Cordilleras. Its multiple use and low-input requirements make sweet potato an ideal backyard crop for a growing number of urban poor. A participatory approach to technology development was employed starting with a workshop designed primarily to raise consciousness regarding the importance of home gardening and the role of the sweet potato. All the participants were women from selected neighbourhoods where home gardening was being practised.

An advantage of using a participatory approach in this project was the opportunity for researchers to do more in a very short span of time: introduce different sweet potato varieties, processed products and recipes; document selection and maintenance; document traditional management practices; assess the performance of introduced varieties in varied garden micro-niches; and assess the impact of the project. It proved both a learning process for the researchers and the home gardeners and provided immediate feedback to the researchers.

On the overall impact, the number of households and the area devoted to sweet potato increased as did the number of sweet potato consumers. The diversity of crops in the garden increased and the backyard became a major source for the sweet potato consumed. Furthermore, the participants gained additional knowledge on its nutritional value and other uses. But most of all, representatives from the Departments of Health, Education, Agriculture, and the City Nutrition Council, Municipal office and University Extension Department all recommended an expansion of the project to other communities and schools. A trainer's training for field health workers was suggested and Benguet State University's role as source of planting materials and technology was reinforced (Gayao *et al.*, 1993).

11.9.6 Soil fertility management in an upland community

A series of diagnostic studies conducted in an upland community in Quezon Province, the Philippines, showed declining soil fertility, increasing use of inorganic fertilizer without increase in yield and a growing need for credit to finance such a production system. A historical transect, constructed on the basis of information from knowledgeable elders, revealed changing cropping patterns through time. From subsistence rice production, farmers shifted to vegetables such as carrots and cabbage when they saw neighbouring villages making money from these crops in the 1970s. When the price of vegetables dropped due to oversupply, they shifted in the 1980s to maize, sweet potato and peanuts. Along with the change in cropping patterns came changes in soil fertility management. In the 1960s, fallowing was the most common practice; in the 1970s, crop rotation and fallowing were the dominant practices. Some even applied animal manure, particularly where farm lots were near houses where animals were kept. In the 1980s, farmers started to apply inorganic fertilizer as they continuously planted sweet potato as a monocrop for the market (Bagalanon, 1992).

When the results of the diagnostic studies were presented to farmers and they were asked to rank their priority problems, soil degradation was not in the list, but the need to reduce the amount of inorganic fertilizer was included. When asked why they did not include soil degradation, most farmers replied that they expected it to occur as a consequence of their unabated use of the land. This process could not be halted and fertilizer application was a measure to control its negative effects, even if it only maintained and did not increase yield. On the other hand, use of organic fertilizer was not attractive because it was laborious and the manure was not sufficient to cover the whole farm area.

Farmers also admitted that they had sacrificed soil fertility maintenance for economic gain. However, if information on the long-term effects of inorganic fertilizer use had been available, they might have had second thoughts about using it. Given alternatives, they professed willingness to change their practices if this meant prolonging soil fertility. An inquiry into practices which did contribute to the maintenance of soil fertility revealed several local practices, such as (see also Campilan, 1995):

- alternate planting of corn and sweet potato;
- planting of sweet potato prior to the peak rainfall months so that the crop can form a canopy and reduce soil erosion. This canopy is maintained by staggered harvesting;
- use of 'gen-gen': the trunks of bananas together with weeds are placed in a shallow trench (gen-gen) dug along the contour. When covered with soil, these break the run-off flow and provide compost which helps produce big tubers;
- use of grass strips along the contours;
- bananas planted in a strip at the edge of the sweet potato area.

Given the recognition of declining soil fertility as a problem and knowing that farmers wanted to reduce their dependence on inorganic fertilizers, the project organized a technical working group composed of sectoral organizations in the community and representatives from various government and non-government agencies working in the area, including soils, biotechnology, and communication specialists from the University of the Philippines at Los Baños.

The project was divided into three phases. The first phase was to concentrate on increasing community awareness and understanding of soil fertility management issues. Community meetings, radio schools, information campaigns, and training in soil fertility management were to be conducted along with multi-sectoral consultations. Phase Two focused on organizing separate local groups into working teams with the involvement of the technical people working in the area. In essence, this is a networking mechanism to reach out and diffuse information to more people in the community through the sectoral/organizational representatives. In this phase, participatory consultation was to be regularly carried out through meetings and farm tour activities sponsored by the group. Phase Three involved the establishment of demonstration farms for participative on-farm trials on the use of biofertilizers and SALT (sloping agricultural land technology) and the establishment of a community enterprise which was to serve as the outlet for bio-fertilizer.

Since the women in the area are not engaged in farm work because the farms are far from their homes, the project enlisted them to lead the sale and distribution of bio-fertilizers. Because of the role they were expected to play, women participated in training programmes and other activities to become knowledgeable about soil fertility management. In the on-farm trials, indigenous soil fertility management techniques such as grass stripping, staggered harvesting, and crop rotation techniques are monitored by the farmers themselves. Cross-farm visits were planned to enhance information and technology dissemination. All these activities were meant to further encourage farmers to adopt technologies which would improve soil fertility management and to work together on a common problem (Bagalanon, 1995).

Most inspiring in this project was a song about 'Holding on to fertile soil' (*Pagpigil sa Tabang Lupa*), composed by a female farmer based on her own soil management practices.

11.10 Conclusions

Food security, disease control, and trade relations require as much investment in social relations and community organization as in production systems. Sustainable agriculture typically implies sustainable social organization.

The phenomenon of 'people plenty – labour short' places a heavy demand on the design and development of technologies. Low external input technologies should not be labour intensive.

Farmers in unfavourable environments tend to be cross-ecosystem farmers and often have a very complicated system to manage. Local knowledge, therefore, is indispensable in the entire research–action process.

Livelihood calendars are valuable tools for identifying problems, prospects, and potential sustainability of likely interventions.

Understanding seed systems is a 'must' as the seeds of sustainability are usually own-farm produced. Improvements in the quantity and quality of seeds are much needed and, in this regard, indigenous knowledge and modern science must meet.

Macro policies are often credited or blamed for what does, or does not, happen at the community and household level, but efforts to trace the micro-impacts of macro-policy are rather rare and should be more deliberately pursued.

The technology development process, even in the new research paradigm, also needs science, because coaxing productivity and sustainability from fragile lands is asking a great deal from mother nature, but as the examples given earlier suggest, when 'hard' scientists pursue the philosophy embodied in the new research paradigm, the results are indeed very encouraging.

Finally, there is nothing easy nor self-evident in the application of the new research paradigm. The diagnostic phase may be rapid, participatory, etc. and perhaps exciting in process and in the identification of problems and the promise of results, but the design and development of social and technological solutions are complex, difficult and demanding of the human intellect and spirit. How many of us are prepared to devote our lives to seeing the 'fruits of springtime'? There are too many who stop at analysis and diagnosis and do little with it thereafter.

'New style' agricultural systems research has a significant 'process' dimension, bur is rather elusive when it comes to capturing and documenting and therefore contributing to the 'soft' image of its underpinnings. The requirements of rigour, which makes science scientific, and of relevance, which gives science a human purpose, demands a refinement of methods, of technique, and of analysis. The new research paradigm is rich in relevance but perhaps a little lacking in rigour. Relevance is not a substitute for rigour. We have, therefore, a great deal of work ahead of us.

11.11 References

Abosetugn, S. B. (1993). Survey of iron and boron status and study of the effect of or on supplementation on growth and tuber yield of two potato cultivars. *SAPPRAD Newsletter*, **VIII**, (2), December 1993, pp. 20–1.

Amihan-Vega, B. (1994). Conservation, indigenous knowledge, and the market: experiences from the sweet potato project in Lanao, Philippines. Paper presented at the international seminar on indigenous knowledge at Savoy Homann, Bandung, Indonesia, 11–15 July 1994.

Aromin, F.B. (1994). Highland Filipino farmer 'discovers' the use of borax on potato. *SAPPRAD Newsletter*, **VIII**, (1), 19–20.

Aromin, F.B. & Rasco, E.T. Jr. (1992). In-ground potato storage by Filipino farmers: a critical analysis. *SAPPRAD Newsletter*, **VI**, (2), 11, 12.

Bagalanon, C.L. (1995). Status report on diffusion and sustainability of bio-organic fertilizer use and indigenous fertility management practices, UPWARD.

Bagalanon, C.L. & Santos, T. (1992). Farmers' knowledge on soil fertility management in two regions of the Philippines. Paper prepared for the Asian Farming Systems Symposium, 2–5 November 1992, Colombo, Sri Lanka.

Bhuktan, J.P. (1990). Patterns of rural household system organization in relation to farming system under different environmental conditions in Nepal, unpublished PhD thesis, University of the Philippines at Los Baños.

Campilan, D.M. (1995). Learning to change and changing to learn: managing natural resources for sustainable agriculture in the Philippine uplands. Thesis, Wageningen Agricultural University.

Cassman, K.G. & Pingali, P.L. (1994). Extrapolating trends from long-term experiments to farmers' fields: the case of irrigated rice systems in Asia. In *Agricultural Sustainability in Economic Environmental, and Statistical Terms.* Chapter 5. Chichester: John Wiley.

Castillo, G.T. (1995). Rice, rice and beyond: understanding fragile lives in fragile lands. Paper prepared for the International Rice Research Conference, International Rice Research Institute, Los Baños, Laguna, Philippines, 13–17 February 1995).

Crissman, C.C. (1989*a*). Seed potato systems in the Philippines: a case study. CIP and PCARRD.

Crissman, C.C. (1989*b*). Evaluation, choice and use of potato varieties in Kenya, CIP International Potato Center Annual Report 1986–87, October 1989.

Crissman, C.C. & Uquillas, J.E. (1989). Seed potato systems in Peru: a case study, CIP and FUNDAGRO, 1989.

Dwarakinath, R. (1994). Redirecting the extension effort in agriculture. A draft note placed before the sub-group on agriculture.

Escalada, M.M. & Heong, K.L. (1992). Human and social constraints to the implementation of IPM programmes. Presented at the 15th Session of the FAO/UNEP Panel of Experts on Integrated Pest Control, 31 August–4 September 1992, FAO, Rome.

Garzon, E. & Arocena-Francisco, H. (1992). Institutions in the Philippine Sweet Potato Subsystem: their implications to farming systems research and extension. Paper presented during the Second Farming Systems Symposium, 2–5 November 1992, Colombo, Sri Lanka.

Gayao, B.T., Alupias, E.B., Sim, J.M., Quindara, H.L., Gonzales, I.C. & Badol, E.O. (1993). Sweet potato household gardening development. Report, Northern Philippines Rootcrops Research and Training Center, Benguet State University and User's Perspective with Agricultural Research and Development, International Potato Center, Los Baños, Laguna, Philippines.

Goodell, G.E., Litsinger, J.A. & Kenmore, P.E. (1990). Evaluating integrated pest management technology through interdisciplinary research at the farmer level. Conference on future trends of integrated pest management, International Organization for Biological Control of Noxious Animals and Plants, Bellagio, Italy, 30 May – 4 June, 1990.

Indonesian National Integrated Pest Management Program IPM (1993). *Farmer Training: The Indonesian Case*. FAO-IPM Secretariat, Yogyakarta, Indonesia.

Keasberry, I. & Rimmelzwaan, M. (1993). *Socioeconomic Relations Within Household Food Security: The Transition from Subsistence Farming to Cash-Crop Farming Among the Ifugao People in the Small Philippine Village of Balog*. Department of Household and Consumer Studies, Wageningen, The Netherlands and UPWARD, Los Baños, Laguna, Philippines 1993.

Kenmore, P.E. (1980). Ecology and outbreaks of a tropical insect pest in the green revolution: the rice brown planthopper, *Nilaparvata ugens* (Stal). Berkeley: University of California, Graduate Division, PhD dissertation.

Kenmore, P.E. (1991). *Getting Policies Right, Keeping Policies Right; Indonesia's Integrated Pest Management Policy, Production and Environment*. Colombo, Sri Lanka.

Kenmore, P.E., Gallagher, K.D. & Ooi, P.A.C. (1995). *Empowering Farmers: Experiences with Integrated Pest Management FAO Inter-Country Programme for Integrated Pest Control in Rice for South and Southeast Asia*. Metro Manila, Philippines.

Mula, R.P. (1992). Farmers' indigenous knowledge of sweet potato production and utilization in the Cordillera region. *UPWARD Working Paper Series No. 1*, User's Perspective With Agricultural Research and Development, Los Baños, Laguna, Philippines, December.

Mula, R.P. & Gayao, B.T. (1991). *Urban and Rural Homegardens in the Highlands of Northern Philippines: The Case of Sweet potato.* Benguet State University, La Trinidad, Benguet, Philippines.

Mula, R.P., Fang-Asan, L.D. & Bennette, N.C. (1994). The dynamics of seed supply and variety maintenance of sweet potato in Bayyo, Mountain Province. Paper prepared for UPWARD Workshop, March.

Pradhanang, P.M., Dhital, B.K., Ghimire, S.R., Gurung, T.B. & Gurung, K.J. (1994). Community approach to manage bacterial wilt of potato in the western hills of Nepal. Paper prepared for UPWARD Planning Workshop held at Tall Vista Lodge, Tagaytay City, Philippines, April 5–8.

Prain, G.D. & Piniero, M.C. (1994). Community curatorship of plant genetic resources in Southern Philippines: preliminary findings. In *Proceedings of an International Workshop on User Participation in Plant Genetic Resources Research and Development*, Ed. G.P. Prain & C.P. Bagalanon Alaminos, Pangasinan, Philippines, May 4–8, 1992 published by UPWARD, Los Baños Laguna, 1994, pp. 191–220.

Prain, G.D.R. & Triangles, D. (1992). Local expertise, social science and genetic resources research. Paper presented at the Seminar on local knowledge and agricultural research at the Eastern Highlands of Zimbabwe, 28 September–October 2, 1992.

Rasco, E.T. Jr. (1992). SAPPRAD research shows need for location-specific variety recommendation for sweet potato. *SAPPRAD Newsletter*, **VI**, (2), 12–13.

Rasco, E.T. (1993). New sweet potato breeding scheme regards farmers as partners. *SAPPRAD Newsletter*, **VII**, (1), June 1993, p. 12.

Rasco, E.T. Jr. (1994*a*). Coordinator's Report, SAPPRAD on the third year of phase III, Southeast Asian Program for Potato Research and Development.

Rasco, E.T. Jr. (1994*b*). The theoretical basis and methods of sweet potato variety evaluation Vol. 2: Background papers and SAPPRAD country reports, SAPPRAD, pp. 1–13.

Redfern, V.P., Daink, F. & Woruba, M. (1991). Indigenous conservation farming practices in Papua New Guinea: two selected examples: the Giu technique and compost mounding, Reproduced in *SAPPRAD Newsletter* in the article 'Compost mounding of sweet potato in PNG highlands', **V**, (1), December 1991, p. 12.

Rentian, Z., Xianjie, D. & Jianhua, W. (1994). Potato marketing witnesses farmers' experiences in poverty alleviation. Paper prepared for the UPWARD Planning Workshop, 5–8 April. Tagaytay City, Philippines.

Rola, A. & Pingali, P.L. (1993). *Pesticides, Rice Productivity, and Farmers Health: An Economic Assessment.* IRRI, Los Baños, Laguna.

Rosamma, C.A. (1995). Evolution of rice varieties for tropical humid west coast of India. Paper presented at the International Rice Research Conference, 13–17 February, IRRI, College, Laguna.

Sandoval, V. N. (1994). Memory banking: the conservation of cultural and genetic diversity in sweet potato production, local knowledge, global science, and plant genetic resources towards a partnership. In *Proceedings of an International Workshop on User Participation in Plant Genetic Resources Research and Development*, Ed. G.P. Prain & C.P. Bagalanon Alaminos, Pangasinan, Philippines, May 4–8, 1992) published by UPWARD, Los Baños Laguna, 1994, pp. 23–55.

SAPPRAD Newsletter (1983*a*). Editorial: Learn from Farmers, **VII**, (1), 10.

SAPPRAD Newsletter (1983*b*). Sri Lankan farmer copes with high seed cost, **VII**, (1), 1, 18.

Thongjiem, M., KewKaew, S. & Rasco, E.T. Jr. (1992). Seed cutting and planting potatoes using cut seed in Thailand: procedure and critical analysis. *SAPPRAD Newsletter*, **VI**, (1), June 1992, pp. 8–9 and 15.

Van de Fliert, E. (1993). *Integrated Pest Management. Farmer Field Schools Generate Sustainable Practices: A Case Study in Central Java Evaluating IPM Training.* Wageningen: Agricultural University, WU Papers 93–3, published doctoral dissertation.

Verdonk, I. & Vrieswijk, B. (1992). *Sweet Potato Consumption in Two Municipalities in the Cordillera.* UPWARD, Los Baños, Laguna, Philippines, July 1992.

Part IV: Platforms for agricultural resource use negotiation

12 Integrated farming systems: a sustainable agriculture learning community in the USA

JOHN W. FISK, ORAN B. HESTERMAN AND THOMAS L. THORBURN

12.1 Introduction

One way to create change in food and farming systems in the USA is to build a learning community around the values of sustainable agriculture. The W.K. Kellogg Foundation (WKKF), through its integrated farming systems (IFS) initiative, is fostering a learning community from which is emerging both the sustainable technologies and social constructions needed to transform our current system. This learning community is made up of individuals, organizations and institutions that participate in the 18 IFS projects located in geographically representative regions of the USA. These projects are tied together in the IFS Network which facilitates the sharing of experiences and information as well as increasing leverage for catalysing policy change.

The goals of these IFS projects are to help farmers develop and adopt more-sustainable farming practices and systems, and to help people and their communities identify and overcome the barriers to sustainable agriculture. The strategy used to achieve these goals is to support community-based demonstration projects that innovatively address the issues of agricultural viability and productivity and environmental protection. Within this strategy is a focus on building and strengthening collaborations and true partnerships, even across traditionally opposing groups, and the development of leadership capacity among community members.

The first part of this chapter discusses the concept of a learning community and how it is being used to catalyse change to a more sustainable agriculture. The development of the IFS initiative will then be described, including the comprehensive approach of the IFS projects which ensures that social as well as technological barriers are addressed. This will be followed by a description of the IFS network, which constitutes the whole of the IFS learning community. Results from the first 4 years of the initiative will be discussed in terms of strategies that are facilitating success in building a learning community and lessons that are being learned about how to create and maintain collaborative relationships. Lastly, we will describe the next phase in the IFS initiative, which builds on what has been learned so that policy and systems wide change can be achieved.

12.2 The concept of a learning community

The learning community is emerging as a powerful approach to addressing complex social issues by operationalising the concepts of systems thinking, true collaboration, and effective partnerships. Ecology has taught us that actions and events that make up life, both common and momentous, do not exist in isolation; they are intricately enmeshed in the fabric of society and culture from which they come. And so solutions to complex social problems must emerge from within the affected society and not be imposed upon it. A community, made up of multiple stakeholders, focused on a common issue and involved in learning relationships, can be an effective means to create and enact needed solutions.

For this community to act as a learning community, there must be something holding them together at a fundamental level which encourages them to expand together. A common vision held by all serves to bind them together, encouraging trust and collaborative effort as well as guiding the community. A learning community focuses on the process of learning and change as well as on the content of the issue at hand.

Peter Senge (1990), in his discussion of a learning organisation, provides some ground work for how to go about creating a learning community. He defines a learning organization as 'an organization that is continually expanding its capacity to create its future' (Senge, 1990). According to Senge, in order for an organization to become a learning organization, five disciplines must be practised by the participating individuals, including: personal mastery, awareness of mental models, building shared vision, team learning, and systems thinking. Each is a developmental path leading to new skills and capacities.

Personal mastery means personal growth and commitment, it is about living life in the service of our highest aspirations, and taking responsibility for our life. The practice of personal mastery by the members of an organization or community determines its capacity for learning. Mental models are our internal pictures of the world, how we stereotype people, how we filter reality to fit our preconceptions. Organizations and communities can have commonly held mental models, holding a shared conception of who they are, who their partners are, as well as their opposition. By holding onto these mental models we limit our learning. By bringing awareness to them, we can choose to see new possibilities. Building shared vision is the practice of finding commonalty among individuals' visions of the future. 'Shared vision is vital for the learning organization (community) because it provides the focus and energy for learning' (Senge, 1990). The discipline of team-learning acknowledges that most activity is not done alone, but with others, and that more can be achieved by working as a team than alone. However, learning and working as a team requires different skills to when learning or working alone. Effective team members practise the skills of understanding individual leadership styles, effective listening, negotiation and conflict management, and maintaining a systems perspective. Systems thinking allows us to see the whole picture, giving insight into how events are interconnected, which can help us see how to change them more effectively. 'At the heart of the learning organisation is a shift of mind – from seeing ourselves as separate from the world to connected to the world . . .' (Senge). Although these five disciplines are described individually, they are themselves a system, each to be developed in concert with the others.

An effective learning organization or community supports the expansion of individual consciousness and effectiveness as well as the consciousness and effectiveness of the organization or community.

Senge's discussion is directed towards an organization such as a corporation, whose ultimate goal is to produce profit, thus sustaining income to its employees and owners. A community can be a much broader group including individuals, organizations, and institutions that cut across cultures, age groups, and geographic locations. A community may also embrace a larger vision of social change. By considering IFS as a learning community, we are elevating Senge's ideas to another level, which brings different challenges and opportunities. However, the five disciplines still serve effectively to guide the community.

There are several reasons why we believe the development of a learning community is a strategy well suited to shifting our current food and agricultural systems towards sustainability. First, no activity is more fundamental to society than its food and agricultural system. Because of this, creating change in that system requires social as well as technological change. True social change occurs when individuals and institutions consciously choose to modify behaviour. In this case the choice may be to eat foods grown more locally, to eat foods produced under certain production practices, or for farmers to choose to prioritize soil health along with high yields, or develop alternative marketing strategies. To encourage and sustain these choices, changes are needed at policy levels which remove barriers from or provide incentives for the desired change. In order for the system to shift as a whole, component choices need to be aligned. This requires information, communication and concurrent action which are inherent in a learning community. Secondly, a learning community is suited to creating a sustainable agriculture, because we really do not know what form our agricultural systems need to take to be sustainable and to contribute to the quality of life of those it serves. One first step in a learning community is to admit that we don't know these answers, that 'experts' don't have the answers, and that it is going to take a systems approach and creative input from many sources to find solutions. In effect, people who care, who have knowledge from other sources or means, who come from all sectors of society are invited to participate in shaping the future. 'Achieving a sustainable agriculture will require integrated farming systems that involve many diverse individuals and institutions in rural communities' (Hesterman and Thorburn, 1994).

The W.K. Kellogg Foundation is a philanthropic foundation located in Battle Creek, Michigan, USA. The Foundation has a long history of supporting rural development and empowering individuals and organizations in the area of agriculture. Through the IFS Initiative, WKKF is creating a learning community to address the critical need for change in agricultural production systems and rural communities throughout the USA.

12.3 Evolution of integrated farming systems

The current 'conventional' farming systems in the USA and elsewhere have serious problems associated with them. Many contribute to the loss of topsoil and contamination of surface and groundwater with fertilizers and pesticides. Larger farms and greater yields have not consistently resulted in economic viability for farmers. Instead, there has been a steady decrease in farm and rural population, a lack of opportunity for young and entry-level farmers, and an overall economic and social demise of many rural communities.

The vision of IFS is to contribute to a paradigm shift which will help bring into being a system of agriculture that is more sustainable (see Box 12.1). Having and holding a shared vision of a sustainable farming system is the root of this learning community. Building that shared vision is part of the forming process of the community of stakeholders, and is an activity itself which requires collaboration.

Box 12.1 Shifting paradigms in learning

The IFS approach to learning and catalysing change is different from conventional approaches which rely upon a positivist world view. In a positivist world view, knowledge and information are generated by experts and transferred to users in a unidirectional learning process. The IFS embraces a constructionist learning process. In essence, IFS is creating a forum for a new round of socially negotiated agreements about farming and agricultural systems. The group of experts recognised as valid participants in this round go beyond scientists and policy makers and is inclusive of many stakeholders who will have to live with these agreements. Utilizing diverse sources of input and acknowledging multiple ways of knowing and learning will contribute to generating social 'constructions appropriate for survival' (Röling and Wagemakers, 1997, this volume). Within the IFS framework, scientists work in collaboration with farmers, policy makers, consumers, environmental and rural activists, acknowledging the unique contribution each is able to make as a professional and as a fellow community member. This interaction modifies the truth which scientists pursue (i.e. the what and why of research) and influences the perspective on reality which they share with the rest of society.

Those who initiated IFS wanted to provide a conceptual framework around which potential grantees could form their projects. It was strongly felt that sustainable agriculture is not a defined set of practices but an evolution of practices, strategies and ways of thinking that are dependent on the context of the production system. In fact, it was felt that creating a specific definition of sustainable agriculture would be counterproductive, serving only to generate disagreement and to put an end point on the learning horizon. Instead, conditions were created to allow IFS to evolve in accordance with an emerging vision.

The concept of farming systems evolving towards sustainability is illustrated in Fig. 12.1. A conventional system is identified with characteristics such as high yields, specialization, and monoculture. As the system evolves towards a higher degree of integration and sustainability, additional characteristics are included reflecting the broadened awareness of the relationships agriculture has to the environment, community, and human and animal health. There is a growing

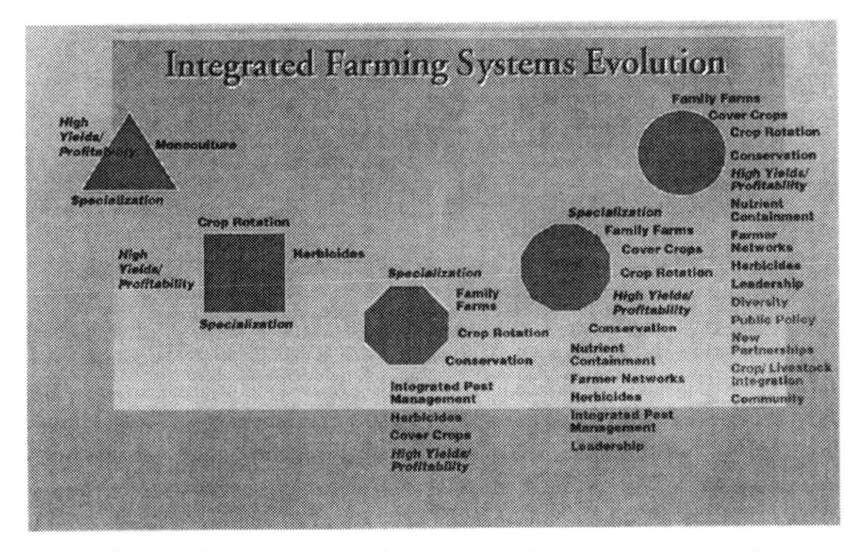

Fig. 12.1. The concept of farming systems evolving towards sustainability.

recognition of the effect agricultural policy has on the farm and rural community, a shift in how information is generated and communicated, and an increasing recognition of the value of new partnerships and diversity of people and ideas.

An important aspect in the initiation of IFS was the identification of characteristics of farming systems that a diverse array of stakeholders would find desirable and could use as part of their common vision, such as resource efficiency, productivity and profitability, protection of the environment and of personal health, support for rural communities, and increased economic opportunities.

As desirable as these system characteristics may be, it is also recognized that there are barriers to actualizing them. Much of the work of IFS projects involves identifying and reducing barriers which stand in the way of developing and using integrated approaches in food and farming systems.

12.4 The IFS learning community: translating vision into action

The learning community concept provides a framework with which to translate the IFS vision into action. Building collaborative relationships and partnerships around the goals of IFS encourages partnerships among organisations, institutions and communities which ultimately shape the future. This happens at two levels. The first is *within* each of the 18 projects, which are focused on regionally relevant issues. The second is *among* the 18 projects, in the form of a national network which is increasing leverage to affect policy issues. At both levels, connections are being made among farmers and scientists, policy makers and those who are impacted by policy, bankers and distributors, producers and consumers. It is these stakeholders who are building shared vision, challenging themselves to practise personal mastery and to look beyond their mental models, to learn as a team and work with the whole picture in mind.

12.4.1 IFS project level

The goals of IFS are to help farmers find and adopt more sustainable farming practices and systems; and to help people and their communities identify and overcome the barriers to adopting more sustainable farming practices and systems. The strategy to achieve these goals is to support community-based demonstration projects that innovatively address the issues of agricultural viability and productivity and environmental protection.

Priority has been given to projects that are comprehensive in their approach. This creates the potential for all stakeholders in the system to be involved in framing the issues and envisioning creative solutions. Solutions fashioned in this manner are more fully supported within rural communities and by the institutions with which they interact (Hesterman & Thorburn, 1994). To encourage projects to be comprehensive in scope each project is encouraged to:

- develop, test, and validate technologies that support the development of more resource efficient integrated farming systems (see Box 12.2);
- include an innovative educational component to encourage adoption of these technologies;
- explicitly address the challenge of enabling effective communication and responsible decision-making among stakeholders in our agricultural communities;
- formulate a strategy to develop leadership capacity within farm families and businesses and then use that leadership to shape the dialogue between farmers and ranchers and the non-farm members of the community;
- create an ongoing evaluation process to continually guide the project.

The Foundation encourages project participation from a variety of stakeholder groups including farmers, citizens, community organizations, land grant universities and other mainstream agricultural and rural institutions.

Box 12.2 Changes in farming practices and systems

Many alternative farming technologies are being developed or validated in the IFS projects. Farmers have found that using intensive rotational grazing as an alternative to confinement livestock systems has cut costs, reduced disease incidence, and often increased net income. In addition, farmers report that a grazing system allows for greater family involvement in the operation, allowing them to spend more time together.

Other projects have focused on the use of alternative crop rotations to create more sustainable production systems. Many alternative rotations include nitrogen-rich legumes as a cover crop or as pasture before a grain crop. Research conducted on-farm is being complemented with on-station research to increase the applicability of the information produced. For example, rotations have to be environmentally sound and productive, but they also need to be achievable, within the limits of common farm equipment, and maintain farm profitability.

Many soils are susceptible to erosion and require conservation tillage to remain productive, yet, low tillage systems rely more heavily upon herbicides. In some IFS projects, conservation tillage has been coupled with an effort to reduce herbicide use by testing the effectiveness of band spraying and reduced rates in no or low tillage systems.

Other production technologies being explored in IFS projects include integrated pest management, alternative manure and nutrient management, soil analysis to guide fertilizer application, and drip irrigation.

In order for many of these production practices to be effective, farmers have to be pro-active in their adoption process and may require whole farm system changes. Farmers are learning through on-farm trials and demonstrations, participation in local innovation groups, attendance at pasture walks and field days, and project newsletters. Extension agents and others are also learning about alternatives through training efforts featuring project farms and farmers.

Each project is a community-based demonstration of the potential of integrated farming systems. Because it is recognized that many barriers to IFS are beyond the farm gate, projects are not limited to production challenges. Each project is developing technologies and assisting others to learn and adopt them. While technologies are being developed, attention is also placed on communication among stakeholders who influence the rate of adoption and on developing the leadership skills among farmers and others to extend the impact of the technologies. This comprehensive approach supports each project to address several barriers in the movement towards sustainability (see Box 12.3).

Box 12.3 The Darby project

Ohio's Darby Creeks are home to one of the most important assemblages of fish and mussels in America. As the use of the watershed land is 80% agricultural, farmers play a critical role in the protection of this ecosystem. The Darby Project, funded through an integrated farming system grant by the W.K. Kellogg Foundation, provides support for an innovative farmer organisation called the Operation Future Association (OFA). In collaboration with The Nature Conservancy and Ohio State University, OFA promotes a production system which enhances protection of this natural resource and net farm income. Desired outcomes of this effort include: a community of farmers connected by the Darby Creeks; increased capacity of individuals to create their own destiny and sustain their community; and a clear example that control of agricultural nonpoint source pollution can be accomplished through a voluntary approach.

The mission of the OFA is to link economic and environmental soundness to improve the quality of life in the agricultural community of the Darby Creek

Watershed, while overcoming the barriers of economic cost of change, historical adherence to conventional practices, and loss of farmland to urban development. It is a non-profit organization with an elected board of directors and a membership that includes more than 20% of all of the farmers in this 370 000 acre watershed.

Activities of The Darby project include:

- Demonstrations of on-farm conservation practices: 'Farm of the future' field days feature water quality protection measures as well as state of the art crop production technology such as prescription farming, reduced rate applications of herbicides in no-till, and the planting of beneficial forages;
- Leadership and community development: OFA enhances leadership capacity through an approach which facilitates the interaction of downstream and upstream perspectives; agricultural, environmental, and residential perspectives; and local, regional, and national perspectives. These experiences are supplemented with training in many of the general skills required for effective leadership;
- Collaboration: OFA members have become township trustees and commissioners, speakers to their community and to national audiences, and members of the Darby Partnership – a group of more than 60 federal, state, and local agencies and private organizations working together to protect the Darbys;
- Innovative educational programming: Environmental education is an important condition for change to occur on the farm. An appreciation for the creeks is fostered through canoe trips that show off the aquatic diversity of the stream.

The Darby Project has implemented an array of programmes designed to address the key threats to the streams, promote compatible economic development within the watershed, and sustain the human and non-human communities. Because of the success of the partnership's work, more than nine million dollars of Environmental Protection Agency Water Pollution Control Loan Fund monies have been made available to the farmers of the watershed for use in installing best management practices. Federal and state agencies have changed their policies to work at a 'watershed' level. Thirty-six thousand critical corridor acres have come under Resource Management Systems Plans. More than one-third of the farms in the watershed have applied highly erodible land conservation plans. Eighteen new wetlands have been created, and a reduction in overland transport of 35 000 tonnes of sediment per year have been recorded. Best of all, no reduction in either the streams' rare and endangered species or the economic bottom line of the farmers has occurred.

12.4.2 IFS network level

Even though each project is addressing issues relevant to farming systems in their region of the country, there are strong commonalties among the projects. Each project reflects a common vision of agricultural sustainability and each was selected for its ability to meet the IFS goals. On a more practical level, each project is building relationships and forming groups. People who share values and hold a common vision of a more sustainable agricultural system are connected and provided a forum for working together.

Just as important are the relationships built across diverse viewpoints, histories, cultural identities and gender. These relationships are an important source for change. Secondly, each project focuses not just on farming but on contributing to the development of the community of which farming is a part. This includes the interface between the rural community and any nearby urban areas. Thirdly, each project is building the capacities of individuals, organizations and networks. Increased capacity arises from more knowledge and inspiration, stronger relationships, new skills, and more and stronger voices committed to sustainable agriculture systems and strengthening communities.

In addition to supporting the 18 IFS projects, WKKF also nurtures the formation of a larger learning community made up of all IFS projects across the nation, known as the IFS Network. Within the IFS Network, information and experiences are shared, enabling more leverage to be created in overcoming barriers than is possible by individual projects. Networking conferences are held half yearly, each hosted by an individual project and attended by several members from each project. Each conference focuses the available time on three objectives; leadership development, informational networking, and evaluation of progress across projects. Participants in the networking conferences bring back information, new skills, and inspiration to be shared within their project.

There are several ways in which the move towards learning communities is being facilitated in IFS. At the project level, the selection criteria of broad-based collaboration and inclusiveness helps to create the learning community. In addition, strong and direct relationships between project leadership and WKKF programme staff ensures that projects are continually challenged and coached in this area. At the IFS Network level, much time is spent at networking meetings helping project participants build the skills necessary for engaging in community learning (e.g. understanding individual leadership styles, effective listening, conflict management, maintaining a systems perspective, etc.).

12.5 What has been learned from the IFS initiative?

From the initiation of IFS, WKKF has contracted with a cluster evaluation team to create cross-project indicators. The evaluation team, in collaboration with project leaders, are using these indicators to draw out and illustrate common challenges encountered and successful breakthroughs by the projects, as well as to discover themes which cut across all projects.

From this cluster evaluation, we have found that the body of evidence in support of sustainable farming practices and farming systems is increasing, while individuals are increasing their capacities and organizations are becoming more effective. This section will focus on what has facilitated these accomplishments, and what capacities are necessary for individuals and organizations to work as a learning community in moving us towards a sustainable agriculture.

12.5.1 Strategies that facilitate success in building a learning community around sustainable agriculture

1. Building a support structure for sustainable farmers and community voices has been crucial. Groups, networks, and associations all act to provide support and empowerment as well as create a conduit for information exchange and technical innovation. People who are creating an alternative to currently accepted ways of doing or thinking need to be in relationship with others doing the same thing to provide emotional support, continued motivation and inspiration. What begins as a group of individuals or separate organizations, becomes a community where friendships are made, experience is shared, and learning and change are catalysed and accelerated.

   A prime example of this are farmer-driven research and education organizations. These groups, of which there are several among IFS projects, conduct on-farm research, educate other farmers and community members, and illustrate the potential of integrated farming systems to policy makers and others. The day-to-day operation and activities of these groups provide laboratories in leadership development for farmers and community members. By being conscious of the need to strengthen leadership skills, ongoing responsibilities such as event organization, information outreach and public relations can be rotated among members.
2. Building relationships across diversity is a valuable means of increasing leverage for change. However, it also presents a challenge in that it takes time to truly understand others and manage conflicts which often do not arise with less diversity. A high level of attention on group process has been rewarding to some projects and frustrating to others. Most projects spend considerable effort attempting to balance the effort placed on group introspection and time spent on task accomplishment.
3. Having a broad collaborative base strongly encourages systems thinking. Within a diverse group, multiple perspectives arise forcing the group to maintain a systems perspective. A more complete view of the barriers to sustainable agriculture guides the activities and energy of many projects. As a result, these projects have evolved from their initial focus of on-farm practice to include, for example, the development of a marketing model.
4. An effective strategy for engaging large institutions is to find the constituency they most listen to and place priority on that relationship. For example, Land Grant Universities may be more responsive to farmers and citizens than to non-profit advocates. If the project has a broad, collaborative base representing a diversity of ideas and cultures, the constituency most listened to will be involved.
5. A successful means of finding farmers and others who can provide leadership in establishing more-sustainable systems, is to partner with existing community-based organizations with a sustainable agriculture mission.

Sustainable agriculture non-government organizations (NGOs) already exist throughout the regions where the 18 IFS projects are active. Farmers and others interested in alternative systems are increasingly finding assistance they need from these groups.

6. Funding provided by WKKF through IFS was issued primarily to NGOs. This proved to be very valuable in terms of increasing the credibility of the NGOs and in strengthening their hand in negotiating project specific decisions with their partners.
7. The IFS Network has facilitated relationships, served to transfer information, helped to develop leadership skills, and served to catalyse action on national policy issues. With the combination of project level work and national level community, the IFS Network has the potential to create major shifts in food and farming systems in the USA. For example, a recently initiated farming systems research programme at the US Department of Agriculture was modelled after one of the IFS projects and was catalysed by the IFS Network.
8. Formative evaluation at all levels can be used to facilitate learning and maintain focus. The use of an evaluation process during the course of IFS (both at the project level and network level) helps reveal what is working, who or what needs attention, how people feel, what impact the project is having, and what steps need to be taken in the future. The role of evaluators becomes twofold; they help document the project as the project is ongoing and, as a result, can help guide the project to success.

12.5.2 Lessons learned about creating collaborative relationships

One of the most important features of the IFS learning community is its highly collaborative projects. We believe that it is only through broad-based collaboration that the barriers to IFS can be overcome. We have therefore asked our cluster evaluators to focus attention on this aspect of the IFS projects. The following is a list of emerging lessons about creating collaborative relationships.

1. At the onset of collaborative relationships, often trust is lacking and needs to be established before much else can be accomplished. With steady involvement over time, trust among collaborators usually increases. Trust is earned as partners demonstrate an ability to listen, follow through on tasks, keep commitments, share credit, share authority, and nurture others. Many IFS projects have demonstrated that building effective relationships takes time and requires disclosure both through task accomplishment and through other kinds of self-revelation and sharing.
2. Decision making among collaborators can be a challenge, and attention should be paid to assuring the process is perceived to be fair. Several approaches have been used by the IFS projects: writing down agreed upon rules for decision making; hiring a consultant to be a neutral

meeting facilitator; use of an expanded executive committee to balance out representation; and establishment of clearly defined oversight committees.

3. Expect conflict, both on an individual basis and at the organization level. In our experience, tensions that existed between collaborators prior to IFS, have been reduced. However, in several projects the tensions have persisted. Some projects have experienced overt institutional conflict during their start-up year over such things as how funds were to be spent and commitment level by various institutions. Working with conflict in creative ways is a continual challenge for all projects.
4. Relationships do not always come easily, especially among individuals or organizations which hold diverse viewpoints. It is important to spend time together in ways that facilitate being open minded. Time spent socializing during meals, occasional retreats, group exercises that take participants out of there familiar surroundings, both mentally and physically, along with presentations and discussions have been valuable in supporting project members in going beyond their mental models and building new relationships.
5. Some have found it hard to make the transition from mentor/mentee relationships to peer–peer relationships. The learning community approach encourages all individuals to be both teachers and learners. This has been a challenge to those who are accustomed to being the experts as well as those who are accustomed to being the passive receivers of information. However, there are some traditional mentors making the transition and using their knowledge to facilitate activity and encourage other people's initiative. Work in leadership development can assist this process by addressing both how to be an effective leader and an effective follower.
6. Active development of stronger group skills such as listening, negotiating, making group decisions, and making room for others to take on leadership responsibilities contribute to the projects and the IFS Network being able to create movement towards more-sustainable agriculture systems.

The work with IFS has illuminated successful strategies which facilitate learning and catalyse change. Difficulties in the process of collaboration have also been identified. Perhaps the universal lesson being learned is that building this type of learning community takes time. Many of us are accustomed to focusing on tangible outcomes such as the task of developing a production technology, crafting a marketing strategy, or evaluating economic performance. In a learning community the focus is on both the task and the process of learning and creating change. Time is spent on forming relationships, learning to collaborate, and teaching ourselves and others. This may initially slow down the task. However, in the long run, the synergy resulting from this work facilitates greater change with deeper roots in society.

12.6 Translating community-based learning into system change

An important step needed to catalyse continued systems change toward agricultural sustainability is to translate community-based experience and success into long-term policy change. Policies which effectively remove barriers to change toward integrated food and farming systems and provide incentives for their creation need to be crafted. Agricultural policy in the USA is currently in transition, with many subsidy programmes being reduced or eliminated and environmental protection programs under evaluation. In order to guide future policy decisions, policy makers are looking increasingly towards grass roots organizations for feedback on the impact that policy has at the farm and community level. There are relatively few organizations providing this feedback related to integrated food, agriculture, environment, and rural communities in a comprehensive way. As a result, there exists a window of opportunity for community-based collaborations, such as the 18 IFS projects and the IFS Network, to have an impact on the policy process in food and agriculture.

The community-based demonstration models of IFS have created a foundation upon which an initiative to catalyse systems and policy change can be built. Policy makers recognize the incompleteness of information that comes from many special interest groups and are seeking more detailed input on proposed legislation and regulation and more comprehensive background information on which to base proposed legislation. Individuals working on IFS projects have experience with integrated food and farm systems and can speak to the barriers present and the encouragements needed. Together, the collaborators have created demonstrations of integrated food and farming systems which can be powerful tools in educating policy makers. A major contribution which can be made by IFS is the linking of those explicitly involved in policy and those who are affected by policy.

The next phase of IFS is supporting projects in policy development and education that examine IFS policy issues comprehensively. This phase involves a diversity of those who are affected by policy, and engages people at various levels in the policy process while drawing upon the experiences of the 18 projects in the existing IFS Network. The newly funded projects are helping to create capacity to overcome policy and other systemic barriers to the adoption of integrated farming systems on a national scale. The major goals of this phase of IFS include the following.

1. Developing and extending the IFS network

Continued support for networking of the 18 community-based IFS projects will provide policy analysts and decision makers an identified source for input and information. Grass roots experiences from the IFS projects will be used in the policy formulation process. As a national network, IFS is expanding to include people and organizations working in sustainable agriculture who have not been part of the 18 WKKF-funded projects. The IFS Network is likely to become a major voice in setting and informing future agricultural policy as an independent, self-supporting organization that serves as a strong base for the future of the sustainable agriculture movement.

2. Communicating and disseminating lessons learned

We will continue to facilitate communication and dissemination of lessons learned to targeted audiences that have been identified as a critical part of the strategic policy making process. The development of specific communication activities and products will be pursued to reach targeted audiences with messages of grass roots success in sustainable agriculture through print and broadcast media. As a result of these activities, we expect key decision makers and opinion leaders in the agricultural policy arena to become acutely aware of the success of community-based efforts in IFS and to use that awareness as they make decisions about future policy direction. We also expect many more practitioners at the community and farm level to replicate successful IFS strategies and practices because of the lessons to which they have been exposed.

3. Strengthening leadership capacity

We will continue to build capacity of grass roots practitioners, policy analysts, and policy decision makers to engage collaboratively in policy formulation. While we build on leadership capacities that have been gained during the first phase of the IFS initiative, we will place additional emphasis on building the capacity to catalyse larger systems change. Components of these leadership-building projects will include development of communication and conflict resolution skills, increased knowledge of our agricultural and food system and the policy-making process, the ability to motivate others and expand the circle of influence, and experience that instils courage and creates a sense of hope that their participation is valid and meaningful. As a result of these activities, we expect a cadre of community-based spokespeople (farmers and others) who can effectively carry the messages about needed policy change to decision makers and other opinion leaders in their own communities and in centres of policy activity.

4. Policy visioning, analysis, and design

We will support public policy visioning, analysis, and design in IFS. Catalysing the creation of a broad-based learning community is only one ingredient necessary for policy and systems change. It is also important to identify those policies that would, if enacted, reduce barriers to and create incentives for sustainable agriculture. Policy analysts must work hand in hand with community leaders in charting policy pathways for a more sustainable future.

These pathways will have been created by a broad array of stakeholders and will be supported by the various sectors of people involved in our agricultural system. These pathways will be communicated clearly and effectively to policy decision makers and will result in the creation of new food and agricultural policy that institutionalizes the sustainable strategies and practices that are now operational on just a few farms and in only a few communities.

12.7 Summary

In 1992 the WKKF initiated Integrated Farming Systems (IFS), which is founded on the concept that a sustainable agriculture is an evolution of practices, strategies, and ways of thinking. To catalyse this process of evolution, the Foundation is fostering a learning community made up of the participants in eighteen IFS projects across the USA, and an IFS Network, which ties these projects together. The IFS learning community focuses on the process of learning and change as well as on the technical challenges in creating a sustainable food and farming system. Each project addresses several barriers to the implementation of a sustainable agriculture by developing or validating a resource efficient technology, including an innovative educational component to encourage adoption of the technology, addressing the challenge of effective communication and decision making, formulating a strategy to develop leadership capacity, and creating an evaluation process to provide project guidance.

IFS is bringing together people, organizations and institutions into collaborative relationships and new partnerships, both at the project level and at the network level. As a result, WKKF is creating the potential for all of the stakeholders in the system to be involved in framing the issues and envisioning creative solutions. In this chapter, we have presented insights from the experience of building a learning community around the values of sustainable agriculture and lessons learned about creating and maintaining collaborative relationships. In addition, we have described the next phase of IFS which continues project and network support, and initiates a program to create long-term policy changes.

12.8 References

Hesterman, O.B. & Thorburn, T.L. (1994). A comprehensive approach to sustainable agriculture: W. K. Kellogg's Integrated Farming Systems Initiative. *Journal of Production Agriculture*, **7**, 132–4.

Senge, P.M. (1990). *The Fifth Discipline: The Art and Practice of The Learning Organisation*. pp. 12, 14, 206. New York: Doubleday-Currency.

13 Fomenting synergy: experiences with facilitating Landcare in Australia

ANDREW CAMPBELL

13.1 Introduction

Problems confronting (post)modern societies can be more effectively resolved by affected and effective people coming together, sharing a common table or platform, learning from their multiple perspectives, world views and value systems, and finding a collective path forward which is inherently more sustainable than those presently prescribed by the traditional technocentric troika of scientists, bureaucrats and politicians, who are leading society headlong towards ecological meltdown.

This is a crude précis of a seductive emerging idea which finds expression in the communicative rationality of Habermas (1990), the post-normal science of Funtowicz and Ravetz (1993) and the participative approaches to agricultural extension and research developed in the Farmer First (Chambers, Pacey & Thrupp, 1989) and Beyond Farmer First (Scoones & Thompson, 1994) literature. Seductive, because it is clear that current paradigms – positivist experimental science, neo-classical economics and the liberal democratic nation state – have not delivered and seem incapable of reconciling and delivering social equity, economic efficiency and ecological integrity; and because it also seems clear that many relevant actors are excluded from traditional problem-solving and decision-making processes. Accordingly, there is an urgent need to get a much wider range of people involved in seeking, developing and implementing solutions to environmental (and consequently economic and political) problems.

If only it were so simple. Consensus, synergy, and rational, equitable outcomes do not just happen, they are never automatic. Situations where individuals participate as equals are very much the ideal, an ideal which seems unrealistic in practice. Resource use negotiations almost inevitably involve clashes and struggles between conflicting interests, between alternative worldviews, between those comfortable with the status quo and those trying to change it, between diverse actors with differential power and access to resources. Getting the appropriate actors around the table to improve resource management and hopefully sustainability is difficult enough, as already-powerful actors are rarely interested in sharing power. Engendering a situation where actors feel they can participate as equals would seem even harder.

Australian experience emerging from the Landcare movement (Campbell, 1994*a*,*b*, 1995) suggests further dilemmas. Creating a platform for resource use negotiation is not as simple as delineating the problem on the ground and designing a participatory structure and process accordingly. Given the usual mismatch between the scale of environmental issues and the remit of conventional social institutions, a key challenge is to develop processes for establishing agency at levels of social aggregation appropriate to each environmental issue. According to Röling and Jiggins (this volume), this involves creating shared perceptions of reality, highlighting interdependence among participating actors, developing processes for collective decision-making and conflict resolution, and integrated structures for rationalizing access to resources.

In developing new ways of tackling environmental issues, we are therefore confronted with the need to develop new skills and competencies and new types of professionals (Pretty & Chambers, 1994). The word 'new' should perhaps be construed as new for fields such as agricultural research and extension – as many of the approaches and techniques championed in the Beyond Farmer First literature find parallels in the community development movements of the sixties and seventies.

This chapter explores some of the issues involved in developing and sustaining new social institutions to resolve environmental problems and improve natural resource management, concentrating on the process of facilitation and the role of the facilitator. This is not a theoretical discussion, rather some reflections from the field, the field in this case being the Australian Landcare movement.

13.2 An Australian experience

Rural Australia suffers from the same inexorable decline in terms of trade that afflict farmers elsewhere, but without the insulation of agricultural subsidies that protect farmers in North America, Europe, Scandinavia and Japan. Many farm families are leaving the land, and those remaining are running larger farms with fewer people and higher levels of debt and stress, with declining access to social services such as schools and hospitals, as rural towns wither and rural people get older. The ancient land, for which many of the introduced European farming practices and species have been found to be profoundly inappropriate, is similarly stressed, with deepening crises in depletion and degradation of water, soils, and unique native flora and fauna.

Against this gloomy background, an exciting phenomenon emerged in the 1980s. It is called Landcare, and at its core are about 2700 Landcare groups, which now involve more than one-third of Australian farm families. Landcare groups are classic examples of what Niels Röling calls 'platforms' or fora for resource use negotiation, i.e. new social institutions which have formed at the interface between societies and the natural resources from which they live, in response to perceived sustainability problems. Landcare receives financial, technical and administrative support from all levels of government, business, and key lobby groups such as farmers and greens. The Landcare movement is described in depth elsewhere (Campbell, 1994*a*,*b*, 1995), but its bare bones are sketched here in order to set the context.

13.3 Introducing Landcare

The Australian Landcare movement is highly differentiated. Hence, one cannot describe a 'typical' Landcare group except in broad terms, as a voluntary group of (usually rural) people working together to develop more sustainable systems of land management. Groups usually involve less than 100 members (often 20–30), covering areas ranging from 500 to 15 million hectares. Common activities of Landcare groups include field days and farm walks; demonstration projects – usually land degradation rehabilitation works; development of a catchment or district plan which sets out a coordinated approach towards sustainability; facilitating the development of individual property plans within the context of the catchment plan – employing consultants, running workshops, coordinating incentives and resources; active involvement in natural resource monitoring programmes, often in conjunction with schools and government scientists; providing land conservation equipment for hire to members and other land users; research and development trials with state agencies, universities, agribusiness, CSIRO; and production of educational materials.

Landcare groups can potentially, through co-operative, coordinated approaches, solve problems on a district scale which cannot be effectively tackled at the individual property level – especially water-related issues, such as salinity, erosion, waterlogging, water quality decline and irrigation management; nature conservation, in particular, preservation of biodiversity; and management of vertebrate pests and weeds. Landcare groups create collective social pressure in favour of developing more sustainable farming systems, reinforcing and supporting the efforts of individual farmers already having a go, and exerting others to become more involved, or at least better informing them of the issues (Cock, 1992). Landcare groups thus generate commitment to the goal of sustainability on an individual and community scale, and they play an increasingly important role in gathering and managing information, in education and raising awareness (Campbell, 1992). There is some evidence that Landcare groups have enabled new practices to be tried which would have been unacceptable in the past, and have ensured faster and wider dissemination of results and learnings (Curtis, Tracey & De Lacy, 1993). Landcare groups re-establish a community focus, creating networks for social support, for sharing the stress of, and doing something constructive about rural decline (Carr, 1994). Landcare groups provide a useful structure, on an ecologically and socially sensible scale, for more efficient and effective use of government, private and community resources.

13.4 The institutional context

Australia is a federation of former English colonies. Under the century-old Australian constitution, natural resource management is the responsibility of the States. While the Commonwealth Government has encouraged a national approach to issues such as land degradation, and can exercise some influence, implementa-

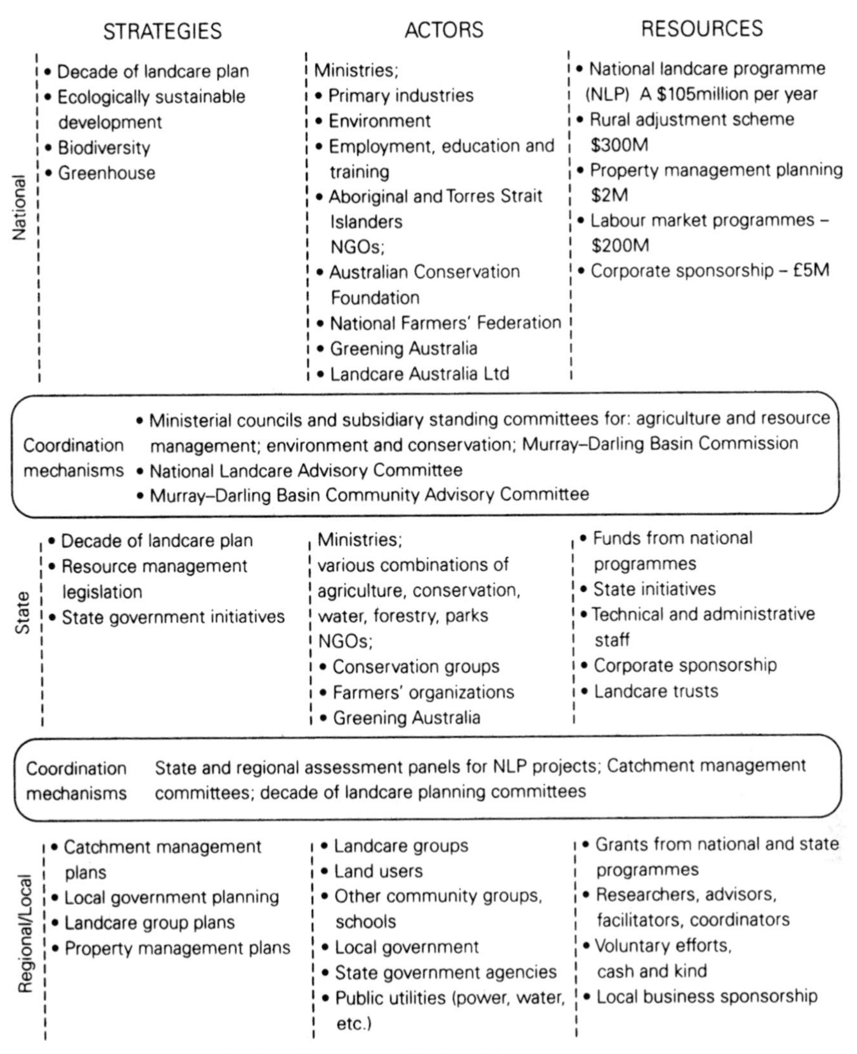

Fig. 13.1. Institutional framework for landcare.

tion of policies influencing land management must occur through state governments and their various agencies. There is no guarantee that the Australian States will take a consistent or coordinated approach to any issue – that would be the exception, rather than the rule. Doug Cocks (1992) likens a state border to 'a line on the ground impermeable to the flow of ideas'. Each state has different land management legislation and the development and support for landcare differs substantially between states.

Fig. 13.1 outlines the institutional context in which Landcare operates. It is impossible to depict Landcare as an organizational flowchart, with discrete institutions joined by lines and arrows representing flows of information, influence and/or resources. Landcare is simply not structured in an organizational sense, and is better thought of as a voluntary land conservation movement, with 2700 community groups at its core, which attracts resources from government and other sources to the extent that its activities overlap with their agendas.

Most of the grant monies available to Landcare groups are provided through the National Landcare Programme (NLP), whereas most of the technical support for Landcare groups comes from state government land management ministries. Most of the people funded by the NLP (including Landcare facilitators and coordinators) are located within state agencies, with the NLP providing a salary and some operating expenses and the state contributing infrastructure, training (usually NLP funded) and administrative support. National Landcare Program funding is provided on a competitive project basis, with submissions assessed at national, state and sometimes regional level by panels comprising a mix of stakeholders: Landcare groups, government and community interests.

These projects are assessed according to guidelines developed jointly by national, state and community interests, ostensibly consistent with plans, policies and strategies for the Decade of Landcare, Biodiversity, Greenhouse, Water Quality, Coastal Decline and Forest Use. The overarching strategy is the National Strategy For Ecologically Sustainable Development.

There is a great deal of frustration within the Landcare movement at present, particularly at community group level, with a confusing array of institutions, committees and strategies, all consuming time and energy and generating paperwork for Landcare group secretaries. The focus of funding is generally on education and raising awareness, research and development, monitoring and planning. Implementing 'works on the ground' is seen to be the responsibility of individual land users, encouraged by tax deductions[1] but not directly subsidized by the state, at least while current ways of analysing public and private benefits and costs prevail. The focus of funding programs and strategies has been on achieving a national 'Landcare ethic' and on ensuring that land managers have the appropriate knowledge, skills and technologies to manage their own land in a sustainable way. From the perspective of community groups, it seems that, once they have gone through the phases of becoming aware of the nature of their problems and potential solutions, identifying options and resources, and developing strategic plans on a farm and catchment basis, they can get money for everything *except* actually implementing solutions to the problems that are their key concern.

Landcare groups are voluntary groups of people trying to develop and implement more sustainable land use systems in an environment characterized by severe rural economic decline, exposure to worsening terms of trade in agricultural commodities, lack of practical and profitable solutions to many land degradation

[1] Given that the average Australian broadacre farmer now makes a farm business loss in the average year, tax deductions are as handy as an ashtray on a motorbike – useless as an incentive to undertake land conservation works.

problems, government 'rationalization' of many rural services and declining investment in infrastructure, and a general policy climate dominated by economic rationalist thinking – a smaller, less interventionist state, privatization of public assets and services, user pays, and faith in the market. A new breed of actors has (re)emerged to work with these voluntary groups, in a role loosely called facilitation.

13.5 Landcare group facilitation

This discussion is grounded in experience with Landcare in Australia, and is centred on community group facilitation. It should not be construed as suggesting that facilitation is the only or even the most important factor in developing and sustaining effective platforms for resource use negotiation. Nor should lessons from Australia be taken literally and transposed elsewhere without an acute awareness of Australian conditions, which may not apply in other contexts. Among the key features are, for example: relative cultural homogeneity and narrow power differentials among and between facilitators and facilitated; starkly obvious environmental problems demanding cooperative approaches; young farming systems introduced from Europe only a century or so ago, evolving rapidly to suit old landscapes; general farmer acceptance of the need for change and the need to work together on scales greater than the farm; relatively autonomous farmers with low expectations of financial support from the state, comfortable with an ethos of self-help; stable political climate, well-educated population, highly developed technological capacity; low population pressure; and a commitment to community-based approaches, expressed at the apex of political power. The fact that Landcare facilitators work with groups of people who have already perceived a need for co-operative, collective action is fundamental – a great start for effective platform facilitation.

Landcare has evolved and grown very quickly. Supporting and sustaining Landcare groups is very different from conventional agricultural extension and research. The National Landcare Program has a large investment in facilitators and coordinators, whose role is described in more depth elsewhere (Campbell, 1994*a*, *b*, 1995). Facilitators tend to operate at the regional level within state land-management agencies, working with a number of Landcare groups simultaneously, concentrating on group processes, especially with emerging and struggling groups. Coordinators tend to be employed by active Landcare groups (usually with NLP funds), with a more practical focus on helping groups to develop and implement projects, although this often requires the application of group facilitation skills. Coordinators are often former group leaders, local people who work from home with their own car and phone and are paid on a part-time basis to do in a more professional way what they used to try to cope with voluntarily.

Facilitators usually have some formal qualifications, but not necessarily in natural resource management. Both facilitators and coordinators are selected mainly on the basis of their energy, enthusiasm and ability to work with people. Their subsequent training in group facilitation techniques is at least as important as any prior qualifications. This training is one of the NLP's most important investments. It is usually carried out in short modules at state and regional level with a combination of state and NLP funding for trainers comprising state agency specialists and outside consultants, and using material such as Spencer (1989) and Chamala and Mortiss (1990, 1992). Like most aspects of Landcare, there is considerable variation in delivery across the continent. A crucial innovation in some regions is that training is not just provided for professional staff and funded coordinators, but also for Landcare group leaders and community and local government representatives. In this way, training itself becomes an important mechanism for platform construction, not just a means to acquire or disseminate skills.

Among the 91 NLP-funded Landcare facilitators and 135 coordinators (1994 figures), there are farmers, small business managers, teachers, agricultural scientists, foresters, engineers, geographers, archaeologists and journalists, about one-third of whom are women. There are many more 'traditional' extension staff working with Landcare groups who are also now receiving training in group skills. They tend to be mainly graduates in agriculture, and until recently, most state agency advisory staff were male (Reeve, Patterson & Lees, 1988).

As the title of this chapter suggests, landcare facilitation is about fomenting group synergy. Fomenting, because its connotations of fostering and nurturing, as with a plant or a child, are accompanied by an element of blending ingredients in such a way that something develops which is much more than the sum of its parts. In this case the synergy comes from a community group being able to achieve more or differently than aggregate individual contributions, in particular with respect to improving collective management of natural resources with an ultimate aim of enhancing rural sustainability. For many Landcare groups in Australia, this is self-evident and the trendy terms are superfluous. If you are trying to prevent dryland salinity due to a rising saline water table, then it is obvious that you need to co-operate with your neighbours on a catchment basis to have any meaningful long-term effect.

In a practical sense, landcare facilitation involves helping groups to make best use of the human resources available, by acting as a link person within the group and the local community, and also between the group and outside sources of information and assistance. Facilitation in the Landcare context usually also means helping to develop a shared sense of direction among the relevant actors. This requires a sufficient insight into group processes to be able to assist groups to find and set direction, to identify factors preventing the group from reaching its potential, and the skills to work through these issues with the group, dealing with conflict, apathy, collective decision making and action without imposing direction upon the group.

Facilitation is much more a matter of skilled listening, asking the right questions of the right people at the right time, than it is of delivery of technical information. This can mean challenging farmers to open their minds to new possibilities, to new ways of looking at their situation, their resources and the options open to them. Facilitators are often 'providers of occasions', organizers of encounters designed to stimulate new ideas, new ways of thinking, new perspectives or new liaisons between groups and sources of assistance. The art of fostering group synergy is delicate. It involves knowing when to lead and when to wait. It also requires empathy with the target stakeholders.

The word 'target' may jar in a discussion about something as apparently non-threatening as facilitation, but facilitation should be seen for what it is – a strategic intervention for a more or less well-defined purpose. Effective facilitators do not just sit and wait for things to happen. They think, they anticipate, they plan and they act, with the emphasis on action, rather than reaction. Although most Landcare facilitators in Australia are submerged under demands from many groups over large areas, they best cope with this pressure by being even more strategic in their interventions.

Anna Carr (1994) notes that facilitators were a critical factor in the success of Landcare groups with whom she carried out detailed case studies, not least by acting as 'bureaucracy busters'. Seeking technical information, clarifying regulations, diminishing the administrative workload of the group, linking groups with schools, advising on project submissions, providing a pathway for government to enlist group support in activities such as farm planning workshops or developing a rural strategy, were all important contributions made by facilitators. Based on her case studies, Carr listed the following skills as being necessary for 'good' facilitation; communication, animating, conflict resolution, public relations, networking, planning, educating and organizing.

There is a fine line between helping a group to get themselves organized, and becoming pivotal in the achievement of the group to the point where if the facilitator leaves, the group fizzles out. Good facilitators tend to work themselves out of a job, withdrawing as groups become self-reliant. As facilitators' performance improves, their interventions become more subtle, and the groups with which they work become more independent.

Relations between the new breed of facilitators and traditional agricultural extension staff are not always smooth and complementary. From the perspective of the archetypal extension worker (a 45-year-old agricultural scientist, ex-footballer with a beer belly, well known for his ribald jokes and detailed knowledge of crop varieties and herbicide mixes), Landcare is a fuzzy, mildly threatening notion. Initially seen as yet another reflection of trendy environmental concerns in the cities, and cynically interpreted as political vote buying, Landcare for traditional extension staff can be a crystallization of unwelcome change.

Picture the new, young, nationally funded Landcare facilitator, with her training in geography (and unabashed ignorance of herbicides – 'she can't even drive a tractor!'); her new nationally funded car and relatively generous travel budget; her membership of the Wilderness Society; her feminism; and her frequent trips away for training courses in 'group dynamics', 'conflict resolution', 'media skills' and 'community consultative processes'. When she arrives in their region, traditional extension staff are rarely indifferent. They may welcome their new colleague as a valuable addition to meet an urgent need to service community expectations, appreciating the different skills required and the need for fresh approaches.

Or, they may resent the resources allocated to Landcare and its implied criticism of the traditional focus on agricultural production, disparaging the lack of experience, technical knowledge and practical capabilities of the new breed called facilitators.

This contrast is a caricature, of course, highlighting two extremes of a well-populated spectrum. There are many government extension staff willing to acknowledge the poor environmental record of conventional agriculture, who recognize that effectively tackling sustainability problems on a regional scale means more than transferring technology to leading farmers. Even for these staff, however, the contrasts between their own training and skills and the skills and insights required to help a diverse community group understand a complex environmental issue, to develop ownership of this issue and to take collective responsibility to try to resolve it, can be unsettling. For staff who prefer to regard environmental concerns in agriculture as peripheral, it is easier to criticise Landcare as 'populist' and 'touchy–feely' than to confront the changes required to adopt new extension approaches, let alone to question extension itself, or, heaven forbid, the role and nature of science.

13.6 Issues confronting would-be platform builders

There are certainly circumstances in the Landcare facilitation role which are uniquely Australian. However, landcare facilitators also face everyday issues which are of universal relevance to people wishing to encourage and sustain new social institutions for improving resource management. Who are you, with whom are you working, and for whom/what are you working? These are some fundamental starting questions and on-going concerns for anyone working in a platform facilitation role. In any situation where natural resource management is seen to be problematic there are likely to be a range of stakeholders with multiple, partial perspectives. Clashes of values and interests between stakeholders with a vested interest in the status quo and those agitating for change are inevitable. Furthermore, in situations where the would-be platform facilitator is employed and/or funded by an institution, there are potential conflicts between that institution and other stakeholders.

One of the first challenges for facilitators is to be clear, first to themselves and then to other stakeholders, about who they are working for, and why. It should then be possible to carry out an initial stakeholder analysis, to identify those social actors who need to be involved in the platform if it is to be successful in improving natural resource management. This process is never finished, but in its initial phase it should start to scratch the surface of conflicting values and interests, which leads to the next thorny question.

What do you do when you realize that one of the powerful vested interests in a resource use conflict is the agency which is employing/funding you? And the reason it is funding you is to work towards its own ends under the rubric of 'bottom-up', 'community-based', or 'grass-roots driven' approaches – more sophisticated, less heavy-handed intervention than traditional methods which have proven to be ineffective, but intervention nevertheless. The interests of other stakeholders, landless farmers or non-agricultural land users for example, may be inimical to those perceived by the ministry or NGO funding the facilitator, yet the rhetoric of the contract/project may glibly refer to 'community empowerment' as a goal. When push comes to shove, whose views prevail, whose interest does the facilitator take up? Who gets to sit around the table?

There is no such thing as a neutral, detached, value-free facilitator. It is ridiculous to pretend otherwise. One must make one's values explicit. Anything less undermines the chances of facilitating meaningful dialogue among participants, let alone subsequent changes on the ground. If the values and agenda of the facilitator are not reasonably coherent with the ultimate goal of more sustainable natural resource management or a more equitable sharing of resources or whatever, then the notion of platform facilitation tends more towards manipulation of stakeholders, or, as one Australian farmer put it: 'You already know where you want us. All this talk of facilitation and self-determination is just designed to help us decide to do what you wanted us to do all along. I reckon it should be called *facipulation*, not facilitation'.

A facilitator needs then to be quite open and frank in this situation. Most stakeholders accept that everyone has an agenda of some sort, and it is best to make this explicit. If an actor is willing to be a facilitator, or to sponsor a facilitator, it is always for a reason. In Australia, some Landcare facilitators describe themselves as 'the voice of the catchment', speaking and acting explicitly in the public good interests of sustainable use of land, water and biodiversity, while accepting that the farmers in the catchment will be acting in the interests firstly of their individual farms and their families, before altruistically considering other people, other species or future generations. Where these interests clash, or where knowledge is clearly inadequate, facilitators are likely to be more credible if everyone is clear where they stand. For facilitators who are 'insiders' rather than people imposed from an external institution, it is even less feasible to act, and be seen to act, in a neutral, value-free way.

13.7 Technical competence and/or people skills

An issue which has emerged in Australia and will probably do so in other contexts, is whether the facilitator can afford to concentrate solely on group process, on resource use negotiation and conflict resolution processes, without any technical background in natural resource management. In other words, can one survive on facilitation process skills alone, or is some content expertise necessary? Australian experience suggests that it is very difficult for facilitators to conceive or create occasions or encounters during which interaction, learning or synergy among stakeholders might occur, if the facilitator's technical appreciation of context is not sufficiently fine to recognise or contrive opportunities. Similarly, a stagnant or stalemated situation can often be overcome with the introduction of information, expertise or assistance from outside, and it helps if the facilitator has sufficient technical nous to know what to ask for, where to look for it, and how to distinguish between useful information/people and clever packaging.

It is quite feasible to develop facilitation competencies among people trained in natural resource management, just as it is possible for people trained in the social sciences and community development to become ecologically literate in a particular context, as long as facilitators are humble about their expertise, aware of gaps in their knowledge and keen to learn. It is no more tenable to assert that 'I am an expert facilitator, I have no need to know anything about farming systems/ soil conservation/ agroforestry', as it is to say 'I am highly qualified to develop solutions to water quality/ soil erosion/ wildlife habitat issues, so I am competent to work with community groups in this field.' It is important, albeit difficult, to maintain a balance. Anna Carr notes that, where facilitators were expected to provide technical support as well as facilitating community groups, the former role tended to take over. The facilitator was drawn by their technical role into giving one-to-one advice to individual group members, thus losing sight of the quality of relationships *among* group members. When the facilitator moved on, so did group cohesion (Carr, 1994).

13.8 Facilitation is not always the most appropriate intervention

Individual facilitators and institutions, such as natural resource management agencies which may employ facilitators, both need to be acutely aware of the nature of facilitation as one policy instrument among many, the effectiveness of which depends on the context in which it is applied. Appreciating the limitations of facilitation and being able to relate these to particular contexts is a prerequisite for knowing when to withdraw or to choose alternative policy instruments.

Australia has more than ten years' experience with community-based participatory planning processes for natural resource management on private land, for example, through the Victorian Salinity Program, in which dozens of salinity management planning committees[2] were formed in the early 1980s with government support and encouragement and a firm commitment to a bottom-up approach. This programme was predicated on the assumptions that the complex problem of salinity demands a coordinated approach over large areas, that those closest to the land have a key role in implementing salinity management plans, and therefore their involvement and support is critical. Reviewing this experience, Roger Wilkinson and Neil Barr (1993) suggest that there are two key factors which point to the likely effectiveness of a participatory community planning process: the nature of the mea-

[2] These were classic platforms in the sense of new social institutions designed around a resource management issue. The basis for delineating the catchments were groundwater aquifers, rather than surface drainage basins and thus the salinity management groups bore only coincidental relation to existing social communities or administrative units, and were not even as intuitively accessible as the idea of co-operation among all the actors in a particular valley, or along a particular river. To 'belong' to such a committee required a leap of faith that the hydrogeologists knew what they were talking about, that to manage surface salinity required integrated planning and coordinated action over groundwater provinces, units which cannot be readily seen or appreciated by the average land user or voter.

Table 13.1 *Factors influencing the likely effectiveness of participatory planning*

	Solutions	
Problems	Voluntary	Compulsory
Salient	The 'dream run' plan. Participatory planning is likely to be constructive	Usually a 'grand plan' which most people desire, but not everyone will agree with the outcomes, especially if significant costs are involved
Not salient	'Someone else's plan' most people unlikely to participate without material incentives	The 'Danger Plan', very risky, likely to go off the rails, community support is unlikely, regulation or market-based incentives may be more effective

Source: From Wilkinson & Barr, 1993.

sures arising from the process, whether voluntary (up to the individual land user to adopt or not), or compulsory (one in, all in, such as community drainage schemes); and the nature of the environmental threat – is it seen as a salient personal management problem by local people or is it mainly a worry for the experts and cognoscenti? In combination, these criteria provide useful insights into the influence of context on the potential outcome of participatory planning processes, as outlined in Table 13.1.

There are close parallels here with platform facilitation. Participatory approaches are context dependent. As much as possible should be known about the context, as early as possible in the process, especially by anyone intending to facilitate resource use negotiation among relevant stakeholders. The salience of problems and the nature of possible solutions are not the only criteria for judging whether a participatory approach involving platform facilitation may work. Pretty and Chambers (1994) contend that more is required than the adoption of more participatory approaches for lasting improvements to occur. They suggest that three areas need to be improved and interwoven: supportive institutional contexts, new interactive learning environments for professionals and rural people to develop capacities, and new methods for partnerships, dialogue, participatory analysis and sharing.

Pretty (1993) cautions that the term 'participation' has been used to justify extending the control of the state and to justify external decisions, just as it has been used to devolve power and for interactive analysis. Participation (or at least the rhetoric of it) has often been used in the past, in particular during 100 years of soil and water conservation projects, in most of which farmers have been advised, paid and/or enforced to adopt conserving practices. However, this 'controlling participation', despite impressive physical results in the short term, has a sorry long-term record, as local people usually have no stake in, or commitment to, long-term maintenance and management of soil conservation works. Guijt (1991), reviewing the efforts of 230 rural development institutions employing 30 000 staff in 41 African countries, found that external agencies rarely permitted local groups to work alone. They usually controlled all the funding, and participation was more likely to mean simply having discussions or providing information to external agencies. Functional or interactive participation (Pretty, 1993) was rare.

Facilitators are often caught between what seems to be an irresistible force and an immovable object. Where facilitators perceive that their own values and interests, or those of key stakeholders in a problematic natural resource issue, are at odds with the institution which employs them, they have fairly clear choices: (i) ignore the incompatibilities between their employer and its 'clients' and try to manage as best as possible – the 'ostrich option'; (ii) try to tease out and articulate the agenda of the various parties (including the facilitator), identifying common ground, negotiable areas and irreconcilable differences, then start from common ground to try to facilitate change *within* the employing institution in order to foment synergy among other stakeholders, consistent with the expressed goals of the facilitator; or (iii) withdraw.

Option (i) is a recipe for mounting stress and conflict and is inherently unsustainable. Option (ii) is preferable in terms of the potential for improving a problematic situation, but it is also guaranteed to generate stress and conflict, albeit with more chance of an occasional win than the ostrich option. Option (iii) may well be the most sustainable option and should not be dismissed lightly. The trick is to be able to tell the difference between the latter two situations, to know when facilitation is unlikely to be effective (it is by no means appropriate in every situation) and to be able to retire gracefully, in good shape to fight another day.

13.9 Dealing with burnout – tools of the trade and where to find help

A would-be platform builder has decided that their facilitation skills and practical savvy can help a collection of stakeholders to achieve synergy in improving natural resource management, and has become committed to this end. An issue that will almost surely emerge is that of 'burnout'. Burnout occurs when someone has given of themselves to the extent that their physical and mental well-being, and their relationships with family and friends suffer. This is a major issue among Landcare group leaders and facilitators in Australia. Maybe a good facilitator should be able to facilitate themselves out of burnout situations, but this is rarely the case. Like the mechanic's car or the builder's house, facilitators' stress levels are often not their own best advertisement.

In coping with burnout, just as in cases where facilitators feel an inconsistency between their employer's goals and their facilitation role, horizontal networks and contact with other people who are or have been in similar situations can be a lifeline. Facilitators are unlikely to find much sympathy in the formal hierarchies of government or even NGOs. They are more likely to get support and valuable feedback and ideas from their peers – people working in similar situations. Network maintenance is a crucial competence for facilitators, both for dealing with crises, and for improving everyday effectiveness, learning tips and contacts from people working in similar roles.

While many facilitators feel that they are breaking new ground, pioneering a new role for which there are no guidebooks, in fact, many people have struggled along similar paths before. Many of the facilitation techniques used by Landcare people in Australia are derived from community empowerment experience in the ghettos of Chicago in the 1960s (Spencer, 1989). The community development movement of the 1960s and 1970s has many lessons for participatory approaches to resolving the environmental issues of the 1990s (Warner, 1989, cited in Carr, 1994). Finally, the *Farmer First* (Chambers *et al.*, 1989) and *Beyond Farmer First* (Scoones & Thompson, 1994) movements in agricultural extension and research have stimulated and fostered a body of published experiences, case studies, and theoretical and methodological discussion. The RRA Notes (now PLA notes – Participatory Learning and Action) published by the International Institute of Environment and Development (IIED) in London are a valuable, accessible source of insights and practical tips from practitioners all over the world.

13.10 Old wine in new bottles?

A brief aside to emphasize the point that few things are really new. The Transfer of Technology (ToT) model of rural extension has taken a battering in recent years from Farmer First advocates and their descendants, starting perhaps with Chambers and Jiggins (1987). This model is widely seen to be derived from diffusion of technology theory arising from American rural sociology, notably Rogers (1962). Yet, American rural extension in the 1960s already had a participatory discourse which finds many echoes in the 1990s. At the 1962 Australian Agricultural Extension Conference in Melbourne, A.H. Maunder, then Chief of the Foreign Education Branch in the Division of Extension Research and Training of the United States Department of Agriculture Extension Service, quoted the following definition of extension education:

> The process of extension education is one of working with people, not for them: of helping people become self-reliant, not dependent on others; of making people become the central actors in the drama, not the stage hands or spectators.

Maunder (1963) went on the describe how the US Cooperative Extension Service had started with top-down (ToT) extension, but had found that farmer adoption of their recommendations was patchy. Then,

> Thousands of farmers and their wives gathered in small groups around kitchen tables and in school houses to tell the extension workers of their wants and needs. This was a significant step forward. It aroused the interest of great numbers of people never before reached.

This sounds very like some expressions of participatory inquiry and has echoes of Landcare. Maunder notes that this approach after some years led to burn-out, to extension staff spread too thinly, with many unrelated projects, responding to local needs but lacking a strategic focus. The pendulum then swung back to a more instrumentally rational approach, before moving on to other emphases (such as marketing, differences between rural and urban living, world competition, demographic trends, technological change) in response to the times (Maunder, 1963). Finally, Maunder summed up the state of the art of agricultural extension as he saw it in 1962:

> [through extension programmes] it is hoped to assist rural people to adjust to the ever-changing situation. Agricultural adjustment is becoming an increasingly important element of extension education. The progressive, democratic approach assumes that the soundest programmes result from a fusion of the ideas of many people – extension workers, technical specialists, farmers, and many others. It recognises that programmes, to be effective, must be geared to the roots of local problems and that no one can know the true character of those problems better than the one who has them. It implies going on the offensive and it involves leadership in pointing the way. It recognises that professional people must lead their clientele to make up their own minds on the most useful targets or objectives and not make their decisions for them.

Not much seems to have changed in 30 years. Comparing this quote with modern tenets of participatory inquiry (for example, Pretty, 1993), it certainly picks up the need for seeking multiple perspectives in local contexts to tackle local problems. The role of the professional as someone helping local people to help themselves is also common to the view from the early 1960s and the 1990s, although the former is stronger on leadership and the latter emphasizes facilitation. Viewed in context, however, they probably mean the same thing, as facilitation did not have the currency in 1962 that it enjoys today. As to the need for the defined methodology and systematic learning process identified by Pretty (1993), Maunder (1963) elaborates a detailed planning framework including criteria for acceptance of jointly developed solutions. He lists 12 propositions about effective extension programme building, including: 1. The central purpose of extension programmes is to promote socially significant learning, 7. People representing major economic and social levels must be deeply involved in analysing and arriving at decisions about the problems, needs and interests in an area, 8. To define a rural problem, one must bring to bear both scientific and folk knowledge, and 9. The programme building process is a teaching technique. Phrases such as 'socially significant learning', the point about combining 'scientific and folk knowledge' and the notion of the process as a learning exercise in and of itself, are all entirely consistent with modern interpretations of participatory inquiry and platform facilitation.

13.11 Final points

Responding to environmental issues is a rapidly evolving field of human endeavour, a sure bet growth industry for the 21st century, and both 'soft' and 'hard' technologies will turn over often. The above quotes are a reminder that there is much to be learned from the past, and that which seems radical or cutting edge today may well have been old hat somewhere else, some other time.

Facilitation is a critical process to establish and sustain platforms for resource use negotiation and to capture some of Niels Röling's postulated synergy. It must be stressed again that good facilitation is a necessary, but not a sufficient condition for platform effectiveness. Involving the community can be time-consuming and frustrating, and it is scary for people who are not naturally disposed to dealing with people and/or have not had relevant training (Campbell, 1992). Seen through the prism of technocratic institutional cultures, involving a range of stakeholders in an ill-defined, open-ended facilitation process is tedious, its outcomes are often intangible and its cost/benefits debatable. The complexities of developing new ways of using the land which meet environmental, social and economic objectives, however, mean that genuine stakeholder participation in generating, using and exchanging knowledge, in decision making, and in resource negotiation, simply cannot be side-stepped or fudged.

The role of the facilitator is critical in bringing relevant actors to the platform, in attempting to develop a shared problem appreciation among different actors, in exposing and building areas of agreement and resolving or mediating conflict, in sustaining involvement and assisting processes for marshalling information, negotiating and making decisions. This is clearly a value-laden, political role. Facilitators need to make explicit where they stand on the issues at stake. Being a successful facilitator often means being able to bring about changes in attitudes, processes and organizational cultures within the employing institution.

Over the longer term, survival as a facilitator necessitates a fine understanding of power. It also demands support networks of peers and reference material to help facilitators deal with the inevitable onset of burn-out. Ultimately, facilitation of multiple stakeholders for natural resource management negotiation is likely to be most effective in a sympathetic institutional context, in harness with complementary policy instruments such as regulation, economic (including market-based) incentives, education and training, in situations where problems are salient and consensus solutions feasible. As noted by Wilkinson and Barr (1993), there is no magic button for eliminating conflict. Facilitation may enhance stakeholders' appreciation of the perspectives and values of others, and may open the way for genuine dialogue and new responses to environmental issues. This is a worthwhile start.

13.12 References

Campbell, C.A. (1992). *Taking the Long View in Tough Times – Landcare in Australia.* p. 148. Canberra: National Landcare Program.

Campbell, C.A. (1994*a*). *Landcare – Communities Shaping the Land and the Future.* p.344. Sydney: Allen and Unwin.

Campbell, C.A. (1994*b*). Participatory inquiry – beyond research and extension in the sustainability era. In *Proceedings of the International Symposium on Systems-Oriented Research in Agriculture and Rural Development.* Montpellier.

Campbell, C.A. (1995). Landcare – participative Australian approaches to inquiry and learning for sustainability. *Journal of Soil and Water Conservation*, Iowa. March/April: 125–31.

Carr, A.J.L. (1994). Grass-roots and green tape: community-based environmental management in Australia. PhD thesis, Australian National University, Canberra.

Chamala, S. & Mortiss, P.D. (1990). *Working Together for Land Care*, University of Queensland and Queensland Department of Primary Industries. Brisbane; Australian Academic Press.

Chamala, S. & Mortiss, P.D. (1992). *Training the Trainers: A Landcare Training Manual*. Brisbane: Australian Academic Press.

Chambers, R. & Jiggins, J.L.S. (1987). Agricultural research for resource-poor farmers; Parts I & II. *Agricultural Administration and Extension*, **27**, 35–52, 109–28.

Chambers, R., Pacey, A. & Thrupp, L. (eds.) (1989). *Farmer First – Farmer Innovation and Agricultural Research*. London, UK: Intermediate Technology Publications.

Cock, P. (1992). Cooperative land management for ecological and social Sustainability. In *Agriculture, Environment and Society: Contemporary Issues for Australia*, ed. G.Lawrence, F.M. Vanclay & B.Furze, pp. 304–28. South Melbourne: Macmillan.

Cocks, K.D. (1992). *Use With Care: Managing Australia's Natural Resources in the Twenty-first Century*. Kensington, Australia: New South Wales University Press.

Curtis, A., Tracey, P. & De Lacy, T. (1993). *Landcare in Victoria: Getting the Job Done*. The Johnstone Centre of Parks, Recreation and Heritage, Charles Sturt University, Albury, Australia.

Funtowicz, S.O. & Ravetz, J.R. (1993). Science for the post-normal age. *Futures*, **25**, 739–55.

Guijt, I. (1991). *Perspectives on Participation*. London: International Institute for Environment & Development.

Habermas, J. (1990). *Moral Consciousness and Communicative Action*, Translated by C. Lenhardt & S. Weber Nicholson. London: Polity.

Maunder, A.H. (1963). Extension programme development. *Proceedings of the Australian Agricultural Extension Conference, Melbourne 1962*, pp.206–17. Melbourne: CSIRO.

Pretty, J.N. (1993). Participatory inquiry for sustainable agriculture. Unpublished monograph. London: IIED.

Pretty, J.N. & Chambers, R. (1994). Towards a learning paradigm: new professionalism and institutions for a sustainable agriculture. In *Beyond Farmer First – Rural Peoples' Knowledge, Agricultural Research and Extension Practice,* ed. I. Scoones & J. Thompson. pp. 182–202. London: IT Publications.

Reeve, I.J., Patterson, R.A. & Lees, J.W. (1988). *Land Resources: Training Towards 2000*. Armidale: Rural Development Centre, University of New England.

Rogers, E.M. (1962). *Diffusion of Innovations*. New York: The Free Press.

Scoones, I. & Thompson, J. (eds.) (1994) *Beyond Farmer First – Rural Peoples'*

Knowledge, Agricultural Research and Extension Practice, pp. 301. London: IT Publications.

Spencer, L. (1989). *Winning Through Participation*. Iowa: Kendall-Hunt.

Warner, P.D. (1989). Professional development roles. In *Community Development in Perspective*. ed. J.E. Christenson & J.W. Robinson Jr. Ames: Iowa State University Press.

Wilkinson, R. & Barr, N. (1993). *Community Involvement in Catchment Management: An Evaluation of Community Planning and Consultation in the Victorian Salinity Program*. Melbourne: Department of Food and Agriculture.

14 The implementation of nature policy in the Netherlands: platforms designed to fail

M. WAGEMANS AND J. BOERMA[1]

14.1 Introduction

'God created the world, but the Dutch created Holland'. Many readers will be familiar with this saying, but they might not have thought of its implications for the country's 'natural heritage'. In Holland, nature is also man-made. The seaboard marshes of the Rhine and Meuse deltas have been drained and under cultivation since the Middle Ages. On the slightly higher land to the east and south, the last virgin forest was cut down in the late eighteenth century. What's more, the natural ecology of flood plains, river meadows, silt deposits, etc., that is, the kind of nature which 'belongs', can regenerate fairly rapidly. One of the largest nature reserves, a pristine wetland with colonies of spoon bills and other rare birds, is the result of draining the Yssel-lake in the 1950s. In other words, nature, 'real' or 'wild', exists only by design.

Most Dutchmen have a conception of 'nature', however, which features meadows, farm lands, hedges, woodlots, drainage canals, old (pollarded) willows, all richly populated with meadow birds, otters, storks, song birds, amphibians and masses of wild flowers. Many cannot imagine nature without black and white cows. In other words, for most Dutchmen, nature is what the countryside used to look like until about 1960.

Modern farming has destroyed all that. There is no need to go into the depressing details here. It is easy enough to imagine what a circle mower going at 30 km an hour through the high grass in early May will do to the nests of peewees, curlews, godwits, redshanks, ruffs, reeves, and larks, and what a good dose of herbicide and a surplus of nitrates and phosphates leaching from over-fertilized fields will do to the erst-while teeming wildlife in a field ditch. Add the good work of the Land Development Agency, which has invested millions of tax guilders into straightening canals and roads, enlarging fields, getting rid of old trees and other impediments to modern farming, and one can easily imagine the virtually complete destruction of what most Dutch, taught to see by Rembrandt, de Hooch, Ruysdael and many other painters, consider 'nature', but what really was an intimate, centuries-old farmscape (see also Van Woerkum and Aarts, this volume).

It is little wonder that an army of determined naturalists, ecologists, bird watchers and others have arisen who, duly armed with a degree in biology or other

[1] The authors wish to thank Mr Lambert van Gils, formerly of the Ministry of Agriculture, Nature and Fisheries for his substantial contribution to their understanding of the Fens case.

qualifications, are now 'manning the planning' in voluntary agencies involved in nature conservation, and in area-planning departments of national, provincial and local government. They have joined battle with the interests of agriculture. However, the latter are not prepared to give up easily. After all, the farmers who are still left have survived a market-driven gruelling selection-of-the-fittest contest which has seen their numbers dwindle from 250 000 in the early 1960s, to perhaps 60 000 full-time stayers at present.

The nature lovers, however, together with urban recreational interests, have made substantial progress and the political climate has shifted from unquestioning support for agricultural development in the late 1970s to present animosity. The Nature Policy Plan, ratified by Parliament, calls for the creation of 'nature', roughly defined as 'no farming' or severely restricted farming, on thousands of hectares of Dutch countryside in order to establish the 'ecological infrastructure' to link scattered biotopes and to provide space for bio-diversity. To this end, certain areas have been designated as special area-based planning projects, with dedicated budgets and a special decision making structure which is supposed to allow for the involvement of local stakeholders and for voluntary change. This set-up has basically led to an impasse, with farmers on the one side, and nature, environmental and recreational interests on the other, locked in interminable negotiations (van Woerkum and Aarts, this volume). One can look at the various fora which have been created for the planning exercise as 'platforms for resource use negotiation' (Woodhill and Röling, this volume). But unlike Landcare in Australia (Campbell, this volume), the conditions for success seem absent.

The first part of the chapter, then, describes a typical area, the Fens (de Venen), and the issues, stakeholders and battles involved. The second part, from paragraph 14.8 onwards, provides an analysis, based on intimate experience with, and responsibility for, such an area.

14.2 The project 'the Fens'

The project area was a fresh-water swamp in the Middle Ages, well protected from the sea by a row of dunes. Since the 11th century, it has been drained by numerous windmills and reclaimed as a large pasture area of ± 20 000 hectares suitable for dairy farming. The area, now encircled by a fringe of cities (Rotterdam, the Hague, Utrecht, Amsterdam and others) is an indelible part of a number of large-scale strategic national plans: the Green Heart of the urban conglomeration, the Blue Network for water recreation, the Nature Policy Plan, the Fourth Extra Spatial Planning Paper, and the National Environmental Policy Plan. In all of them, the Fens are designated as a nature development area. The cost of the project is estimated at 350 000 000 guilders (US $200 million) over 25 years (Stuurgroep Groene Hart, 1992; Ministerie LNV, 1992; Tweede Kamer, 1989/1990).

The project plan calls for a radical change in the function of the area, with a large extension of some 3500 hectares (17.5%) of the area for nature development to be used for recreational purposes. This natural area will connect the Vecht lakes with the Nieuwkoop lakes into one large reserve. Over the 3500 ha, the water level will be raised so as to allow the land to revert to swamp.

In a buffer zone around this new nature area, as well as in other parts of the Fens, farmers are asked to sign a pasture management contract. They may choose among different 'management packages' (from heavy to light), e.g. not fertilizing the sides of ditches, or postponing the first cut of grass until young birds are able to escape. Depending on the package, farmers are compensated at standardised rates for their loss of revenue. Water levels in the buffer zone will be raised by 10 cm. High-water tables are essential for meadow birds, but are expected to also make the area more attractive for recreation.

14.3 Some essential characteristics of the area

The project area has some specific attributes which have consequences for the use of the land in the past, now and in the future. The soil consists mainly of peat, the remains of vegetation deposited in wetland anaerobic conditions over thousands of years. Peat has been used for various purposes since Roman times. Properly dried and cut, it was an important source of energy for heating and cooking. As a result, the peat has been removed down to the clay in a number of areas. In others, the land was reclaimed for farming by lowering the water table by windmills. In such areas, especially when the lower soil is sandy, as in the Fens, the peat layer has gradually diminished (probably by some 2.5 metres since the 11th century) through mineralization as a result of exposure to the air (Driessen & Dudal, 1989). Mineralization releases nutrients making peat soils very fertile.

Mineralization, however, also means a lowering of actual ground levels. Since the water table has to be kept at 60 cm below ground level for optimal modern dairy farming, the Water Management Associations, which until recently served only farmers' interests, have systematically had to lower the level at which water tables are maintained. It takes increasing amounts of energy to pump the excess water out. Nowadays, wind mills would not be enough. The ground level in the polder (water management area) Groot Mijdrecht Noord, for example, which is part of the Fens, is now 6 metres below sea level. This means that the pumps must have a high capacity. Since the peat soils here lie over salty sands, the groundwater contains salt which has a negative influence on the vegetation. The future use of these lands is thus endangered by high energy costs and low soil quality. The Central Board for Land Reclamation has now decided that the government will no longer subsidize the lowering of water tables in peat soils.

14.4 The dominant types of land use

The project area has several uses. The two most important, dairy farming and 'green' open-air recreation for the urban agglomeration, are strongly inter-connected. The meadows for the cows provide virtually all of the green space.

As we noted before, the ideal water table for dairy farming on peat soil is

60 centimetres below ground level. It gives the best quality grass and the soil structure will not be destroyed by grazing, mowing or fertilizing. Yields can be very high because of the high soil fertility. The farmers have no drought problem in the summer, but wet peat soil is vulnerable. Once the upper layer with grass has been destroyed, walking on it becomes difficult. During autumn, winter and early spring, cows are kept inside. In spring, after two or three mowings to gather grass for the next autumn and winter, and when the upper layer is dry enough, the cows are brought to graze outside. When the water table is not low enough, the heavy tractors which the farmers use for mowing and fertilizer application destroy the soil. That means they have to wait for the water table to fall, which might cost them a cut of grass. Thus the introduction of heavy machinery since the 1960s has acerbated the problem by requiring lower water tables than before and hence causing more rapid mineralization of the peat. The system is clearly unsustainable.

The intensification of agriculture in the area, and especially the very intensive use of manure and nitrogen fertilisers, has led to a mono-culture of a few varieties of grass and a sharp reduction in bio-diversity. There is clearly something 'wrong' with a farmer if one can observe flowers in his meadows.

The Fens have an important 'green' recreational function for the population of the urban fringe. Bicycles and boats are basic means of transport at weekends, holidays and long summer nights. The views across this flat, but intimate and still diverse landscape are very pleasing. The grasslands are dotted with lakes, small rivers and canals, some of which are already protected nature reserves. As wetlands in a delta with a Europe-wide eco-system function, they are of great importance for birds and other wildlife.

The two different functions of the area are coupled to different water tables. Farmers need a low water table, whereas wetlands need a high one. It is therefore impossible to serve both functions in one polder.

14.5 Stakeholders' perspectives

14.5.1 The farmers' perspective

The ground water table is clearly a central concern of the farmers. Productivity depends on it. Yet a farmer has no individual control but is dependent on the Water Management Association running his/her polder. These Associations are among the oldest structures for governance in the Netherlands. Farmers for many hundreds of years ran their own water affairs, deciding on their own levies, the necessary pumps, the required water table, etc. Farmers elected the members of the boards of the Associations. Now, the composition of the Associations has been changed in recognition of the multiple purposes of farm land. Since 1993, all inhabitants of a polder pay for the services and elect the board members. The farmers have lost their monopoly. What's more, the new set-up of the Associations allows nature conservation organizations also to have a say. Farmers feel that they are no longer the managers of their own land.

When the decision is made to raise the water table so as to create more nature, farming with present techniques is no longer possible. Farmers will have no choice but to sell their land. Given that such drastic measures are proposed, farmers cannot believe in voluntary change and participation. In addition to solidarity with their colleagues threatened by inundation, farmers who will be adjacent to the new swamp areas are afraid of damage by foxes, coots and weeds. In all, 'participation' in this project has been reduced to a formality, and voluntary selling of land to a farce in the change process.

The government is proving an unreliable partner. The way it is carrying out this project has reinforced its unreliable reputation. It did not involve the farmers, the most direct stakeholders, in the design of measures, nor did it try to share the project goals with them. The possibility for collaboration on nature development was ruined at the start of the project, when it turned out that not only the goals, but also the concrete measures had already been determined. Briefly, the government determined that 3500 hectares were to be turned into new nature. But can nature be expressed in a certain number of hectares? The project documentation does not clarify what is meant by 'nature'. Is it a swamp? Or does a bird's nest on a farmer's land also qualify?

The farmers argue that they can produce 'nature', for example, by protecting bird's nests, if they are paid to do so (see also Roux and Blum, this volume). To encourage farmers to produce nature is a far cheaper policy option than a huge project like this one, especially if one takes the maintenance costs into consideration. In addition, if swamps are not well maintained, they can become an eyesore, and cause annoyance to the farmers adjacent to the project.

The farmers in the project area feel that their interests are not defended strongly enough by the traditional farmers' organizations. The latter have learned to negotiate with national, provincial and local authorities in terms of slowing down or cutting concrete measures, while accepting, or at least not discussing, the goals, and not obstructing projects or developing entirely new plans.

The farmers and others who live in the area have taken matters into their own hands and are formulating goals and measures that oppose those of the project. They have organized into two pressure groups, 'the Agrarian Platform' and 'the Viable Fens'. These two parties are important channels for defending the interests of the direct stakeholders (see also Section 14.6.2).

14.5.2 Local government perspective

Local authorities have less uniform interests. They are represented on official committees involved in implementing the plans (see below), but are obliged to carry out national government policies. Their perspective is based on budgetary considerations and on local politics (local governments are elected). What is important for them is the impact on their own area of jurisdiction and the financial compensation.

Local authorities have not been involved in setting project goals, but they have signed the letter of intent, expressing willingness to help realize the policy's goals. They have little influence on the proposed measures, yet the impact of these measures on council elections can be significant.

The various community authorities are co-operating in the negotiations, although their interests seem similar. The reason for this is that the 3500 hectares are to be divided over the whole project area. Most of the villages are trying to get as few 'nature hectares' as possible included in their jurisdiction. Each focuses on his own territory. But, as representatives, they have little authority to negotiate, and they have no voice in the definition of nature. They remain focused on what happens in their own village and what financial arrangements can be obtained. This stance is a block to the formulation of new ideas and solutions.

14.5.3 Provincial government perspective

The provincial authorities seem to have chosen to support nature development and conservation moves. The need to develop more nature in this part of the Netherlands has become for them an unshakeable conviction. It arises from the rationale that underlies national policy, the ecological infrastructure and the Green Heart. The provincial authorities look at the project from the national perspective, shaped by the need for nature conservation in this densely populated country which straddles the Rhine Delta with its international wetlands protection function.

The provincial governments thus have a stake in the measures which are to be taken to develop nature. This means that there is not much tolerance among them for accommodation with other perspectives. Several actors have the feeling that provincial governments are not providing all available information. They suspect them of trying to avoid the commotion that could result from providing 'premature' answers to stakeholders' questions. The local stakeholders are only informed about the final results of the project staff's internal negotiations. The working groups, in which local stakeholders had some chance to defend their interests, were given some say in the implementation of the measures. The measures themselves, however, had already been decided and were not subject to advice. However, the provincial authorities underestimated the active resistance that the proposals have created.

Clearly, the presumed collective long-term national interests are in conflict with interests as perceived by farmers and other local stakeholders. Provincial governments, having chosen a defined position, are hardly in a position to foster a compromise among the different parties.

14.5.4 Environmentalists' perspective

The space for nature development in the Netherlands is very small and fragmented. There is some doubt about the effectiveness of scattered small reserves. However, the Fens, as part of the ecological infrastructure, seems to offer an opportunity to divert a significant amount of land from agriculture into a large reserve which would incorporate several existing smaller ones. Quite a few environmentalists blame modern agriculture for the loss of bio-diversity, and only see solutions which exclude the farmers. Given the present excess capacity of agricultural production, they claim that there are 'too many' farmers. It is in the collective national interest to get rid of some in favour of nature development.

Others have a more sophisticated view. The present excesses of high intensity farming are not merely a consequence of the farmers' decision making, but also of the consumers' desire for cheap agricultural products. Solutions to the present problems should therefore be sought in consultation with farmers. They know the area best and should be involved in any nature development on their lands. In the eyes of these environmentalists, nature can be more than an area from which all humans are excluded. In other words, even among the naturalists, the definition of 'nature' is not unequivocal.

14.6 The 'platform' for resource use negotiation

The 3500 hectares, the buffer zones and the funds are supposed to be divided over the whole project area in a process of change that follows the procedures of a society that prides itself on its attention to civic liberties, due democratic process, etc. To work out the plans at the provincial and regional levels and translate them to the working level, a three-tier project structure has been established.

14.6.1 The formal structure

The three levels have different functions and executive power. The *Steering Committee* is the core. Its members discuss and make all final decisions about the project, be it that these are to be ratified by the elected provincial and community councils. The Committee also has to create the social and managerial basis for change. Only government authorities are represented on this committee (three ministries, provincial and local authorities and the water authorities). No non-government organizations (NGOs) are represented. They have professional and political influence but no decision making power. The main argument for this is that NGOs have no executive power, whereas the public authorities run public services which are responsible for the public domain. Nowadays, this kind of argument is subject to political debate, as government is increasingly searching for ways to privatize its activities (Ministerie LNV, 1994). Several organizations openly lament that they are not represented on the Steering Committee (Agrarisch Platform De Venen, 1993) as they are in some area planning projects elsewhere in the country (Tatenhove, 1993).

Directly under the steering committee is the Project Committee. Its members coordinate all project activities and prepare the propositions, 'scenarios' and/or alternative solutions to be discussed and decided upon by the Steering Committee. In the Project Committee all relevant non-government organizations are represented, as well as representatives from all the communities, provinces and three ministries concerned. A Professional Unit, composed of engineers and other specialists from the public service agencies, prepares everything for the Project Committee. This unit draws up the final plans for action, based on its own professional work, the advice of the working groups, and consultations with the Project and the Steering Committee.

Under the Project Committee are three working groups, agriculture, environment and recreation. They advise on the use of the natural resources in the project area from their own sectoral point of view. Public organizations are represented in these working groups. What happens to their advice depends on what the Professional Unit is able to use and integrate into its proposals, on their representation in the Project Committee, and on their political influence.

14.6.2 The informal structure

In addition to the formal structure, several other committees and non-government organizations play an important informal role in the Fens. One of them is the 'Agrarian Platform', established by agricultural organizations and fed by the farmers' resistance to the project. It has become an important central contact for farmers and the formal project structure. The Agrarian Platform has developed alternative plans and proposals for the Fens (Agrarisch Platform De Venen, 1993).

The Viable Fens Association involves all inhabitants of the polder Groot Mijdrecht Noord. According to the project plans, this polder is to be turned into a swamp in its entirety, since it is considered to be economically and ecologically unfeasible to maintain it for agricultural use. The inhabitants do not agree.

Another important actor is the Department of Environmental Biology of Leiden University. Its thesis is that nature conservation is the work of man (Kruk, 1993). It opposes the way things are being handled in the Fens since people do not seem to have a say there. It is therefore helping the Agrarian Platform.

Landscape Conservation Utrecht is a foundation that, together with farmers, is trying to create more nature on farm land. It is interested in what farmers consider important and in what nature means to them.

The Nature Monuments foundation and the government Forest Management Service both expect to manage part of the projected nature reserve. Before the start of the project, the former published a report on the future of the Fens claiming that 7000 hectares of farm land should become a swamp. Nature Monuments never contacted any of the farmers of the area before publishing the report. Although the farmers cannot stand Nature Monuments' higher officials, the contact between the farmers and the field officers is good. According to the farmers these people know what is going on.

The Regional Agricultural Research Centre Zegveld specializes in dairy-farming on peat soils, and farmers from the region can come there with specific problems. The station has become an important link between policy and action.

14.7 The process

Because of the many uncertainties about financing the eventual management and raising of the water level of the 3500 hectares, and other key issues, the officials involved thought it better to avoid unrest by withholding information from local stakeholders. Answering their questions seemed a worse option than any potential resistance that the final plan might create. The project has taken the view that stakeholders would be informed in due course when things were certain. At that moment there would still be enough room for participation.

The lack of information and participation, however, has already caused much unrest and distrust, especially in the farming community. Farmers do not believe in the promises that the sale of their farmland will be voluntary, or in their right to participate in the project. These beliefs are reinforced by the importance of the water level. Once this is raised, the sale of their land will be the only option and not a voluntary action. Several farmers have already sold out and moved to a safer place. Others have become afraid that the area will impoverished long before they are forced to sell up. Individual farmers feel that they have little choice but to sell out now (Agrarisch Platform De Venen, 1993).

As it turns out, not informing the stakeholders has endangered the project's progress. It was a mistake to think that everything could be kept quiet until certainty existed. It was also rather risky not to guarantee the financial capacity to buy the land of all farmers involved and allow their resettlement. Much mistrust has also been caused by the fact that the representatives of local stakeholders were not allowed a say in formal decision making procedures. Admittedly, in a project of such a scale, the officials already form a huge group. 'Adding more participants would cost far too much time', according to some. 'Only government authorities are able to make the kind of decisions the committees have to make', argued others, expressing fear that due national electoral process would be endangered if local interests could alter decisions ratified by Parliament.

The net effect of the procedure chosen, however, is that opposition has created an impasse, the only option left open to other perspectives given the project structure. The impasse means that no reasonable agreement can be reached and that the project will have to follow the institutional route. Farmers can and probably will use the many options they have in that process to further delay project implementation. If one considers that the 3500 hectares offers no more than 50 farmers a viable farm, one begins to realize the immense power local stakeholders have against the combined forces of those representing the national interest. Van Woerkum and Aarts (this volume) have drawn important conclusions for policy making from this type of experience. Below we try to analyse the processes involved in more detail with a view to identifying those factors that impede arriving at 'agency at a higher level of social aggregation' (Röling 1994*a*,*b*). What happened to 'fomenting synergy'? (Campbell, this volume).

14.8 The Dutch agricultural policy actor network in shambles

Only 15 years ago, the Dutch agricultural policy network was considered both effective and efficient, and especially complimented for the close relationship between practice and policy and between practice and research, for a high degree of organisation, and for intensive consultation among the constituting actors. Foreign delegations still visit Holland to learn how it can be done, but today, the network is in shambles, as suggested by the Fens case presented above. Below, we analyse what has gone wrong and make recommendations.

Intensive interaction has always been one of the hallmarks of the policy network. Important policy decisions were concluded only after detailed consultation with, and approval from, agricultural organizations. Issues concerning research, education and information were also discussed with such bodies. The network was sustained by broad and intensive rounds of consultation at the national, regional and local levels. The degree of organization and organizational redundancy was high. The success of the sector in terms of goals, which were unquestioned at the time, i.e. productivity and competitiveness, was attributed, to a large extent, to the quality of the policy network.

Now there is increased discontent. Tensions have developed even within agricultural organizations; members do not recognize themselves any more in the positions taken by their own organization. New organizations are being formed. The network is at the point of restructuring.

To explain what has happened, we must examine the changed societal meaning of agriculture. Directly after the Second World War, the most important policy objective was food security. Access to sufficient and cheap food supported industrial development by keeping wages low. The period that followed was characterized by an enormous rise in productivity, both in a technical sense (output per unit input), and in an economic sense (lower costs of production). New technologies allowed for scale increase and for intensification. Over time, the economic function of agriculture became increasingly important, especially with respect to its large contribution to national exports and the balance of payments.

All this has changed. The success has turned into a problem. Dutch agriculture is faced with a laborious adaptation to market conditions. Societal willingness to finance chronic overproduction has evaporated and production must be cut back to conform with demand. Secondly, intensive methods of production have been accompanied by increasing damage to nature and the environment. This is no longer accepted. Nature and environmental interests carry increasing weight in shaping public opinion and political decisions. One result has been an increasing number of regulations which condition and constrain farm management.

These developments also have caused a reorientation of agricultural and nature policy and of ideas about the role of the public sector. The new government strategy emphasizes the integration of sector policies (e.g. nature, environment and agriculture), decentralized policy development and participation of the organizations and citizens involved.

A relevant question is whether the changes described ought also to have consequences for the policy network. Is it only a question of adapting policy content (targets), or must traditional policy processes and consultation structures also change? In other words: is the existing network capable of supporting the change, or is a revision of the network itself required? Until now, the actor network has not changed fundamentally and continues to function in its splendid self-evidence. It has been reproached for catching on much too late to changed public opinion, but this is a superficial explanation of why the network has reacted insufficiently to the need for change. There is, in short, a need for analysis.

14.9 Taking a constructivist perspective

Our starting point is that people act on the basis of sense-making. We ascribe meaning to the world around us, to the behaviour of others and act on the basis of definitions of reality that we construct in interaction with others. In so far as we speak of a reality, it is a reality that is interpreted. Sense once made implies reduction: it is only one interpretation of a world that is interpretable in multiple ways. At the same time such reduction cannot be avoided in action.

The constructivist perspective is important for understanding the policy network and the realisation of policy. Policy interventions are based on a concept of a problem, on a reduced perception of reality. That raises two questions.

1. What can be said about the nature of the reduction wrought by the dominant sense-making paradigm within the agricultural policy network?
2. How are differences in situational definitions between the policy network and its environment dealt with?

Both questions will be considered in relation to experiences in various regional area planning processes in the Netherlands.

14.9.1 The nature of the reduction

In the first place, the sense-making paradigm within the network is to a great extent, sectoral. This has an historical background. Until about 1975, the Ministry of Agriculture had a narrowly agricultural focus. Within the agricultural policy network, there was a great degree of agreement about goals and interests. Moreover, agriculture was (and is) sectorally organized. The result of this is that policy, and the instruments of policy, also were (and are) to a large degree sectoral.

The decision to broaden the Ministry's policy field from agriculture to an integral approach to the rural area, inclusive of nature and environmental interests, did not lead to fundamental change. Essentially, the broader mandate meant that two policy networks functioned alongside one another, agriculture, and nature and environment, both of which have a sectoral origin and orientation. Integration did not take place and was hindered by the fact that sectoral sense-making paradigms were employed: a sectoral policy network meets the world sectorally. Questions are reduced to sectoral questions, and policy is consequently to a large extent the summation of a sectoral agricultural policy and a sectoral nature policy (Box 14.1).

Box 14.1 Integrated policy review from a sectoral perspective

A policy brief in which an integral approach for a certain area was worked out, was reviewed from sectoral perspectives. What was favourable for the sector in question could count on approval, what had neutral impact was acceptable, and what had an unfavourable impact was rejected.

The sense-making paradigms that dominate agricultural and nature policy differ markedly. Agriculture is seen traditionally as an economic activity in which income must be earned through the market. The objectives of nature, on the other hand, can only be realized through intensive government interference. But they differ not only in terms of the driving mechanism and the role of the government. The interests, and thus the definitions of reality, also diverge (Box 14.2).

Box 14.2

As agriculture became more intensive, the negative effects for nature and the environment increased. What is favourable for nature often leads to losses in yield for agriculture. The sectoral perception of agriculture is that biodiversity means weeds, pests and disorder. The sectoral perception of the environment is that productivity means habitat destruction, loss of diversity and pollution.

The role and position of government presume a world that can be fashioned or at least regulated. Without the power to intervene, a government can fulfil its function only with difficulty. Every government intervention is thus also based on the power to influence. Further, every government intervention in its turn must satisfy a large number of precisely detailed rules. Intervention can not be arbitrary; citizens in comparable circumstances deserve to be treated equally. These are the principles of decent administration (Box 14.3).

Box 14.3 How interventions reduce reality

The high standards that government interventions must meet have consequences for the content of regulations. On the one hand, they must produce the intended effect, but at the same time, they must satisfy requirements which have been precisely formulated. One consequence is that the degree of detail increases. Thus, any given regulation begins with a large number of definitions which exactly define 'what is meant by this regulation . . .'. In other words, each regulation defines reality. In so far as reality is not regulatable, it is defined to *be* regulatable, and thus reduced.

Within the policy network, consultation with the different actors traditionally plays an important role. Policy measures originate in close consultation with representatives of the organisations involved. The assumption here is that agreement with representatives automatically implies supportive public opinion. This assumption no longer holds. Representatives of organizations have gradually become part of the policy network, within which formal definitions of reality are dominant. They have been absorbed, as it were, by the policy circuit which has become highly institutionalized. This means that representation no longer automatically implies that perceptions of those represented will be considered.

The consequence of this is that citizens' definitions of reality not only differ in principle from the definitions within the policy network, but that differences between both definitions are not, or are hardly ever, the subject of discussion within the formal circuit.

14.9.2 How the differences in situational definitions are dealt with

The ways in which citizens define their situation are rather different from the definitions that are current within the formal policy arena. This is more than merely a difference in use of language. What a citizen does is not always significant within the policy network. Because of this, a communication problem arises. How can both sides ever communicate 'sensibly' with each other when they give divergent meanings to situations? The situation threatens to become one of permanent misunderstanding. In practice, this problem is 'resolved' because the dominant formal sense-making paradigm tips the scales when situational definitions differ.

This has consequences for communication between citizen and government. If citizens wants to attract the attention of a government organization(s), they must reformulate problems in such a way that it has significance within the formal sense-making paradigm. This is an annoying exercise for which the individual citizen is usually not prepared. That is, the formal sense-making paradigm is complicated, and as a result of a continuing flood of legislation, subject to continual change. An extra handicap is that the formal paradigm is centralized. Thus, there is a strong emphasis on national regulation. Consequently there is little room for regional differentiation and diversity.

Moreover, the complexity of the formal structure is increasing. In drafting new regulations, making them accord with already existing regulations demands increasing energy. This is a logical consequence of the fact that the density of policy and rules has markedly increased. Each new regulation calls for the need to define new concepts. Such definitions must agree with, or at least not contradict, the descriptions of concepts that are recorded in existing regulations (Box 14.4).

As a consequence, the policy network becomes increasingly self-referential. The already weak orientation to its surroundings thus declines still further, while, at the same time, the threshold for citizens to enter the policy arena becomes still higher. The formal paradigm is not receptive to citizens' perceptions. Communication between government and citizen thus for a significant part takes on the character of a confrontation between divergent sense-making. What is regarded as relevant by citizens has no 'significance' within the formal paradigm.

Box 14.4 the definition of a poplar meadow

In official circles, an intensive discussion recently emerged about the question of whether a poplar meadow should be considered a pasture, be it with a number of trees, or a forest, be it with room for grazing. This discussion, which appears theoretical at first, was important because defining the poplar pasture as a 'forest' has consequences under the Forest Act. Defining it as a 'pasture' could open up possibilities for farmers to take advantage of applicable subsidies.

The opposite is also true. Formal conceptions of planning are not understood by citizens; at least it is not clear to them whether and how plans contribute to a solution to what they experience as a problem. The communication between the two sides thus becomes static and polarised, to the point where they experience the other's information and proposed interventions as meaningless. The upshot is that a sense-making framework, within which the tension between agriculture, nature and environment could be conceptualized, is absent.

14.10 Consequences for policy development

14.10.1 Limited receptiveness of the policy network to societal problems

The dominance of the formal paradigm means that problems, as experienced by citizens, do not come up for consideration, or else receive no attention within the formal policy process. The formal paradigm is interested in societal developments and problems only in so far as they can be thematized within it. The opposite is also true: one peers out of the policy arena through 'formal spectacles', and everything that has no significance within the formal paradigm is ignored, and everything that cannot be understood runs the risk of being shoved aside as irrelevant. This has radical consequences for the relationship between government and society. The character of this relationship was originally that of service: the government was to generate and apply solutions which were beyond the willingness or capability of other parties. This relationship has been turned around: society serves in the final assessment only as a confirmation of formal policy. The government determines from existing regulations what is relevant. The existence of societal problems, however urgent, is in itself no guarantee for government intervention. A societal problem might not be a 'problem in the sense of the regulation'.

14.10.2 Competency and procedures

Though the policy network is not very open to problems as experienced by citizens, it is very open to 'formal' problems. A typical concern is the accommodation among the competencies of different government departments; another is the question of which procedures the policy process ought to follow. Within the formal arena, much significance is given to such questions. In a way, they even have a pre-emptive character: in government organizations, one can actively collaborate in policy development only provided that there is clarity concerning formal competence and procedures. When policy development involves an issue that falls undisputedly within the competence of one government organization, and which is a matter of clear, existing procedures, there is no problem. This is not the case when a number of government organizations are involved (Box 14.5).

Box 14.5 Unclear competency

A number of government organizations are involved in the integral regional approaches. Not only do national, regional and local government play a part, but at each administrative level, several institutions are also involved. For example, in a certain area three national ministries are involved with mutually unclear demarcations of competence. Thus the administrative responsibility for agriculture and nature is vested in the Ministry of Agriculture, Nature Management and Fisheries, but that is only partly the case for environmental policy, for which the Ministry of Housing, Physical Planning and Environmental Management is also responsible. Coming to clear, concise agreements in these matters caused considerable delay at the beginning of the project.

14.10.3 Judicial limitations

In addition to problems with competency and procedures, judicial questions score high. The marked increase in the density of policy at the national, provincial and local levels has led to an impressive body of regulations. This results in a flow of rights, privileges and obligations which hinders the formulation of new agreements concerning a joint approach. For judicial reasons alone, one cannot make a fresh start without first carefully exploring its effects and taking adequate anticipatory measures (Box 14.6).

Box 14.6 The momentum of existing rules, permits and regulations

Municipalities have given permits to farmers in past years as part of environmental policy. These permits usually establish the maximum number of animals that may be kept on a given farm in great detail, as well as the composition of the livestock population, the layout of the pens, and numerous other stipulations. Such a permit has the character of a complex set of limits and conditions. One can, however, also derive rights from it. As long as the permit holder operates in accordance with the permit, he has the right to keep a given number of animals, irrespective of what would be desirable under a new area plan.

In short, the relationship between government and farmer is determined in a judicial framework. This means that the government bodies concerned are also bound by it. This limits the room for manoeuvre in policy. In the event that there are convincing arguments to change environmental policy or to sharpen it, obligations that are imposed on farmers by permits are to an increasing degree interpreted by farmers as rights with which they can defend themselves against policy changes.

14.10.4 Little room for creativity

As a result of the sectoral orientations of the parties, it is hard to predict at the beginning of a policy process what the outcome will be, which measures will result from it, or whether one's interests will be positively or negatively affected. This leads to uncertainty, which has an influence on the positions taken by the different parties. It makes them search for certainty. Uncertainty is dealt with by building-in certainty beforehand. One agrees to participate in policy processes only provided that one is promised, beforehand, that certain effects that would work out unfavourably for the sector will be excluded. This is deadly for creativity. It hems in the space for thinking about solutions. In the extreme, it leads to a situation in which the policy process generates only solutions that are already known. The actors force each other to repeat their moves (Box 14.7).

Box 14.7 Reducing the number of livestock

The increase in scale and specialization has resulted in a large increase in the number of livestock and, consequently, in an unacceptable level of acidification and a gigantic liquid manure surplus. For many, the only solution seem to be massive reduction in the number of animals kept. The agricultural sector was initially fiercely opposed to this. When the problem was first taken up, the agricultural organizations would only co-operate in solving the manure and acidification problem, *provided that* the number of animals were not reduced. There could be no question of 'new policy' on this point.

Policy is looked on as a threat. The energy of the different parties involved is more likely to be directed at impeding and destroying efforts to arrive at a policy. Communication is static and calls to mind the image of trenches. History has taught that those are not the places best suited to creativity.

The fact that citizens' definitions of situations do not, or only to a limited extent, come into consideration within the policy process also has consequences for the public support of policy. The formal conceptual framework and the process of policy realization are complex and difficult to fathom. This leads to alienation between government and citizen; citizens are confronted with the consequences of policy measures, even when they cannot understand the underlying policy processes. There is, therefore, little to connect them to, or involve them in, policy measures as outcomes of such processes. The measures are, for example, solutions to problems that citizens do not experience as problems. The willingness to contribute to the execution of those measures is thus limited.

14.10.5 What is needed for change

A radical modification of the existing policy context is required. Policy implementation must once again be experienced as a chance to make possible solutions to experienced problems.

This does not happen by itself. One condition is that the definitions of all those involved in the policy process expressly come up for consideration. At present, communication is dominated by the formal sense-making paradigm. In particular, differences in situational definitions need attention here. A policy process must do justice to the premise that reality is multi-interpretable. It is exactly when one situation or one problem is allowed to be perceived in divergent ways by different people, that creativity is engendered. More strongly, one could say that making sense of a situation in another way than the conventional one is an important source of creativity.

There is an additional aspect. Planning assumes some collective process. This assumption is often not met. As a consequence of the domination of the formal paradigm, but also as a result of the sectoral orientation of the different parties, the accentuation of *differences* receives much attention. There is little interest in the perceptions of other parties.

This attitude can only be overcome when policy development is no longer experienced as threatening, but as a way to tackle problems. Trust must develop in the policy process and in its outcomes. This is possible by bringing into consideration, at an early stage of the process, the uncertainties that all parties experience, and by making them an explicit subject of discussion (Box 14.8).

Box 14.8 Creating common concerns

The position taken by the farmers is principally dictated by the fear that the planning process will lead to new constraints. In a situation in which farm incomes are already under a great deal of pressure, such a position is indeed understandable, but at the same time a formidable obstacle in the search for solutions. The income concerns ought to be placed on the table at the beginning of the process. This might lead to the agreement that solutions thought of during the course of the process are examined in terms of their consequences for the incomes of individual farmers. A plan to restructure a region which does not make explicit the effects on incomes is, of course, not complete. To put it another way, it is a *common* interest that sufficient income-generating power within a region is retained. Thus the pre-occupation of an individual or of a professional group is reformulated into a joint concern. This favours the progress of the planning process. For example, proposals to give nature interests more weight will not then be shoved aside beforehand by agriculture as 'not negotiable'.

Bringing the situational definitions of citizens up for consideration has yet another advantage. The formal policy process is, characteristically, rooted in formal sources of knowledge. The knowledge of the residents in a region has, until now, played a subordinate role in policy processes, and moreover has a low status, compared to the knowledge of research institutions, for example. This is often not fair. Residents are more often than not perfectly capable of describing what an area looked like several decades ago, and what changes have taken place. People also appear to experience it as rewarding when this knowledge is acknowledged as meaningful. People feel stimulated to actively think along in the search for solutions. Within the conventional planning process, willingness to acknowledge local knowledge is scant because solutions must satisfy the condition that they fit into the existing formal policy instrumentation.

Present tensions are principally directed at changes in policy content and targets, while essentially a change in policy structure is what is needed. What is further needed is that centralization, which is characteristic of the existing policy structure, is reduced and that more room is created for policy development at the regional and local levels. Policy instruments must *facilitate* solutions, while the current unchangeable regulations regularly stand in the way of solutions. The necessary changes in the policy context may be summed up in Table 14.1.

Table 14.1. *Comparison between actual and desirable policy context*

Existing policy context	New policy context
Policy is viewed as threatening	Policy is viewed as an option
Static; policy is fixed	Development; policy is process
Confrontation between different sense-making paradigms	Search for a communal sense-making paradigm
Energy is directed at accentuating differences	Search for collectivity comes first
Inspiration is not there; the emphasis is on the defence of one's own interests	Inspiration comes in doing it together
Energy is directed at destruction or obstruction	Energy is directed at creation
Existing policy frameworks must be respected	Policy frameworks are respected only in so far as they contribute to solutions to problems
The policy network is inwardly oriented	The policy network is outwardly orientated

14.11 The management of change

The foremost consideration ought to be that the sense-making paradigms of all involved be explored and considered early in the policy process. This means a form of participative planning. The system of indirect representation, such as is customary within politics, as well as within sectoral organizations, has not been able to prevent estrangement between government and citizen. A broader representation, or a somewhat altered composition of the present consultative mechanisms, therefore, does not appear to be the solution either. An alternative is to involve citizens more directly in the planning (Box 14.9).

Box 14.9 Citizen involvement

During the preparation of a Rural Zoning Development Plan, not only formal policy briefs and research reports ought to be consulted, but the citizens themselves must also be directly involved. It is important that communication with citizens is not dominated from the outset by formal paradigms, but that citizens be invited and stimulated to give their own situational definitions. This can happen, for example, by asking them to describe in their own words their environment, so as to indicate positive and negative points; the developments which have occurred in past years; whether their environment has improved, or whether it is in fact in decline; what they would like to be different; what would be needed for this; and what opportunities and constraints they see.

Secondly, it is of importance that those involved come to see policy development as a joint process. Therefore, it is necessary that differing situational definitions not be shoved aside, but that people get to know and understand just how and why others come to different sense-making. Until now that has hardly happened (Box 14.10).

Box 14.10 Understanding each other's sense-making

It often seems that farmers and nature protectionists have little knowledge of, or insight into, each other's concerns, ambitions or motives. They know each other's sense-making paradigms only imperfectly. They have, nevertheless, mutually delineated images of each other. Thus agriculturalists assert that nature protectionists have eyes only for nature interests and are blind to agricultural business interests. 'They prefer plants and animals to people.' Conversely, nature protectionists hold the image that agriculturalists are only out for economic gain and subordinate flora and fauna to money. The mutually imperfect knowledge that they have of each other does not keep them from having strong and outspoken images of each other.

It is important that people's mutual images be exchanged and tested at an early stage. This is possible through asking the different parties to put themselves in each other's position, and to describe reality through the spectacles of the other. Such interventions are set up to come to an understanding, and where appropriate an acceptance of, each other's definitions of reality. This means essentially that people accept that reality has diverse meanings, dependent on the perceptions of the actors in question. Differing situational definitions thus no longer need to be contested, but must be accepted as part of reality.

Acceptance of differences naturally does not quite mean responsibility for bridging them. To do this, more is needed. Collectivity means that people ultimately accept solutions only when diverse interests and concerns are sufficiently met. This demands trust from everyone that one's own interests and concerns will come up for sufficient consideration in the policy process, and that one will not oneself suffer from having taken an interest in, or been involved in, the interests of others (Box 14.11).

Box 14.11

The strategy of farmers is often defensive. Proposals for new policy are obstructed because constraints to business management could result from it. Suppose that an intensive agricultural practice in an area conflicts with natural elements, whether existing or to be developed. One possibility is to invite the farmers as well as the nature protectionists to develop a farm management model which takes into account what is necessary from the point of view of nature. In doing so, the parties concentrate on the technical consequences for farm management. Thus, it may be necessary to reduce the use of chemical pesticides for the sake of natural resources in the area and to convert to mechanical weed control. The starting question is then: suppose that one wants to exclude negative effects on natural resources, but still wants to maintain a viable agriculture, what type of farm and what farm management model would then be appropriate?

Prior to going through this exercise, it would need to be agreed that the model ultimately developed would also have to be assessed from an economic point of view. Moreover, it would need to be agreed that, in the event of the operation leading to considerable economic loss, this would be accepted as a mutual problem and that the loss would not be borne solely by the farmer. Ultimately it is of mutual interest for a sustainable farm management model in an area to be discussed in technical, ecological, as well as in economic respects.

Such an approach can contribute to a situation in which nature protectionists no longer feel responsible only for nature, while disregarding the farm management consequences. For farmers, the conviction can grow that it is also in their interest that the farm management model to be developed integrates natural and environmental interests in a way which is acceptable to everyone. This differs considerably from the existing form of consultation, in which the attention of different parties is directed expressly at safeguarding their own sectoral interests, and in which people have no, or only marginal, interest in the effects on other functions.

The above makes high demands on the planning process. We are talking of a process in which the stakeholders are intensely involved. This has implications for the size of the area. In comparison with current practice, in which land use planning projects can cover a size of 10 000 ha (the Fens cover 20 000 ha), it is more logical to select areas of about 1000 ha. This holds particularly for areas in which the tension between the various functions is high. The need for changing the planning context will be small in areas of a purely agrarian character in which nature values are hardly relevant.

14.12 Conclusions

The proposed approach is based on the express enlistment of citizens. What stakeholders consider as a solution, or as a problem, must receive proper attention. This differs considerably from the current approach to planning, in which the operative policy framework is taken as a starting point while deviating conceptions and goals are subordinated to it.

This raises the question of how one is to deal with situations in which citizens' goals do not agree with the policy goals which have been established according to due electoral process. Or to put it another way: what status do nationally agreed upon policy documents have within the new planning context? Is policy established in due process only relevant from now on in so far as stakeholders agree with it?

This, naturally, cannot be the intention. To put it more strongly, policy goals, as established at various government levels, will be an important reason to start the planning processes proposed. This must especially be the case for those areas for which goals and measures established in nature, environmental and agricultural policy, are significant.

The approach proposed is intended expressly to explore in what ways and under what conditions established policy can actually be carried out. It must be clear from the outset that the new approach should not serve to repeat the debate that resulted in the policy in the first place. Experience shows that parties sometimes tend to see a new approach to planning only as an option for getting an established policy changed in their favour.

A couple of points deserve attention in this respect. First of all, the involvement of citizens in the realization of policy briefs has been quite limited to date. This means that a redress is in order. Explanations are needed concerning the starting points and deliberations that have underpinned policy processes in the past; on the alternatives that were discussed; and on the reasons that led to the operative policy. Strictly speaking, this communication should have taken place during the realisation of the policy. As we saw, neither the centralized character of existing policy structures nor the imperfect functioning of the system of representation allowed for such communication.

Secondly, attention within existing policy processes is directed mainly at the establishment of policy targets (acidification must be pushed back by X% in the year 2000; use of pesticides must decrease yearly by Y%; in the year 2000 the surface area of natural areas must be expanded by *Z* ha). What has received much less attention is the question of what the consequences are of implementing the established policy goals. How can these goals be realized? What is needed to do that? What shall we do if undesirable effects arise that have not been, or have only insufficiently been, taken into account? The approach proposed expressly affords room for just such questions to come up for consideration. The outcomes and consequences of established policy are made explicit. This can result in situations in which policy goals are not consistent, or in which the assumed relationships between policy goals and policy instruments are misjudged. In this respect also, there can be occasion to reconsider policy measures as they have been established in the past.

In short, it is not a matter of something different from established policy. The ambition is precisely to realize established policy, be it that problems impeding such a realization must come more expressly on the table. Such problems have received little attention until now, or have been all too easily externalized to a single occupational category, such as the farmers.

14.13 References

Agrarisch Platform De Venen (1993). *Landbouwperspectief De Venen? Ja, natuurlijk!* Utrecht: Agrarisch Platform De Venen.

Driessen, P.M. & Dudal, R. (1989). *Lecture Notes on the Major Soils of the World.* Wageningen: Agricultural University.

Kruk, M. (1993). Meadow bird conservation on modern commercial dairy farms in the western peat district of the Netherlands: possibilities and limitations. Dissertation. Leiden: Rijksuniversiteit Leiden.

Ministerie LNV (1992). *Structuurschema Groene Ruimte. Het Landelijk Gebied de Moeite Waard.* Den Haag: Ministerie LNV.

Ministerie LNV (1994). *Sturing op Maat. Een andere Benadering van Milieu-Problemen in de Land- en Tuinbouw.* Den Haag: Ministerie LNV.

Röling, N. (1994*a*). Communication support for sustainable natural resource use management. *IDS Bulletin*, **25**(2), 125–33.

Röling, N. (1994*b*). Creating human platforms to manage natural resources: first results of a research programme. In *Proceedings of the International Symposium on Systems Oriented Research in Agriculture and Development*, pp. 391–395. Montpellier, France.

Stuurgroep Groene Hart (1992). *Nadere uitwerking Vierde nota. Plan van aanpak ROM-beleid. Groene Hart.* Den Haag: Rijksplanologische Dienst.

Tatenhove, J. van (1993). *Milieubeleid onder dak? Beleidsvoeringsprocessen in het Nederlandse milieubeleid in de periode 1970–1990; nader uitgewerkt voor de Gelderse Vallei.* Wageningen: Agricultural University.

Tweede Kamer (1989/1990). *Natuurbeleidsplan. Regeringsbeslissing.* Den Haag: SDU.

15 Communication between farmers and government over nature: a new approach to policy development

CEES VAN WOERKUM AND NOËLLE AARTS

15.1 Introduction

In 1990, the Dutch parliament accepted the Nature Policy Plan (NPP), designed to conserve and develop 'nature' over the next 30 years. A central element of the Plan is the maintenance of a network of areas of outstanding natural value (the 'ecological super structure'). Farmers in such areas are encouraged to restrict certain farm practices such as mowing and fertilizing, and are compensated for any loss of productivity. They may also sell land, if such land is wanted for nature conservation and production purposes. All decisions are of a voluntary kind, as clearly indicated in the NPP (see also Wagemans and Boerma, this volume).

Implementing the plan, however, has not been easy, and there have been clear indications of resistance from farmers and others. In order to understand better what has been happening since 1990, a government grant was made available to study conditions conducive to (non)acceptance. The report here is on two of the four studies conducted in this field (Aarts & van Woerkum, 1994, 1996). They mark the beginning and end of a transitional process in theory and practice from an instrumentalist to a more interactive approach. The first study examined the empirical situation and reports on factors that block acceptance. The fourth, reports on an experiment in interactive policy making. After outlining the findings, some theoretical reflections are offered on the data. In conclusion, the critical points are stressed for making an interactive approach more effective.

15.2 Non-acceptance of the NPP

Sixty farmers in three areas of the Netherlands were interviewed. The method chosen was qualitative, in the form of taped in-depth interviews, as this allowed for a more detailed probing of the motives and arguments behind particular opinions.

One of the main findings was that farmers perceive nature in a very different way to that expressed in the NPP. For farmers, nature is everything that grows and lives around them, and this includes the plants and animals that are the products of their own practices. They respond negatively to the idea of nature promoted by the plan – so-called 'wild' nature.

Acceptance is a process in which several dimensions can be distinguished: the problem itself must first be seen as serious and intervention by government viewed as inevitable; to be acceptable the main lines of the policy and the measures to implement it must be conceived as (a) effective, (b) realistic and in tune with farming practices, and (c) fair, in terms of what is asked of others besides the farming population concerned.

According to their own perceptions of nature, most farmers did not believe the current state of affairs to be as serious as stated and therefore intervention by government was not seen as necessary. Moreover, they did not see such intervention as isolated from, or in any way different from, the many other government interventions in the area of nature or environment. Farmers expressed, in general, being 'fed up' with regulations. The main lines of the nature policy (more nature, better nature) were not received with enthusiasm.

Most farmers do not think the specific measures to be effective. They doubt that there will be sufficient funds available either now or in the future to pay for what the government wants in terms of nature conservation and production. They experience problems in adapting to the measures demanded of them. Farmers who run modern agricultural enterprises prefer lower ground water levels and regimes for mowing than those recommended within the NPP, and are only willing to sign a contract to participate in the Plan if their land is unsuited to modern farming. They are particularly sceptical of, and opposed to, the voluntary nature of the Plan. If a region is selected for nature conservation/production, they anticipate a lower value for their land on the market, and feel they have been placed 'in the wrong corner' and that government is no longer willing to support their efforts to keep abreast of new standards in modern farming. In farming culture this phenomenon is referred to as '*planologische schaduwwerking*', a kind of shadowboxing.

Finally, farmers do not consider it reasonable that they alone should bear the full burden of creating a better natural environment. Other citizens, they believe, are not asked anything in comparison. They do not see the general principle of 'who spoils, pays', as being applicable to them. In their eyes they are the ones who for generations have been responsible for the beauty of the landscape and nature around them. Now they are to be made the victims of what they themselves created.

Looking at government communication activities after the NPP's assessment and ratification, the general impression is one of a chain of misunderstandings, mainly due to the nature of trust at most of the meetings. Farmers are expected to listen to the information being given out by the government, but are instead strongly motivated to speak about their problems. From both sides communication is one way, resulting in ever more frustration rather than understanding.

Farmers are influenced by the collective opinions on nature policy that are formed within their own circles. They feel united in their common defence. By stereotyping 'the other parties', nature conservationists and government, they become aware of and reinforce their own identity, which is the more important the more they experience their world as unstable and threatening. But individually, when confronted with concrete proposals to change their farming practices for money, or to sell their land, their reactions are much more pragmatic. Then they tend to act as *homo economicus*, not as a member of the group.

The research showed three patterns of reaction to the plan.

1. Farmers who do not accept the NPP as a meaningful reality. They are not involved directly in the plans and/or are not willing to change anything. Such farmers are in the majority.
2. Farmers, fewer in number, who do not accept the NPP as just or necessary, but are willing to 'deal' if they are to get well enough compensated for the requested change.
3. Farmers, a small minority, who accept the NPP, not because it is good for society but as an inevitable development which had best be taken seriously in order to influence the outcome.

With this motive in mind, several so-called 'environmental co-operatives' were established by farmers to deal with the other actors involved in shaping nature policy in their own regions. These initiatives will be focused on later, for they constitute an important element in new processes of interactive policy making.

15.3 Analysis

Behind government policy making for solving societal problems lies a vision that is basically 'instrumentalist': Government sets a goal (the creation or preservation of nature) and constructs a set of means (government tools or measures) to achieve the desired effect.

Instrumental thinking can be analysed as a product of social-technocratic philosophy, that is, handing over the responsibility and capacity for solving societal problems to an institution that is presumed to be equipped to analyse those problems and to act according to the best (technical) alternatives. Research is an important support to this kind of policy making. It can reinforce the tendency to look at societal problems in a particular way. As policy studies have often revealed, governments suffer from 'self-referencing': that is, they tend to interpret the world according to their own problem definitions and politically viable perspectives and solutions. Research is used to underpin their outlook and rationalities outside the policy system are ignored.

Links with society, however, are stronger where effectively organized interest or pressure groups exist to press for change. Especially in the field of environmental issues in the Netherlands are such groups powerful and influential, and they have the attention of both the media and the political parties. They push for new measures to change society in a desirable way, but without much consideration for other groups who might be affected.

The NPP fits this instrumental perspective on many points. It is mainly produced by the Ministry of Agriculture, Fisheries and Nature Management, which has links with the organized nature movement. Contacts with the agricultural community are restricted to its representatives at the top level. The case is thus a perfect example of self-referential thinking. The NPP includes no worked out analysis of the process of change in which the farmers are expected to engage. The data they work from are derived mostly from biological and ecological research, and the rationality of the farmers, and their perceptions of nature and nature policies have not been dealt with in any serious way.

As a consequence, farmers 'label' the NPP as stemming from outsiders (Elias & Scotson, 1965), and its supporters as invading the existing established agricultural order. This results in a process of stigmatization, whereby established orders construct an image of outsiders based on their worst characteristics and compare this image with the most favourable features of their own group. These constructions mirror and justify the aversion that the establishment feels for outsiders. The aim of the stigmatization is to maintain a power position by excluding and attacking the outsider. Cohesion, as an outcome of a common history or shared goals, reinforced by daily interaction, is an important power source. From the beginning, this process of stigmatization has blocked effective communication of the NPP's intentions and has frustrated attempts to gain its acceptance. Farmers are used to seeing themselves as masters of the countryside. They have always felt superior in managing nature. Now nature-minded people appear on all sides, behaving as new neighbours.

Instrumental thinking also affects communication strategies. Many communication specialists tend to over-rate the potency of communication campaigns and continue to use a model that has had enormous impact on their professional activities: a message is formulated and transmitted via a particular medium or channel in order to evoke a defined effect. The outcome of communication is seen as a well-considered combination of source, channel, message and receiver factors (Berlo, 1960). The implication is that communication effects can be 'managed' if these factors are taken seriously into account. The model has been criticized for its 'psychological bias' (Rogers & Kincaid, 1981). It neglects the cultural factor, the rationality of seeing the world according to a group's own interpretative rules. It ignores the fact that people do not change in isolation but in discussion with others who share concerns of common interest and importance.

It also neglects to take into account the autonomy of groups to shape and construct their own messages, messages that will accord with their own world views. Many attempts have been made to label or value the activities of 'receivers', see, for example, Bauer's notion of 'the obstinate audience' (1971), the 'sense-making approach' of Dervin (1981) or, more recently the literature that takes a constructionist approach such as Röling, Kuiper & Janmaat (1994).

Attempts to communicate about the NPP is based on unrealistic communication models. Farmers are unlikely to be convinced by arguments and data that are incompatible with or take no account of their own messages, thematization of the nature problem or perceptions of nature. It is not so easy to manipulate the thinking of people with vested interests. They simply label the source of the information as unreliable since it stems from an out-group, and as therefore not sufficiently important as to be taken seriously. As a result they internalize little or nothing of the message; of what is being said.

Communication about the Plan has followed the so-called DAD-model:

decide, announce, defend (Wolsink 1994). Communication took place only after the plan had been approved. It is informative and motivating, but it does not create the desired level of acceptance.

15.4 Another way of thinking

An important recommendation of our first study was that policy making at regional level should be more interactive. A number of problems could be resolved if authorities were more in contact with the thinking of farmers. The farmers' views on the relationship between farming practice and nature conservation and production ought to play a role in regional nature policy. The way forward must be based on two-way communication, respect for each others' opinions and a readiness to look at common solutions.

The basis of interaction is perceived interdependency. Government plans have met fierce resistance from the agricultural community. Some are prepared to acknowledge the need for change but they believe they should be more actively involved in nature policy, so they can influence the nature of the restrictive measures placed upon farming, and steer future developments. They recognize the need to improve their image and, in the end, their position. Members of the organized environmental movements are also coming to the conclusion that demands and pressure for government intervention is not sufficient to achieve the solutions they hoped for. Direct contact with farmers is an alternative with perhaps more promising prospects.

Within a relatively short period a number of so-called environmental co-operatives were established in the agricultural community, aimed at organizing initiatives in conjunction with other actors to deal with the problems and issues involved in the 'nature' problematic. One of these environmental co-operative organizations was formed in De Peel, a region in the south of the Netherlands of outstanding natural beauty, threatened by a particular kind of agriculture – pig breeding. In this region, full of tensions, young farmers decided to talk instead of using protest and 'wait and see' tactics.

We were given an opportunity, over a 2-year period, of observing the functioning of this cooperation. Our starting point was to see whether interactive policy making could contribute to a more acceptable plan for preserving nature, and to examine which factors furthered or hindered this process. We made use of in-depth interviews with the key co-operating actors as well as participant observation of farmers' meetings.

15.5 The De Peel Environmental Co-operative

Our study of the activities of the De Peel Environmental Co-operative (EC), involved looking at the most critical relationships in its network: government, the various nature and environmental organizations, and the agricultural community.

The government is responsible for implementing the NPP. Another way of policy making can thus affect existing norms and policy instruments. Policy makers must be flexible on some points in order to make room for new regional proposals. They are also responsible for funding any initiatives considered more effective in conserving and improving nature. They play an important role in facilitating the communication and decision-making process as a whole, in bringing people together and linking regional ideas with the national policy process. For this reason a special government team was established, *Nadere Uitwerking Brabant Limburg*, to organize the process in these southern provinces.

An important environmental organisation in the region is the *Werkgroep Behoud de Peel*, the WBP, or Committee for the Preservation of De Peel. Until the setting up of the co-operative, they had fiercely attacked the farming community through, for instance, numerous lawsuits. They have now moved to a more co-operative position.

Regarding the agricultural community, as we mentioned, only a relatively small number of farmers had been willing to engage in a pro-active process of negotiation. Most had clung to their old position of coping and protesting. Gradually, however, with the encouragement of the co-operative, they became more open-minded.

The process of interaction resulted in a common plan, supported by all parties, to develop the region in an ecologically as well as economically viable way. The plan was accepted by the government, though as an 'experiment', not a final programme. The plan will thus proceed but the results will be carefully monitored.

Our study of EC De Peel activities showed that their efforts to organize a network of all the actors involved and to keep this network alive gave them the role of 'network manager' – managing relationships and tuning everybody to the right notes.

In discussions with government they were confronted with bureaucracy and slow decision making. More importantly, they felt uncertain of the role government wanted to play. The new face of government as facilitator is not always visible. Traces of its old face as regulator are still there, and nobody can predict what will be the outcome of workings within the government bureaucracy over preferred policy style. As a result of this, criteria for judging the approach or effectiveness of the EC De Peel are uncertain, and it is not clear whether the experiment will be supported in the long run.

Having dealings with the nature and environment committee (the WBP) relieved a lot of tension. The organization moved from a position of pure antagonism to one of co-operation, with the common plan as the final outcome. They were able to learn of, and respect the views of, others once there was an understanding of the motives behind them. A degree of communication has been achieved with at least a minimum of concern for each others' viewpoints.

However, this relationship is fragile. Old reflexes still exist, as was shown by efforts initiated by the WBP against certain farming practices. Moreover, not all farmers have been involved in co-operative interaction. Many of them hold to their stereotypes of the environmentalists as dangerous strangers from outside. The gulf between the groups as measured by their opinions on nature, as well as by their cultural background, is still enormous.

Contacts with fellow farmers are, of course, crucial. Here, the De Peel EC has to balance its role as negotiator in a situation of conflict with its role as mediator. The first calls for a tough attitude towards other parties, with only a minimum of contact with its own constituency until they can produce some results. The second role demands a friendly outside face and a lot of inside contacts to guarantee that farmers stay involved in the learning and collaborating process.

The different roles were in evidence at two meetings with farmers. In one, the EC De Peel presented itself as the promoter of agricultural interests, with the stress on one-sided communication – speaking, not listening. The other meeting was much more interactive, with two-sided communication, allowing the farmers to engage in a process of thinking about new possibilities, encouraging them to explore another attitude towards nature, environment and the WBP.

We must conclude that the EC De Peel has not yet solved the acceptance problem. But they are 'under way'. Much is achieved, much is still vulnerable. The system is moving, however.

15.6 Analysis

What is striking about the activities of the EC De Peel, is how difficult their task of network manager is. They must not only balance the interests of all actors in the network, but must also make sure that changes in one area, among one group of actors, keeps pace with changes occurring elsewhere. The whole picture is constantly moving, and not always in the same direction and at the same rate.

Discussion is the basic mechanism for learning and negotiations to start moving (Putnam & Roloff, 1992). These discussions, however, tend to be restricted to representatives of all the different interests. In such situations representatives can play two possible roles. They can act to promote particular interests or they can act as change agents, internally for 'their' group, but also externally, for the other party. In the first role, there is a strong tendency to keep quiet during the negotiation process. They 'have nothing to offer'. Everything then remains uncertain. Most labour union negotiators, for instance, act in this way. In the second role, they are, on the contrary, busy from the beginning, trying to involve everyone in the process, making sure each understands the other, saying what they feel is necessary and supporting the outcomes. Acceptance is not based on an evaluation of the outcome alone but on participation in the process from which the outcome itself is produced.

Behind these two role types lies a different conception of the negotiation process. We distinguished two types: 1) distributive negotiation, on the basis of one cake that has to be divided, and 2) integrative negotiation, on the basis of the baking process, in which the cake's consumers are directly involved in producing the kind of cake they want. This distinction has several implications.

Distributive negotiation starts from fixed positions which are held on to as firmly as possible. Often the case is overstated, too much is asked, knowing that something will have to be given up, but as little as possible. Integrative negotiation starts from an interest or an idea about the desired future. People do not overstate in this case. In distributive negotiation, the negotiators keep underlying motives and their background to themselves. They keep quiet about personal feelings. In integrative negotiation, participants are much more open, trying to share their feelings, motives and beliefs. In distributive negotiation, threats are common. The constituency is kept alert, with an energetic spread of images of the enemy, to guarantee, if necessary, an activistic response. In integrative negotiation such threats are minimised. They try to keep functional relationships as amicable as possible. Joint fact finding is a common phenomenon, whereas such activities are impossible in distributive negotiation. There is a concern for the consequences of discussions for the other, and most important, in integrative negotiation people are learning, they learn to see themselves from the position of the other, they learn to be socially reflective. Such learning processes are absent in distributive negotiation.

The more negotiations take place according to the principles of integrative negotiation, the more successful the regional communication process becomes (see also Pruitt & Carnevale, 1993). The De Peel EC has made great efforts to establish for itself an integrative negotiating position. However, the move towards this position is not yet complete. Representatives of EC De Peel behave differently in different situation, for instance, at one meeting with a group of farmers, they acted according to old role models. Their responses are not consistent. Sometimes they 'promote' their beliefs instead of stimulating debate about them. Sometimes, they engage in 'scapegoating'. It is no longer the regional environmental working group, the WBP, that is the enemy, but government. In picturing the government as unwilling and unco-operative they create new distances, now between regional and national levels of government.

15.7 Conclusions

Theoretically, interactive policy might appear to be the obvious answer to a failing system of governance where target groups are 'steered' instrumentally. It has many attractive features: the actors involved learn from each other, they understand better their interdependency, they create together more effective policy plans whose implementation will be supported by the different groups. However, in practice it is not so easy. Old attitudes, belonging to old situations, are strongly embedded in social relations and interactions. The rate at which representatives and their constituencies evolve do not keep equal pace. And representatives are unable to always act and react consistently. Only a conviction of the importance of an interactive way of solving problems and a willingness to carefully monitor the on-going process can prevent difficult confrontations. The outcome of the process in De Peel is promising, but fragile.

15.8 References

Aarts, M.N.C. & Van Woerkum, C.M.J. (1994). *Wat heet natuur? de communicatie tussen overheid en boeren over natuur en natuurbeleid.* Wageningen, Landbouwuniversiteit, Vakgroep Voorlichtingskunde.

Aarts, M.N.C. & Van Woerkum, C.M.J. (1996). *De Peel in gesprek; een analyse van het communicatie-netwerk van Milieucoöperatie de Peel.* Wageningen, Landbouwuniversiteit, Wetenschapswinkel.

Bauer, R. (1971). The obstinate audience. In *The Process and Effects of Mass Communication*, ed. W. Schramm & R. Roberts, pp. 319–28. Urbana: Illinois Press.

Berlo, D.K. (1960). *The Process of Communication.* New York: Rinehart & Winston.

Dervin, B. (1981). Mass communications: changing conceptions of the audience. In *Public Communications Campaigns,* ed. R.E. Rice & W.J. Paisley, pp. 71–87. Beverly Hills: Sage.

Elias, N. & Scotson J.L. (1965). *The Established and the Outsiders.* London: Frank Cass & Co Ltd.

Natuurbeleidsplan (NPP) (1990). Regeringsbeslissing. Gravenhage, SDU Uitgeverij.

Pruitt, D.G. & Carnevale, P.J. (1993). *Negotiation in Social Conflict.* Buckingham: Open University Press.

Putnam, L.L. & Roloff, M.E. (eds.) (1992). *Communication and Negotiation.* Sage Annual Reviews of Communication Research, Vol. 20. Newbury Park London New Delhi: Sage Publications.

Rogers, E.M. & Kincaid, D.L. (1981). *Communication Networks: Towards a New Paradigm for Research.* p. 31. New York/London: The Free Press, Macmillan.

Röling, N.G., Kuiper, D. & Janmaat, R. (1994). *Basisboek Voorlichtingskunde.* Amsterdam/Meppel: Boom.

Veld, R. in 't (1990). *Autopoiesis and Configuration Theory: New Approaches to Societal Steering.* Dordrecht: Kluwer, Academic Publishers Group.

Wolsink, M.P. (1990). *Maatschappelijke acceptatie van windenergie; houdingen en oordelen van de bevolking.* Amsterdam, Universiteit van Amsterdam.

Part V: Synthesis

16 The ecological knowledge system

NIELS G. RÖLING AND JANICE JIGGINS

Learning provides an alternative for crisis (Westley, 1995)

16.1 Introduction

This book looks at efforts to make agriculture more sustainable, both at the farm and higher system levels.[1] Its purpose is not to present technical practices or to provide evidence in support of their feasibility in terms of farm management, marketing, food security or community health. Other recent books have looked at these issues (Reijntjes, Haverkort & Waters-Bayer, 1992; Northwest Area Foundation, 1994, 1995; Pretty, 1995).

If this book makes a contribution of its own, it is because it looks at what ecologically sound practice implies for the human actors involved, especially in terms of their learning and in terms of the facilitation, institutional support and conducive policies required for that learning (Pretty, 1994). That is, it examines the implications of ecologically sound agriculture for land users and other stakeholders as social actors and goes on to analyse how conditions for change can be created. For that is the point of departure of this book: the conditions for sustainable agriculture are created in the socio-sphere, through policy, institutional and behavioural change. We shall call these conditions in the socio-sphere 'the ecological knowledge system' (Fig. 16.1).

This perspective is unusual for two categories of likely readers. First, many of those who are interested in making farming more sustainable are agronomists, soil scientists, and others who are more used to dealing with agricultural practices

[1] The chapter uses the words 'sustainable agriculture' and 'ecologically sound agriculture' and sometimes refers to 'integrated' practices such as IPM or integrated arable farming. As is usual with publications on this subject, some definition of terms is in order, given the confusion which prevails. As explained in Chapter 1, 'sustainability' is used in the constructivist sense as a property which arises out of the interaction among stakeholders in a natural resource. It is not a hard property, such as carrying capacity, of an agro-ecosystem. Instead, sustainability is negotiated. Agreeing on what sustainability is to mean is half the problem (Pretty, 1994). In that sense, sustainable agriculture is not defined in terms of the absence of chemical inputs, but in terms of what is locally agreed to comprise a sustainable form of agriculture, which might include measures to prevent further salination, erosion or other land degradation, measures to curb toxic or polluting emissions, measures to stop unsustainable exploitation of non-renewable resources, such as fossil ground water, or measures to regenerate bio-diversity. Sustainable agriculture and ecologically sound agriculture are used intermittently as equivalents. Such agriculture is not necessarily low external input, depending on the situation. Integrated approaches rely on an integration of chemical, mechanical and biological methods. Ecological agriculture does not use any external chemical inputs and is, therefore, not the same as ecologically sound agriculture.

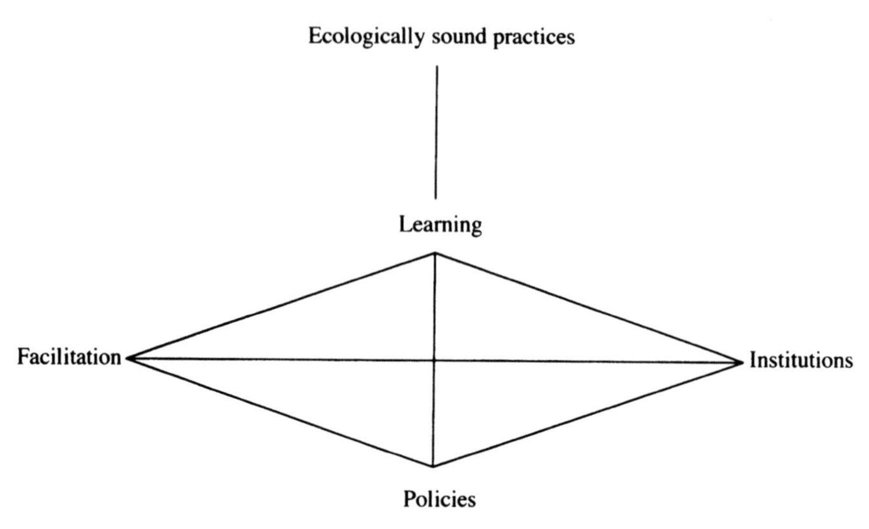

Fig. 16.1. The ecological knowledge system.

than with the social dynamics underpinning them. Secondly, many of those who are interested in managing change focus on market mechanisms, incentive structures, and regulatory policies, rather than on the facilitation of learning. Yet the chapters of the book show that:

- professionalism in ecologically sound agriculture makes much greater demands on understanding social process than professionalism in conventional agriculture; and
- managing change in the direction of ecologically sound agriculture makes much greater demands on the understanding of learning than does the promotion of 'more of the same' within the conventional paradigm.

These lessons require some elaboration before we go into more detailed analysis.

16.2 Professionalism in thinking about people

The book looks at people and environment as distinct constructs and examines their problematic interface. It is useful briefly to make explicit the different ways in which we use the two. Basically, the environment is considered amenable to causal reasoning and instrumental control. It is not itself intentional. It was once, and still is by some, considered to be a given reality, external to the human observer, which can be objectively known through experimental discovery of its secrets, as if it were some sleeping beauty found by a wandering knight.

More people are now inclined to think of the environment as complex, chaotic and inherently unknowable. But we can still construct useful knowledge about it, in the sense of developing 'effective action in the domain of existence'

(Maturana & Varela, 1987). Such effective action 'releases human opportunity' by 'flexible, diverse and redundant regulation, monitoring that leads to corrective responses, and experimental probing of the continually changing reality of the external world' (Holling, 1995).

In such new perspectives, however, the environment is still governed by causal chains, albeit as unpredictable as Lorenz's proverbial butterfly effect (Gleick, 1987). And it is still considered as existing without a purpose, in the sense of deliberate design. Typically, however, it is seen as goal seeking in that, in the case of matter, it moves inexorably towards entropy, and, in the case of life, struggles for survival and reproduction, and evolves towards greater complexity. Such assumptions about goal seeking allow powerful causal analysis and prediction.

People, on the other hand, are essentially intentional. Their goals can never be assumed except, perhaps, in the most abstract economic models. In addition, not causes but reasons are most useful for understanding human behaviour. People are intentional sense-makers. Understanding their activities is, therefore, inherently different from understanding the environment.

One can seek to analyse human activity by observing people as objects so as to attribute meaning and motive to their activities, but one can also interact with people, so as to inter-subjectively construct reasons and objectives for action together with those whose activities one wishes to understand. It is the shift to the latter choice which drives the interest in 'participation'. In the relatively decent society assumed in this book, affecting people's activities is not a question of instrumental control, but of either strategic manoeuvring in exchange or conflict among actors with different objectives, or of consensual decision making to reach shared objectives (after Habermas, 1984, 1987; Röling, 1996). The contributions in this book demonstrate that:

- change to ecologically sound agriculture often implies a shift from strategic manoeuvring to consensual decision making based on negotiated accommodation of interests and on social learning of new shared perspectives.

A final key point about people is that human advance as a historical species does not depend on evolution of genetic traits. Their acting effectively in the environment depends, instead, on their ability to learn, construct and share useful knowledge and technology, and hence on their ability to shape, control and exploit the environment to their purposes.

Criteria for effective action are not determined by in-built goals. People's intentionality is shaped by what is possible, by relative deprivation, and above all, by an insatiable need for novelty, self-presentation, curiosity and enterprise. By their very nature as intentional sense-makers with a considerable capacity to make a difference through action, people behave as autonomous actors in their environment. They have learned to harness a large part of the earth's water, air and biomass for their purposes, and are self-referential to a point where they tend to ignore ecological imperatives when these are incompatible with their designs. A sustainable society is not automatic or God given, but the outcome of a self-willed and learned transformation.

We seem to be in the middle of that transformation, which makes this one of the exciting periods in which to be alive. The environment is no longer a production factor only, much as labour ceased to be a production factor only in the nineteenth century. The transformation potentially leads to as yet unimaginable societal discontinuities, in terms of adapting human wants and technology to perceived ecological imperatives, and in terms of restructuring institutions designed for production and productivity to also serve sustainability. This book is part of that transformation in that it seeks to contribute to our collective understanding of what it means to make the change. The French aptly speak of a 'prise de conscience'.

16.3 The framework for analysis

The main dimensions along which we examine the transformation to ecologically sound agriculture were introduced in Chapter 1. We repeat them here because they form the backbone of this concluding chapter (Box 16.1).

Box 16.1 Dimensions of the ecological knowledge system

1. Ecologically sound practices
2. Learning
3. Facilitation
4. Support institutions and networks
5. Conducive policy contexts

A key insight which emerges from the cases in this book, and especially from the IPM Programme in Indonesia (Röling and van de Fliert, this volume) which has perhaps gone furthest in 'scaling-up' pilot efforts to a significant societal level, is that

- the five dimensions form a mutually interdependent and consistent whole, in that the nature of the ecologically sound practices makes special demands on learning, which, in turn, places special demands on facilitation, institutional support and a conducive policy context. The ecological knowledge system is fundamentally different from a knowledge system to support conventional agriculture.

Ecologically sound agriculture can be seen as a complex system, not only in the sense of complex interactions among soils, crops, animals and farming practices, but also in terms of human knowledge and learning, institutions and policies. Ecologically sound agriculture is not only a 'hard' system of biophysical components and processes which can be modelled in a computer. It is a 'soft' system in that it also, and essentially, includes a 'human factor', which cannot be modelled in a predictive sense, but must by its very nature be constructed interactively and negotiated. This observation has some important preliminary implications.

- The change to ecologically sound agriculture is not only a question of sound scientific claims with respect to its appropriateness and feasibility, but especially also of widely shared learning and social reconstruction of the environment.
- The change to ecological sound farming is not only the outcome of technical intervention, but especially also a negotiated outcome based on accommodation among paradigms, coalitions, institutional interests and politics.
- Ecologically sound agriculture cannot be expected from merely introducing different methods and technologies to individual farmers. Required is a transformation of the entire complex soft system which can be called 'conventional agriculture' to an equally complex but different soft system which can be called 'ecologically sound agriculture'.
- This transition, therefore, requires a management of change which goes beyond providing policy makers with scenario studies based on computer simulation, in that it must forge ways forward which emerge from interaction among stakeholders, based on shared perspectives, shared ways of making visible the state of the environment, shared strategies and collective decision making.

In addition to the five dimensions, also the level of aggregation plays a crucial role in structuring the chapter, as it did in the entire book. We shall basically discuss two levels:

1. the shift to ecologically sound agriculture at the farm level; and
2. the shift to more sustainable management of natural resources at a level higher than the farm, e.g. the watershed, the polder, the ocean or other ecosystem that requires a 'platform for natural resource use negotiation'.

Key observations which arise from the chapters are that

- Ecologically sound agriculture requires change, not only at the farm level, but also at higher agro-ecosystem levels, such as watersheds, biotopes and landscapes.
- Ecologically sound agriculture requires change, not only at the level of the farm household, but also at the level of the institutions in which it is embedded.

16.4 The technological imperative

The cases described in this book show that ecologically sound agriculture represents a way of dealing with natural resources which is different from conventional agriculture. This seems a rather self-evident statement until one tries to specify the nature of the difference. In the various chapters, the following typical observations are made with respect to what constitutes ecologically sound, as compared to conventional agriculture.

- It is *observation intensive*. It relies on observation and monitoring of the state of the crop, the farm and the ecosystem. This typically requires new ways of 'making things visible' and new feedback loops, not only at the farm, but also at the landscape and society levels. Much work still needs to be done in this area. If there is one distinguishing characteristic of ecologically sound agriculture, it is that it is responsive to environmental feedback. It cannot be based on self-referential practice.
- It is *knowledge intensive*. It relies on the ability of the land user and his\her support networks to anticipate, i.e. to infer from observation and timely to take appropriate measures.
- It is *learning intensive*. It relies on monitoring and sensitive probing to maintain adaptive responses to continually changing circumstances. To a large extent, adaptive responses are predicated upon the ability to construct shared fresh perspectives, i.e. on social, as opposed to only individual, learning.
- It is *technology intensive*. Although it explicitly implies decreased use of chemical and other inputs, it does rely on sophisticated technology to manage water, nutrients, energy, genetic resources and pests and diseases. It definitely is not the same as 'traditional', pre-industrial, agriculture. In fact, sometimes it represents 'second-generation', 'post-Green Revolution' technology and seems to provide a window on additional productivity gains now the Green Revolution is running out of steam. It can be intensive and highly productive, depending on the situation (Pretty, this volume).
- It focuses on the *whole farm* rather than on component technologies only and emphasizes *resource development*. The transformation of a farm or larger ecosystem to ecologically sound management might take several years (e.g. for organic matter in the soil to build up, or for populations of natural predators to establish themselves).
- It relies on *local implementation of principles*. Conventional agricultural innovation has emphasized the introduction and widespread adoption of uniform technologies across large recommendation domains which are assumed homogeneous with respect to benefit from the technology, so as to adapt local resources to create the circumstances on the experiment station. Ecologically sound agriculture relies to a much larger extent on the land user's ability to apply principles to locally specific situations and expects them to capture the opportunities provided by diversity. Hence the land user is *of necessity* an expert in his/her own right and not only a user of other people's expertise.
- It usually requires *multi-level management*. Conditions for growing healthy

Table 16.1. *Environment, people and their interface*

Environment	Interface	People
Water, nutrients, energy, ecosystem, pests and diseases, genetic composition, population	Observation, monitoring, making visible, anticipation probing, technology, farm practices, regulation, extraction, intervention	Knowledge, learning, goals, objectives, laws, power, gender, policy, institutions, coalitions

crops and animals and for accessing biomass must usually be created at system levels higher than the farm (e.g. soil and water conservation, habitats for natural predators, bio-diversity conservation, etc.).

- Above all, ecologically sound agriculture relies on *using natural processes*, such as pest control through natural enemies, recycling of nutrients, water retention through mulching, erosion control through grass bunds, etc.

The ecologist Holling (1995) sums up these points very nicely. Human effort to focus narrowly on maximizing the value of one target variable such as crop yield by controlling the ecosystem has led to breakdown and collapse of the ecosystem. It is impossible to control all the feedback loops. The environment is inherently unknowable and continuously changing. It cannot be suppressed to serve a limited goal. Hence Holling suggests a shift to *adaptive management,* which relies on the flexible, diverse and redundant regulation, monitoring for responsiveness, and experimental probing we quoted earlier when we spoke of new ways of perceiving the environment. This repeated quote is not accidental. The chapters suggest that:

- the shift to ecologically sound agriculture coincides with, or is predicated upon, a shift from reductionist, positivist approaches to acceptance of complexity and constructivism. One could go so far as to say that this is part of the Copernican Revolution, which started with the realization that people are not the centre of the universe but rather a temporary and marginal event in an immense space, and is now in the process of creating wide acceptance of multiple equivalent realities, as opposed to one solid objective truth, and of the inherent complexity of nature, as opposed to the certainty that knowledge can be accumulated about it (Tarnas, 1991). Such a shift fits well with a shift from the arrogant focus on the human control of nature to a realization that humankind is inherently part of nature.

In all, the characteristics of the practices which characterize ecologically sound agriculture have far-reaching implications for land users who embrace it and for the social processes in which they are embedded. That is why we called this section 'the technological imperative'.

But we can do better than that. Table 16.1 goes back to our earlier distinction between people and environment. Typically, 'environment' is the domain of the natural scientist. 'People' is the arena for the social scientist. For us, the interface between the two is of crucial interest. For it is the nature of the interface which determines whether the people–environment system can be called ecologically sound. This does not mean that the interface *causes* this outcome. In fact, environmental disasters or political change might drive it. All we are saying is that, in the end, it is the nature of the interface, of the practices, which counts in terms of sustainable outcomes. The book takes off from here and examines what an ecologically sound interface means for the 'people' domain, and hence for learning, its facilitation, supportive institutions and conducive policies.

16.5 Knowledge and learning

For a long time, we have defined learning in agriculture as the adoption process, a farmer's encounter with an external innovation, first becoming aware of it, then gaining additional information and finally becoming convinced to adopt or reject it (Rogers, 1995). The context is thus the delivery of external, usually science-based, innovations, with farmers as potential users. The chapters in this book show that:

- the change to ecologically sound agriculture is not comparable to the adoption of an add-on innovation, but a complex learning process which can take a few years. It has been likened to obtaining a degree.

The transformation, therefore, not only involves a change of the farm, in terms of the establishment of populations of natural enemies, the adaptation of crop rotation practices, the build up of soil organic matter, etc., which might take a number of years in itself, but also, and especially, a transformation of people and, as we shall see, of institutions and policies.

Somers (this volume) has explained the difference between add-on innovation adoption and the shift to ecologically sound agriculture by examining the attributes of ecologically sound farming in the sense used in the diffusion of innovations research tradition. It is clear that its complexity, lack of immediate relative advantage, lack of visibility, etc., all mitigate against its rapid autonomous diffusion.

Of course, certain practices, such as the use of less toxic herbicides or of nematode resistant and tolerant starch potato varieties, do diffuse. This opens the possibility of viewing the shift to ecologically sound agriculture as the sum total of a large number of individual add-on innovations. The case of the Netherlands (e.g. van Weperen *et al*, this volume) is interesting because it throws light on this issue. Dutch farmers have adopted *en masse* a number of technical innovations in order to comply with tough environmental regulations. But this has not been sufficient to reach the targets set in the legislation. The only success has been a reduction by 77% in the use of soil disinfectants, but the use of pesticides and herbicides has actually increased during the last few years after showing a promising reduction at first. The case of fungicides is the most problematic. Instead of the intended reduction by

23%, the quantity used has remained the same (Progress Report on the Multi-Year Crop Protection Plan to the Dutch Parliament, August, 1996). In all, the Dutch experience suggests that

- a substantial reduction of pesticide use, let alone a shift to ecologically sound agriculture, seems not possible by the cumulative adoption of external innovations. It requires a transformation in the manner of farming (Pretty, this volume).

If true, this observation has very important implications for the ways we think about technical change in agriculture. We used to rely on innovation diffusion to a considerable extent. 'Diffusion occurs while you sleep'. In other words, this autonomous process could be relied upon to multiply the efforts of farm advisors and the agricultural press to introduce technical innovations, even if they reached only 10% of the farmers directly. This multiplier effect occurred because of what Cochrane (1958) has called the 'agricultural treadmill'. The mechanism is simple.

The large number of (relative to the size of the total market) small farms ensures that no individual farmer can affect prices. They therefore all face the same price for a given commodity. The early adopters of an innovation, which increases productivity and/or reduces costs, capture windfall profits because their increased output does not affect the market. However, others soon see the relative advantage and follow. Once the diffusion process takes off and large numbers of farmers adopt, the cumulative effect of these individual decisions begins to exert a downward pressure on prices. Farmers who have not adopted must now do so in order to stay in the market place, but the investment now is no longer profitable. At the end of the curve, 'laggards' eventually drop off. In farming, innovation is a condition for staying in the market place. Hence Cochrane's notion of treadmill. The Greek effort to become competitive in a European context (Koutsouris and Papadopoulos, this volume) provides a telling instance.

This model of 'social learning' informs agricultural technology policy in most countries. In industrial countries and 'Green Revolution' areas in the South, it has led to a phenomenal increase in the productivity of agriculture and a rapid reduction in the number of farmers, releasing labour for other purposes. But this market-driven process has led also to the unsustainable type of farming we face today. Of course, this 'second-generation' problem is quite different from the degradation of resource-poor areas in the South. For us it is important to conclude, at this point, that:

- the introduction of ecologically sound agriculture does not seem to occur in 'treadmill manner'. This is largely the consequence of the fact that market forces do not propel technological change towards sustainability as in the case of productivity enhancing innovations. As long as the ecological costs of modern agriculture can be externalized to the future or other sectors, we must conclude that the market fails to ensure the widespread adoption of ecologically sound practices, and that alternative measures must be taken to offset this 'market failure'.

Environmental economists (e.g. Van Ierland, 1996) point out that such market failures require interventions in the form of fiscal measures, price policies or regulation. While such policies can create a conducive context for the transformation to ecologically sound agriculture (Pretty, this volume), the chapters in this book show that:

- the facilitation of learning, in addition to fiscal measures, regulation and price policies, is an important, perhaps essential, though not sufficient, intervention to ensure a change to ecologically sound agriculture. In the social science understanding of most policy makers and scientists, which is usually influenced by economics, learning processes and their facilitation appear often to be ignored.

This is why it is important to explore the nature of learning ecologically sound agriculture. The lessons contributed by the chapter authors appear to be the following.

- Learning and becoming expert in managing the farm or larger eco-system in an ecologically sound manner is intrinsically rewarding and energy giving.
- Learning ecologically sound agriculture substantially benefits from collective learning in small groups or 'field schools' (Röling and van de Fliert, this volume). Discussion plays a key role in concept formation, in the development of theory with which to anticipate on the basis of observation, in establishing what is effective and acceptable action, in assessing external information and in coping with the uncertainty of embarking on a new way of farming (Darré, 1985).
- This collective learning also has a technical dimension. As we have remarked earlier, important conditions for ecologically sound farming cannot be realized at the farm level only. A typical example is rat control in irrigated rice. Creating such conditions at the higher-than-farm level requires collective action and coordination based on shared perspectives and negotiated agreement.

A crucial and exciting aspect of shared learning that emerges from the chapters is what could be called the 'social construction of the ecosystem'. Most farmers, advisors and researchers have taken the farm or the crop as the essential unit around which to create shared knowledge. Scaling up that knowledge to encompass the watershed, landscape or eco-system poses special challenges for learning, and hence for curriculum development, and for 'making things visible'.

An important aspect of learning that emerges from the chapters is the power of 'discovery learning'.

- Carefully structured learning experiences, such as the agro-ecosystems analysis which farmers engage in during IPM field schools, or experiments with the rain simulator in Queensland, seem to have an energizing and mobilizing effect. Discovery learning relies on engaging people in experimentation, observation, measurement and so on which allow people to draw their own conclusions. Creating tools for discovery learning emerges as an important challenge for scientists.

How easy it is to focus on farmers only! But that should not blind us to the fact that, as Wagemans and Boerma demonstrate,

- moving to ecologically sound agriculture requires learning new roles across a wide range of actors.

In many of the cases presented, conventional farm advisors, for example, are struggling to learn new professional roles, or are by-passed altogether by the training and employment of new cadres of professional facilitators. Researchers and research institutes, on the whole, are slow in taking up the challenges posed by ecologically sound farming, and continue to work on the best technical means to achieve maximum productivity and global food security. Policy makers have to learn new approaches to technology policy.

A special case is the learning process of consumers. The experience in Germany (Gerber and Hoffmann, this volume) with biological agriculture clearly brings out its importance. One cannot expect farmers to change their product and add value to it in terms of the sustainability of the mode of production, if the consumers of that product have not learned to appreciate that added value and are not willing to pay for it, and worse, if the consumers continue to be bombarded with advertisements which ignore the added value. In that sense, the present efforts in most countries to introduce ecologically sound agriculture are not based on a holistic perspective and remains a marginal activity, perhaps awaiting more disasters such as mad cow disease, for the body politic to take the need of a shift seriously.

As a result of other actors being slow to come on board, farmers who want to change often find themselves on their own, not only in terms of coping with technical problems, but also in terms of information exchange, finding moral and institutional support, and especially in terms of receiving incentives. The available knowledge about moving to ecologically sound farming is patchy, with huge gaps, not only in terms of the technical practices, but also in terms of knowledge management, technology development, marketing, and promotion.

16.6 Facilitation of learning

In conventional agriculture, facilitation of learning is often equated with extension, and more specifically with transfer of technology (e.g. Chambers & Jiggins, 1987; Röling, 1988). This is a simplification. Especially in highly sophisticated agricultural industries, such as glasshouse vegetable and flower production, mushroom cultivation, intensive animal production, it is perhaps better to speak of a web of information services, consultants, and technical and financial supports in which the producer is embedded, which emphasizes whole farm enterprise development, and pays special attention to farm management and profitability. Across the globe we are witnessing a decline in the (funding of) monolithic public 'unified' or 'integrated' generalist professional public extension agencies which focus on technology transfer while, at the same time, advisory and information support services, which underpin specialized and commercial agricultural industries, are going from strength to strength. The demise of monolithic general extension services does not, therefore, imply a demise of the demand for, or supply of, extension. The highly specialized forms of agricultural production which emerge (also in the South) in response to middle-class demand, increasingly require specialized information support and advice with respect to long-term, strategic and operational decisions. The sophisticated managers in charge of such enterprises assemble the information and advice they require from many different sources and are highly skilled at integrating them in their farm management decisions.

From the chapters, quite a different picture emerges with respect to the facilitation of learning required for the transformation to ecologically sound agriculture. In the first place, the evidence suggests that introducing IPM in irrigated rice in South East Asia (Matteson, Gallagher & Kenmore, 1994, and Röling and van de Fliert, this volume) and Latin America (Agudelo & Kaimowitz, 1989) through the conventional transfer of technology-type extension service simply does not work. The Swiss case (Roux and Blum, this volume) also shows that the same can be said for other forms of sustainable agriculture. The focus on adoption of component technologies through demonstration and other transfer mechanisms seems inadequate to ensure acceptance of the complex of behaviours required for ecologically sound farming. The chapters suggest the following tentative observations about the alternative approaches to the facilitation of learning that seem to work.

- If one distinguishes extension and (non-formal) education by saying that the former focuses on the solution of specific problems while the latter aims at qualification, the type of facilitation required for learning ecologically sound practices seems to resemble the latter more than the former, a fact reflected in the concept of the 'farmer field school' introduced by FAO's IPM Programme. A new way of farming is being dealt with, with different procedures, with replacing reliance on external inputs with reliance on the farmer's own capacity to anticipate and to enhance desirable natural processes. This is a form of qualification with a focus on creating farmers who are expert at managing the locality-specific agro-ecosystem and at applying general principles to create opportunity.
- The fact that the kind of learning required seems to benefit from participating in learning groups, where concepts, norms and acceptable behaviours are collectively established (Darré, 1985), means that an important aspect of facilitation concerns group process. The experience shows that many conventionally trained extension workers find it easy enough to give

a presentation in front of a group, but have real problems with stimulating group learning processes, through fostering discussion, engaging in exercises, creating learning experiences, and so forth.

- Facilitators need to be able to access an arsenal of methods and methodologies, perhaps best illustrated by the experience with IPM in Indonesia, for example, by its use of insect zoos of potted rice plants covered with netting which groups of farmers use to observe the behaviour of various insects. In general, methods and methodologies cover participatory and group methodologies, including group experimentation (e.g. defoliating rice plants and comparing their yields with those of regular plants to establish tolerance for leaf damage by rats and insects).
- One of the key elements in this type of facilitation is fostering discovery learning. Typical for this approach is not to answer farmers' questions on the basis of one's expertise, but to use questions as opportunities for discovery and for working out answers oneself. Increasingly, one sees discovery learning being developed into a method of its own. It involves scientists who deliberately set out to create learning experiences during which farmers can draw their own conclusions. Examples are: learning about water retention and run-off through simulating rain fall on various types of land surfaces (Hagmann, Chuma & Gundani, 1995; Hamilton, 1995, and this volume); 'agro-ecosystem analysis' with an emphasis on counting pests on rice hills (Röling and van de Fliert, this volume, about a methodology developed by Russ Dilts and Kevin Gallagher and their Indonesian colleagues), resource flow mapping (Dufour *et al.*, 1995, using a methodology first developed by Lightfoot (e.g. Lightfoot, Pingali & Harrington, 1993), and water management simulation using a model of an irrigation scheme with farmers (Scheer, 1996). In all these approaches, there is a curriculum for learning, very definitely, but farmers are not told what happens or what to do. The essence is that they discover things for themselves. What they do with it is their own responsibility. The use of this principle emerges very clearly also from the Swiss case (Roux and Blum, this volume).
- Typically, the training of trainers must follow the same principles.

Few lessons have been learned so far with respect to other actors, such as researchers and consumers. The cases in this book mostly refer to training farmers and the training of trainers. Much work still needs to be done.

This holds especially for facilitating collective learning in actor networks. As we have observed, transforming conventional agriculture is not just a question of training farmers, but of social learning in complex interwoven networks of interdependent actors. In most instances, we are not dealing with 'virgin country' but with situations in which highly interwoven actor networks have already evolved around the needs of conventional farming. Saying that not much is known about the facilitation of social learning in such situations is an understatement. Yet, one could point to methodologies such as RAAKS developed by Engel (1995) and by Engel, Salomon and Fernandez (1994), which takes groups of interdependent actors through a mutual learning process by opening various 'windows' on their interaction.

16.7 Support institutions

The institutional dimension of the transformation to a more sustainable society is rapidly emerging as a crucial area of interest. If one recalls Conway's (1994) enumeration of policy goals (production and productivity, equity, sustainability, and stability), one could say that most institutions serve production and productivity (factories), equity (labour unions) or stability (schools, police forces), but only few institutions have been emerged to support sustainability (Christensen, in prep.). Yet careful studies of the management of complex ecosystems (Gunderson, Holling & Light, 1995) or of common property resources (Ostrom, 1992) reveal the immense importance of new institutional arrangements. Overcoming the social dilemmas inherent in managing natural resources in a sustainable manner places heavy demands on trust, on agreed upon procedures, on covenants among stakeholders, and on mechanisms for collective decision making (e.g. Maarleveld, 1996). In this volume a special interest is taken in 'platforms for resource use negotiation', but those will be discussed separately. Here, the focus is on the institutional implications of ecologically sound farming. Implications for the required institutional support for ecologically sound farming emerge from virtually all chapters, and especially from the US case (Fisk, Hesterman and Thorburn, this volume) but much research still needs to be done on the subject. One could speak of a crucial area for further study. We formulate a few tentative conclusions.

- So far, efforts to introduce ecologically sound agriculture focus largely on the transformation of farms and on the activities of farmers. However, these efforts reveal that the institutional frameworks in which farmers are embedded are of crucial importance for the transformation required. This happens especially when pilot efforts are scaled up.
- Existing institutional frameworks mitigate against the emergence of ecologically sound farming practices, for example because marketing and processing institutions do not allow the added value to be realised at the level of retail distribution, or because powerful conventional farmer organizations prevent efforts to make consumers aware of the value added by ecologically sound farming.
- New networks and institutions are rapidly emerging in response to the challenge. Examples are consumer and environmental organizations, and organizations of ecological farmers, from bio-dynamic farmers to integrated arable farmers, usually with a view to creating marketing labels by which the environmental value they have added to their product can be recognized in the market place (Matteson, Den Boer & Proost, 1996).

- No experience has been presented in any of the chapters, of the kind of institutional linkages required for nutrient recycling which is a condition for ecologically sound practice.
- The experience in several countries suggests that ecologically sound farming makes special demands on what Röling and Engel (1991) have called the agricultural knowledge and information systems (AKIS), i.e. the actors who can potentially improve the innovative performance in a given sector or domain if they manage to form a soft system. The knowledge system supporting conventional agriculture typically takes on the character of a set of institutions 'calibrated along the science-practice continuum' (Lionberger & Chang, 1970). The experience reported in the chapters suggests that the ecological knowledge system is characterized typically by learning groups of farmers and networks among such groups which allow the exchanging of experience. These groups are facilitated typically by agents who are highly skilled both technically and in non-formal educational methods. In other words, where a knowledge system designed for technology transfer would deploy its least trained staff as field workers, the ecological knowledge system seems to require a decentralised network of skilled field agents. A special feature of the ecological knowledge system is the possibility for 'farmer-to-farmer extension' or rather the use of farmer trainers, a term which fits our earlier observation that the kind of facilitation required can better be called non-formal education than extension. Once they are qualified, farmers can help their colleagues to learn. What's more, experience in Indonesia's IPM programme shows they are eager to do so.
- Of special interest is the role of agricultural research. Where it is seen as the source of innovations in the conventional knowledge system, the role of research that emerges from the chapters is unclear, except for the crucial contribution of curricula for discovery learning. But other contributions seem not to have crystallized. The old role of developing technologies FOR farmers seems to clash with the logic of ecologically sound farming, while a new role of working WITH farmers seems not to have clearly emerged, at least in the case studies in this book. In fact, there is some evidence, e.g., in Hamilton's chapter, that scientists feel threatened in their traditional role of definers of problems and designers of solutions.
- Scaling up pilot efforts to introduce ecologically sound farming seems to lead to clashes with entrenched interests and established procedures. Therefore, establishing effective programmes requires considerable attention to strategic manoeuvring and to building coalitions. This observation is consistent with social actor network theory (Callon & Law, 1989) which states that scientific paradigms and theories have societal impact, not because of the 'truth value' of their tenets, but because of the emergence of social actor networks which have an interest in creating and maintaining the conditions for their effective operation in society. This seems to

hold for conventional intensive high-input agriculture and for ecologically sound agriculture.

In all, the experience reported in the chapters seems to suggest that the institutional support for ecologically sound agriculture is an important new area for research and development. The introduction of ecologically sound agriculture not only requires agronomic and other technical research, but also participatory action research and field experimentation with alternative institutional arrangements. A typical example of this approach is the experiment supported by W.K. Kellogg Foundation (Fisk *et al.*, this volume) with learning communities that support 'integrated farming systems', but the participatory development and exploration of new institutional arrangements that support ecologically sound agriculture seems to be emerging simultaneously in different places. A typical example is the ILEIA project (see the issues of the ILEIA Newsletter for a regular update) which links local learning groups engaged in participatory technology development in three countries in a global network of information exchange. Still another example is the development of a European network for the participatory design of proto-type ecological farming systems (Vereijken *et al.*, 1994; Vereijken, 1995). Kabourakis (1996) reports on learning groups in Crete which engage in participatory proto-typing of ecological olive production systems. Further research in this area that we are aware of is the work of Glasbergen of the University of Utrecht, and that of Gunderson *et al.* (1995) in the US.

16.8 Conducive policy contexts

Cochrane's (1958) agricultural treadmill and its implication that productivity gains in primary production are passed on to processors, middlemen and consumers, provides a powerful economic theory to explain why even present day's highly capitalized, large scale and incredibly productive survivors of the shake-out during the last decades are still in a vulnerable position, especially if they are highly specialized and have no sources of off-farm income (Knickel, 1994). The nearly perfect market for primary products is self-destructive (Galbraith, 1995).

Yet the prevailing political climate favours the free operation of the market as a beneficial shaper of a desirable society. Public support for agricultural research, extension and education is rapidly decreasing with the dwindling political support for agriculture in general and productivity enhancement in particular. Given the relatively small number of farmers still in business, and given also the decrease in the flow of productivity enhancing innovations, a key mechanism for the most innovative and influential farmers to earn windfall profits is considerably weakened. Where those influential farmers used not to make too much noise about their colleagues who were forced out of the market by the treadmill mechanism, they now face the music themselves. The sector is in crisis throughout the industrial world, and with it those in the South who produce for the market.

The emergent concern about the sustainability of conventional agriculture has, as yet, not led to an adaptation of the societal mechanisms which ensure a more ecologically sound agriculture. While there is widespread recognition that the market fails when it comes to externalized environmental costs, destruction of common property resources or open access resources (Van Ierland, 1995), fiscal policies and/or regulatory measures to off-set market failure have, as yet had limited effect, while public support of farmer learning required for ecologically sound agriculture is generally limited. One could say that the kind of societal mechanisms, such as labour unions, which emerged in response to a greater societal concern with equity as a result of the impact of the industrial revolution, have not as yet emerged in response to the present concern for sustainability.

Incentives for ecologically sound practices can be developed in two ways. Either environmental costs are somehow internalized so that the market automatically ensures ecologically sound practices, or the reward is negotiated among stakeholders (as is the case with labour and employers). Perhaps other ways will emerge which bring to bear ecological imperatives on human behaviour.

What is the 'evidence' that emerges from the chapters? Below we enumerate what appear to be defensible conclusions.

- In the present free market context, profitability is essential for farm survival. Ignoring it, and asking farmers to engage in unrewarding, cost-increasing practices is unrealistic. The margins for change at the farm level are small. The downward pressure on prices forces farmers even to ignore important concerns such as maintaining the long-term productivity of their land for their children.
- The narrow focus on primary production in the search for solutions casts farmers in the role of scapegoats, while consumers and agri-business continue to benefit from unrealistically low prices. It seems impossible to move to ecologically sound agriculture without taking a more holistic perspective which includes other actors in the production column, and, perhaps even more importantly, takes into account the multiple purposes for which land is used. The Swiss case clearly demonstrates what can be achieved when cross-transfers among different land uses (e.g. from tourism to agriculture) are made possible. The German case shows how consumers can be directly involved in agricultural production.
- Internalizing costs by fiscal measures based on mineral book-keeping, monitoring of pesticide use, etc., are difficult to effectuate. Yet, the development of mechanisms to make visible the impact of farming on the wider environment is essential, not only for effective fiscal measures, but especially as a basis for farmer learning and anticipation. Much work is still required in this area.
- Covenants between government and farmers which specify a route of environmental targets to be achieved, while leaving the method of realising the targets to the industry within the prevailing free market conditions, is apparently insufficient to achieve the targets specified in the

absence of transformation subsidies, and public support for facilitation and research.

- The transformation required cannot be limited to primary production only. Agriculture in industrial countries is embedded in 'actor networks' (e.g. Callon & Law, 1989) or 'coalitions of interest' (Biggs, 1995), which have to be taken into account. Actors include processing industries, agricultural co-operatives, advisory services, research establishments, farmers organizations, government agencies, commercial input providers, banks and many other firms and services which form a complex system that supports the prevailing mode of production and is totally dependent on its continuation. In the Netherlands, for example, this actor network has emerged over a period of nearly a century. It is highly inter-connected in that respected farmer leaders hold multiple key positions in co-operatives, boards of research stations, etc. High social pressure is exerted within the network to maintain the coalition of interest. Farmers who deviate are considered a threat to the coalition as a whole. What's more, the logistics of farming often mitigate against a change in the mode of production. For example, the entire production of biological sugar beet in the Netherlands can be processed in one day by the large-scale plants which slowly evolved as Dutch agriculture became more specialized and competitive. The transformation to ecologically sound agriculture cannot be realized without struggle, as new actor networks emerge and undermine the power of others. To be effective, policies will need to address this issue explicitly.
- Policies are often inconsistent with ecologically sound farming, as the chapter by Pretty makes abundantly clear. Sometimes they work in wondrous ways, as described in the Indonesian case where the threat to national food security, and hence political stability, as a result of emergence and resistance of the Brown Plant Hopper, created a conducive context for IPM. But policies that favour ecologically sound practices can be incompatible with the continuation of conventional farming practices. The case of the Netherlands (Van Woerkum and Aarts, this volume) makes clear that various policies duly enacted by Parliament flounder on the intransigence of the conventional farming complex which the market and significant areas of policy continue to sustain.
- An interesting policy issue that emerges is the nature of financing of the ecological knowledge system. It is one thing to finance an extension service, but quite another to finance farmer trainers and farmer meetings. Yet, the latter seem essential ingredients of the ecological knowledge system. Of interest is the considerable funding of such costs by farmers, local communities and local government (reported by Pretty and Röling and Van de Fliert, this volume).

Moral rewards seem hardly at work as yet, allowing large gaps between what people know and what they do (Van Woerkum and Aarts, this volume).

Trade-offs among the policy goals of productivity and sustainability are difficult to operationalize. For the time being, the internally consistent, and highly interwoven complex of policies and policy instruments which supports productivity seems as strong as ever. It is driven by policy theories which emphasize global food security and high chemical input agriculture to ensure that future demands can be met. It goes without saying that such policies are favoured by fertiliser and pesticide lobbies. Furthermore such policies are informed by computer simulations which only take into account biotic factors, emphasize grain production as a proxy to food production, and take the American high fat and high meat diet as an inexorable standard toward which the human food system evolves. As humankind 'progresses', it is expected to 'move up the food chain', and thus become more of a predator.

Such expectations ignore the fact that people's behaviour is not caused but constructed by themselves. Trend is not destiny. In fact, knowledge about trend is a key instrument in ensuring that trend does not become destiny.

An example is the US Food and Drug Administration which has recently, and not without a long and fierce battle with the meat industry, released new recommendations for a healthy diet which feature a sharply reduced animal fat and meat intake and an increased fibre, fruit and vegetable intake. In fact, a vegetarian diet is strongly recommended. The cost of the present US diet in terms of public health is very high.

Jiggins (1994), and Pretty, in his wide-ranging chapter (this volume), have presented a number of alternative perspectives which take the global food security challenge seriously, but point to non-conventional routes to ensure it. It is time to question the scenarios advanced by both the ecological doomsayers such as Lester Brown, or the advocates of renewed investment in conventional agricultural research and extension, such as Norman Borlaug (Borlaug & Dowswell, 1995).

16.9 Platforms for resource use negotiation

Essential conditions for sustainable practices at the farm level are increasingly determined at higher eco-system levels. Parameters at the farm level become variables at higher system levels (Fresco, 1986). Stakeholders using the same natural resources for various purposes, such as drinking water companies, industries, irrigators and other stakeholders in a watershed, are discovering that they are interdependent in that the use one stakeholder makes of the resource affects the outcomes of the others. When such conditions are widely recognized, the situation is ripe for 'scaling up human agency to a level of social aggregation which is commensurate with the level of the eco-system perceived to be in need of interactive management' (Röling, 1996). In other words, such conditions require building human institutions and a capacity for collective learning and decision making about the eco-system perceived to be under threat. Such moves usually require solving social dilemmas, in that people are required either to give more to realize a public good (such as a purification plant) or take less from a common (such as a subterranean aquifer resource) (Ostrom, 1992; Koelen & Röling; 1994; Maarleveld, 1996). The chapters allow us to draw the following preliminary conclusions.

- Sustainable agriculture essentially involves multiple levels of decision making. Even when the challenge is to change practices at the farm level, and when, therefore, there seems no need for collective decision making at first glance, social learning and collective decision making in, e.g. learning communities (Fisk *et al.)*, farmer field schools (Röling and van de Fliert), or learning groups (Hamilton, van Weperen *et al.)* seem essential, not only because learning ecologically sound farming is an interactive process, but also because the shift not only involves the farm but also the whole network of institutions and agencies in which it is embedded. A typical example is collective marketing through packaging, labelling, processing, etc. Somers provides some interesting examples.
- Usually change in practices at the farm level also involves change in practices at a higher system level, e.g. at the level of rat or insect population dynamics, or at the level of the watershed (Campbell).
- Finally, the change required may be entirely at the higher system level, when multiple interdependent stakeholders in an eco-system, with different and often conflicting interests find that they need to scale up their decision making to the higher system level and share in problem definitions, accommodate multiple perspectives and 'rich pictures', and negotiate collective management decisions at the higher level, based on social learning about that higher system level. It is in such situations that we speak of 'platforms for resource use negotiation'.

Wagemans and Boerma provide a compelling example of a situation in which the problem situation could have been improved by using an interactive 'platform approach', but where the central authorities attempted instead to use a heavy-handed regulatory approach. The result is an impasse which emerges out of the many opportunities where stakeholders in a modern and complex society have to resist and sabotage centrally imposed legislation. 'Area based planning' does seem to require a consensual approach, in that ways forward need to emerge from interaction among stakeholders (also Glasbergen, 1996). This leads to interesting implications for subsidiarity, in that decisions taken by platforms of stakeholders might not be consistent with those arrived at on the basis of electoral politics. Van Woerkum and Aarts provide an interesting case in which 'interactive plan formation' at the local level was effective in breaking through an impasse.

The Australian Landcare movement provides an example of a consensual approach (Campbell, 1994 and this volume), in that stakeholders share the definition of the problem and learn to deal with it together, often with the help of a facilitator who uses participatory techniques, and focuses on the process of stakeholder interaction (rather than on a predetermined outcome). Landcare has become a classic example for scaling up natural resource management to at least a sub-catchment level and has experimented with many new and exciting methodologies, such as using geographic information system (GIS) software in combination with resource mapping by farmers, using indigenous classification criteria to create resource maps of the catchment in a process designed to help farmers construct a shared perspective on the catchment, and, from the farm level, scale up their concerns to the catchment level.

The platform notion, if anything, is heuristic and intriguing. It captures in one word a widely shared concern and the increasingly common experience that natural resource management at the customary level and by the customary stakeholders is no longer adequate to take into account the ecological imperatives that make themselves felt. Much research is still required, and is in fact on-going in the search for institutional frameworks that reflect the increasing concern with sustainability of natural resource use. From the chapters and other recent research (Van 't Bosch, 1996), further tentative conclusions can be formulated.

- The platform required to adequately manage a given natural resource perceived to be in need of sustainable management, be it a lake, water catchment, forest, or the air over a metropolis, very often does not coincide with existing administrative or economic institutions. In fact, it might require collaboration of people who are normally at loggerheads, such as Indian caste communities occupying the same catchment. Wagemans and Boerma provide an example of the problems caused for platform formation by the existing, and in such an old society as the Netherlands, very complex, institutional arrangements at the local, community, provincial and national levels.
- Platforms can be one-time meetings, elected committees, formally appointed boards or councils or even parastatal or government bodies. An important issue is the representation of the key stakeholders in the resource, and the accountability to constituencies without bringing the platform to a total impasse of immobile positions. In fact, negotiating from explicitly stated interests, instead of from positions, is considered a condition for effective operation.
- The facilitation of platforms is crucial, especially in conflict situations. In the US, 'consensual approaches to the resolution of distributive conflicts' (Susskind and Cruickshank, 1987) have been developed and used by the Environmental Protection Agency. The EPA has provided negotiators to situations in which natural resource use was disputed among various stakeholders, such as conservationists and commercial interests, with the sole purpose of fostering the process of conflict resolution rather than a specific outcome. In situations with less serious conflicts, the facilitation of platform processes can borrow a leaf from soft systems methodology (SSM) (Checkland, 1981; Checkland & Scholes, 1990), which have been designed for fostering collective agreement about action in corporate situations. It takes a number of stakeholders who have a shared perspective on a problem situation through a learning path, which includes creating a

'rich picture' of the multiple perspectives held by the different stakeholders, accommodation of divergent interests, creating a common perspective on the problem and its solutions and, finally, agreeing to take action and identify who should take action. Discussions with Checkland have made it clear that SSM can be considered a suitable methodology for platform facilitation, albeit that the greater complexity of the situation calls for additional prior exploration to understand the nature of the actors and the local history (Leeuwis, 1993). The RAAKS (rapid appraisal of agricultural knowledge systems) methodology developed by Engel *et al.* (1994) takes off from there and provides a method for stakeholders to look interactively at their interaction through different windows.

- The focus on platforms is often on the purely social processes, such as conflict resolution, negotiation, institutional development, leadership, power, etc. However, especially the Australian experience (e.g. Campell, Woodhill and Röling, and Hamilton) makes abundantly clear that collective learning about the higher level eco-system is an essential element in platform building. The stakeholders need to inter-subjectively construct an understanding of the eco-system and its vital complex dynamics. A very interesting area that emerges from the chapters is the nature of this type of social learning and especially the methods of making things visible.

16.10 The ecological knowledge system

Partly as a result of the work of Röling (1988) and Engel (1995), it has become common practice to speak of 'agricultural knowledge systems', i.e. to use a (soft) systems approach for looking at the interaction among the (institutional) actors operating in a 'theatre of agricultural innovation'. Innovation emerges from this interaction, and is no longer seen, as was customary in the 'transfer of technology perspective', as the end-of-pipe product of a sequential process.

The knowledge system perspective looks at the institutional actors, within the arbitrary boundary of what can be considered the theatre of innovation, as *potentially* forming a soft system. A soft system is a social construct in the sense that it does not 'exist'. One cannot, therefore, say that such actors as research, extension and farmers *are* a system. In all likelihood, they are not, in that there is no synergy among their potentially complementary contributions to innovative performance, but by looking at them as potentially forming a soft system, one begins to explore the possibilities of facilitating their collaboration and hence the possibilities for enhancing their synergy and innovative performance. Innovation is an emergent property of a soft system, that is, it emerges from the interaction among the social actors who form a soft system to the extent that they collectively begin to see themselves as forming such a system. Again, the soft system is a deliberate and often facilitated social construction (Checkland, 1981; Checkland & Scholes, 1990; Wilson & Morren, 1990). The RAAKS methodology (Engel *et al.*, 1994) takes the social actors in a theatre of innovation through a deliberate interactive process during which they learn to become a soft system.

We have earlier described five elements: practices, learning, facilitation, institutional supports and conducive policies, which can be usefully considered to constitute the knowledge system, in that they form a consistent whole in accordance with the purpose of the system. Elsewhere (Röling & Jiggins, 1996), we have contrasted three types of knowledge system: transfer of technology, farm management development, and the ecological knowledge system. By way of a summarizing conclusion to this chapter, we briefly contrast these different knowledge system models.

The most commonly recognized knowledge system is the first one. It is a powerful model of innovation that still drives a great many decisions about investment in research, institutional and organisational arrangements, training and deployment of staff, etc. Millions of dollars have, for example, been invested in developing countries in the training and visit system which is totally based on the transfer of technology model. It sees desirable farming practice as using science-based component technologies, farmer learning as the adoption of external innovations, and facilitation as delivery of those innovations. The supportive institutional framework is 'the institutitional calibration of the science-practice continuum'. Conducive policies include financing elaborate public research and extension systems, and subsidies for inputs and farm demonstrations. Using Habermas' (1984 and 1987; Brand, 1990) distinction between instrumental, strategic and communicative rationality, we can say that transfer of technology is driven by instrumental rationality (Röling, 1996).

Farm management development requires a totally different knowledge system. Its aim is to support the practices of the farmer as an entrepreneur engaged in an economic enterprise. The knowledge system has much in common with the model for commercial innovation developed by Kline and Rosenberg (1986). The focus is on the farm as a whole, and practices are seen as the operational and strategic decisions which optimize the profit making capacity of the farm. Farmer learning is largely information processing, based on being exposed to a wide array of sources of market and technical information. Facilitation occurs through advisory work of various types, such as salesmen, private consultants, book keeping firms, banks, professional journals, etc. A supportive institutional framework typically includes a network of actors arranged to perform mutually complementary functions in a certain industry, such as the production of cut flowers, riding horses, mushrooms, wine, seed potatoes, etc. Conducive policies typically include support for the development of export markets, protection of product labels, quality control, etc. The knowledge system for farm management development operates within strategic rationality.

An ecological knowledge system has not, as yet, crystallized as clearly as those described. Yet, the chapters in this book suggest that:

- The ecological knowledge system has distinctive features of its own, which need to be elaborated and specified so as to constitute a 'model' which can guide decisions about investment, the design of organisations and inter-institutional linkages, staff deployment and training, and so forth. Attempts to introduce ecologically-sound agriculture through knowledge systems based on other models seem to have been ineffective.

The practices, nature of learning, facilitation, institutions and policies have been the subject of this last chapter and need not to be repeated here. A few summary points must suffice:

- The main purpose of the knowledge system is to help land users become experts at managing complex eco-systems in a sustainable manner. Such management requires more than efficient resource use but also some concept of 'wholesomeness' which is learned over time as the land user responds adaptively to feedback from eco-systems (e.g. Gwyn & Walter-Toews, 1996). The management involves eco-systems at different system levels and different commensurate levels of social aggregation. The management does not focus on instrumental control of eco-systems to optimise the value of key variables, but on adaptive and responsive management of diversity and complexity to optimize opportunities and outcomes. As Beck, Giddens & Lash (1994) put it: '. . . the expansion and heightening of the intention of control ultimately ends up producing the opposite'.
- A key feature of learning is that it is largely social learning in learning groups, field schools, platforms, etc. One reason is that ecologically sound agriculture requires mutual adjustment and agreement in solving social dilemmas, taking less from the commons or giving more for the public good. In that sense, the rationality driving the ecological knowledge system is what Habermas calls 'communicative rationality'. As Beck says (Beck *et al.*, 1994): 'It is the same everywhere: the demand is for forms and forums of consensus-building co-operation among industry, politics, science and the populace. For that to happen, however, the model of unambiguous instrumental rationality must be abolished'.
- A key feature of facilitation is that it can only be partially based on technical expertise. A major component is the enhancement of interactive processes for social learning, negotiation, accommodation and agreement. This means that facilitators must be well versed in both technical expertise and skills, and in social science expertise and process skills. Their training must pay explicit attention to this hybrid character.
- Institutions which feature in the ecological knowledge system tend to be decentralized. Key elements are the learning groups and platforms at different eco-system levels. Typical are also highly skilled facilitators who are, themselves, part of learning networks in which experience is exchanged. The nature of science linkage is less clear from the chapters. This is perhaps caused by the fact that few researchers and institutions have as yet fully embraced the promotion of ecologically sound agriculture, and/or still support a view of the contribution of science which is more

in accord with the transfer of technology model than the emerging ecological knowledge system model which relies strongly on interaction. Thus scientists tend to focus on learning themselves to generate a product which is useful for others, rather than deploying their knowledge to allow others to learn. Where this happens, e.g. through the development of tools for discovery learning, the contribution of scientists is dramatically effective. In addition, scientists do, of course, have important traditional roles in developing resistant and tolerant cultivars, tools to make things visible, biological controls, theoretical models of complex system interactions, etc.

- Policies focus particularly on preventing the externalization of environmental and ecological costs of agricultural production and on enhancing consumer trust in ecological products.

One of the interesting questions that emerges from the chapters of this book is how to manage change towards ecological knowledge systems. The conflict and struggle involved in that transition seem to be part of a larger more encompassing societal change that Beck *et al.* (1994) call 'reflexive modernization'. It is perhaps fitting to end this chapter and this book with a reference to their wide-ranging perspective, if only because the social learning of sustainable agriculture seems to be part of the larger human challenge they describe. 'The future looks less like the past than ever before and has in some ways become very threatening. As a species we are no longer guaranteed survival, even in the short term – and this is a consequence of our own doings, as collective humanity' (p.iv). We live in a 'risk society' (see also Funtowicz & Ravetz, 1993, 1994). 'Let us call the autonomous, undesired and unseen, transition from industrial to risk society *reflexivity.* Then 'reflexive modernization' means self-confrontation with the effects of risk society that cannot be dealt with and assimilated in the system of industrial society – as measured by the latter's institutionalized standards' (p 6).

16.11 References

Agudelo, L.A. & Kaimowitz, D. (1989). Institutional linkages for different types of agricultural technologies: rice in the Eastern plains of Colombia. The Hague: ISNAR/RTTL, linkages discussion paper 1.

Beck, U. (1994). The reinvention of politics: towards a theory of reflexive modernisation. In *Reflexive Modernisation. Politics, Tradition and Aesthetics in the Modern Social Order*, ed. U. Beck, A. Giddens & S. Lash, pp. 1–55. Stanford: Stanford UP.

Beck, U., Giddens, A. & Lash, S. (1994). *Reflexive Modernisation. Politics, Tradition and Aesthetics in the Modern Social Order.* p. 9. Stanford: Stanford UP.

Biggs, S.D. (1995). Contending coalitions in participatory technology development: challenges for the new orthodoxy. Norwich: University of East Anglia. Expanded version of a paper prepared for the PTD Workshop on 'The Limits to Participation', organized by Intermediate Technology at the Insitute for Education, Bedford Way, London.

Borlaug, N.E. & Dowswell, C.R. (1995). Mobilising science and technology to get agriculture moving in Africa. *Development Policy Review*, **13**, 115–29.

Brand, A. (1990). *The Force of Reason: An Introduction to Habermas' Theory of Communicative Action*. Sydney: Allen and Unwin.

Callon, M. & Law, J. (1989). On the construction of socio-technical networks: content and context revisited. *Knowledge in Society: Studies in the sociology of science past and present*, vol. 8, pp. 57–83. JAI Press.

Campbell, A. (1994). *Landcare. Communities Shaping the Land and the Future*. St Leonards (Australia): Allan and Unwin.

Chambers, R. & Jiggins, J. (1987). Agricultural research for resource-poor farmers. Part I: Transfer-of-technology and farming systems research. Part II: A Parsimonious Paradigm. *Administration and Extension*, **27**, 35–52 (Part I) and **27**, 109–28 (Part II).

Checkland, P. (1981). *Systems Thinking, Systems Practice*. Chichester: John Wiley.

Checkland, P. & Scholes, J. (1990). *Soft Systems Methodology in Action*. Chichester: John Wiley.

Cochrane, W.W. (1958). *Farm Prices, Myth and Reality*. Minneapolis: Univ. of Minnesota Press (especially Chapter 5: The agricultural treadmill).

Conway, G.R. (1994). Sustainability in agricultural development: trade-offs between productivity, stability and equitability. *Journal for Farming Systems Research-Extension*, **4**(2), 1–14.

Darré, J.P. (1985). *La Parole et la Technique. L'Univers de Pensée des Éleveurs du Ternois*. Paris: l'Harmattan.

Dufour, T., Hilhorst, T., Kanté, S. & Diarra, S. (1995). Analysing the diversity of farmers' strategies. *ILEIA Newsletter*, **11**(2), 9–11.

Engel, P. (1995). Facilitating innovation: An action-oriented approach and participatory methodology to improve innovative social practive in agriculture. Wageningen: Agricultural University, published doctoral dissertation. Will be published commercially by the Royal Tropical Institute in Amsterdam, 1997.

Engel, P., Salomon, M. & Fernandez, M. (1994). *Strategic Diagnosis for Improving Performance in Extension. RAAKS Manual version 5.0*. Wageningen: Agricultural University, CTA and International Agricultural Center. Will be published commercially by the Royal Tropical Institute in Amsterdam, 1997.

Fresco, L.O. (1986). *Cassava in Shifting Cultivation. A Systems Approach to Agricultural Technology Development in Africa*. Wageningen: Agricultural University, published doctoral dissertation, Departments of Tropical Crop Production and Extension Science. Published by Royal Tropical Institute, Amsterdam.

Funtowicz, S.O. & Ravetz, J.R. (1993). Science for the post-normal age. *Futures*, **25**, (7), 739–55.

Funtowicz, S.O. & Ravetz, J.R. (1994). The worth of a songbird; ecological economics as a post-normal science. *Ecological Economics*, **10**, 197–207.

Galbraith, J.K. (1995). *The World Economy Since the Wars*. London: Mandarin.

Glasbergen, P. (1996). Learning to manage the environment. In *Bureaucracy and the Environment*, ed. W.M. Laffenby & J. Meadowcroft, pp. 175–93. Cheltenham: Edward Elgar.

Gleick, J. (1987). *Chaos. Making a New Science*. London: Penguin Group, especially Chapter 2, 9–33.

Gunderson, L.H., Holling, C.S. & Light, S.S. (eds.) (1995). *Barriers and Bridges to the Renewal of Ecosystems and Institutions*. New York: Colombia Press.

Gwyn, E. & Walter-Toews, D. (1996). *Community-Based Agroecosystem Health. A Manual for Rural Communities*. Guelph: University of Guelph.

Habermas, J. (1984). *The Theory of Communicative Action. Vol. 1: Reason and the Rationalisation of Society*. Boston: Beacon Press.

Habermas, J. (1987). *The Theory of Communicative Action. Vol. 2: Lifeworld and System. A Critique of Functionalist Reason*. Boston: Beacon Press.

Hagmann, J., Chuma, E. & Gundani, O. (1995). Integrating formal research into a participatory process. *ILEIA Newsletter*, **11**(2), 12–13.

Hamilton, N.A. (1995). *Learning to Learn with Farmers. A Case Study of an Adult Learning Extension Project Conducted in Queensland, 1990–1995*. Wageningen: Agricultural University, published doctoral dissertation.

Holling, C.S. (1995). What Barriers, what Bridges? In *Barriers and Bridges to the Renewal of Ecosystems and Institutions*, ed. L.H. Gunderson, C.S. Holling & S.S. Light, pp. 6–37. New York: Colombia Press.

Jiggins, J.L.S. (1994). *Changing the Boundaries: Woman-centered Perspectives on Population and the Environment*. Washington (DC): Island Press.

Kabourakis, E. (1996). *Prototyping and Dissemination of Ecological Olive Production Systems. A Methodology for Designing and a First Step Towards Validation and Dissemination of Prototype Ecological Olive Production Systems (EOPS) in Crete*. Wageningen: Wageningen Agricultural University. published doctoral dissertation.

Kline, S. & Rosenberg, N. (1986). An overview of innovation. In *The Positive Sum Strategy. Harnessing Technology for Economic Growth*, ed. R. Landau & N. Rosenberg, pp. 275–306. Washington (DC): National Academic Press.

Knickel, K. (1994). *A Systems Approach to Better Understanding of Policy Impact: the vulnerability of family farms in Western Europe*. Proceedings of the International Symposium on Systems-Oriented Research in Agriculture and Rural Development, pp. 966–73. Montpellier: INRA.

Koelen, M. & Röling, N. (1994). Social dilemmas. In *Basisboek Voorlichtingskunde*, ed. N. Röling, D. Kuiper & R. Janmaat. Meppel: Boom (in Dutch).

Leeuwis, C. (1993). *Of Computers, Myths and Modelling. The Social Construction of Diversity, Knowledge, Information and Communication Technologies in Dutch Agriculture and Agricultural Extension*. Wageningen: Agricultural University. Wageningse Sociologische Reeks, published doctoral dissertation.

Lightfoot, C., Pingali, P. & Harrington, L. (1993). *Beyond Romance and Rhetoric: Sustainable Agriculture and Farming Systems Research*. London: ODI, *Agricultural Research and Extension Newsletter*, **28**, 27–31, July 1993.

Lionberger, H. & Chang, C. (1970). *Farm Information for Modernising Agriculture: The Taiwan System*. New York: Praeger.

Maarleveld, M. (1996). Improving participation and cooperation at the local level: lessons from economics and psychology. Paper presented at the 9th

Conference of the International Soil Conservation Organisation (ISCO): Towards Sustainable Land Use: Furthering Cooperation between People and Institutions. Bonn (Germany), August 26–30.

Matteson, P.C., Gallagher, G.D. & Kenmore, P.E. (1994). Extension of integrated pest management for planthoppers in Asian irrigated rice: empowering the user. In *Ecology and Management of Planthoppers*, ed. R.F. Denno & T.J. Perfect, pp. 656–85. London: Chapman & Hall.

Matteson, P.C., Den Boer, L. & Proost, M.D.C. (1996). The Dutch grasp thorny green labelling issues. *Pesticide Campaigner.* Newsletter of the Pesticide Action Network International. December issue 1996.

Maturana, H.R. & Varela, F.J. (1987, 1992). *The Tree of Knowledge, The Biological Roots of Human Understanding.* Boston, MA: Shambala Publications.

Northwest Area Foundation (1994). *A Better Row to Hoe. The Economic, Environmental and Social Impact of Sustainable Agriculture.* St Paul (Minnesota): North West Area Foundation.

Northwest Area Foundation (1995). *Planting the Future: Developing an Agriculture that Sustains Land and Community.* Ames, Iowa: Iowa State University Press.

Ostrom, E. (1990, 1991, 1992). *Governing the Commons. The Evolution of Institutions for Collective Action.* New York: Cambridge University Press.

Pretty, J. (1994). Alternative systems of inquiry for sustainable agriculture. *IDS Bulletin*, **25**(2), 37–49, special issue on 'Knowledge is Power? The use and abuse of information in development.' pp. 37–49.

Pretty, J. (1995). *Regenerating Agriculture. Policies and Practice for Sustainability and Self-Reliance.* London: Earthscan.

Reijntjes, C., Haverkort, B. & Waters-Bayer, A. (1992). *Farming for the Future. An Introduction to Low-External Input and Sustainable Agriculture.* London: Macmillan and Leusden: ILEIA.

Rogers, E.M. (1995). *Diffusion of Innovations.* 4th edn. New York: Free Press.

Röling, N. (1988). Extension science. In *Information Systems in Agricultural Development.* Cambridge: Cambridge University Press.

Röling, N. (1996). Towards an interactive agricultural science. *European Journal of Agricultural Education and Extension*, **2**(4), 35–48.

Röling, N. & Engel, P. (1991). The development of the concept of Agricultural Knowledge and Information Systems (AKIS): implications for extension. In *Agricultural Extension: Worldwide Institutional Evolution and Forces for Change.* ed. W. Rivera & D. Gustafson, pp. 125–39. Amsterdam: Elsevier Science Publishers.

Röling, N. & Jiggins, J. (1996). The ecological knowledge system. Second European Symposium on Rural and Farming Systems Research, Workshop 2, Granada, Spain, March 27 through 29, 1996. To be published in the Proceedings of the Conference.

Scheer, S. (1996). *Communication between Irrigation Engineers and Farmers. The Case of Project Design in North Senegal.* Wageningen: Wageningen Agricultural University, published doctoral dissertation.

Susskind, L. & Cruickshank, J. (1987). *Breaking the Impasse. Consensual Approaches to Resolving Public Disputes. MIT-Harvard Public Disputes Program.* New York: Basic Books Publishers.

Tarnas, R. (1991). *The Passion of the Western Mind. Understanding the Ideas that Have Shaped our World View.* New York: Ballentine Books.

Van 't Bosch, J. (1996). Platforms with Communities: a case study of the LRMG, a community platform for nature resource management in Lockeyer Valley Queensland, Australia, exploring factors that influence the role of a platform in the community. Wageningen Agricultural University, unpublished MSc thesis.

Van Ierland, E.C. (1996). Milieu economie: over ongeprijsde schaarste en klimaatverandering. Wageningen: Wageningen Agricultural University: inaugural address.

Vereijken, P., Wijnands, F., Stol, W. & Visser, R. (1994). *Progress Reports of Research Network on Integrated and Ecological Arable Farming Systems for EU and Associated Countries.* Wageningen: DLO Institute for Agrobiology and Soil Fertility.

Verreijken, P. (1995). *Designing and Testing Prototypes.* Progress Report 2. Wageningen: DLO Institute for Agrobiology and Soil Fertility.

Westley, F (1995). Governing design: the management of systems and ecosystem management. In *Barriers and Bridges to the Renewal of Ecosystems and Institutions,* ed. L.H. Gunderson, C.S. Holling & S.S. Light, pp. 391–428. New York: Colombia Press.

Wilson, K. & Morren, G. (1990). *System Approaches for Improvement in Agriculture and Resource Management.* New York: Macmillan.

Index

CPSIA information can be obtained at www.ICGtesting.com
Printed in the USA
BVOW041141020413

317077BV00002B/138/A

9 780521 794817